THREE COPERNICAN
TREATISES

THREE COPERNICAN TREATISES

The *Commentariolus* of Copernicus
The *Letter against Werner*
The *Narratio prima* of Rheticus

Translated with Introduction
and Notes by

Edward Rosen

Professor of History
City College of New York

Second Edition, Revised
with an
Annotated Copernicus Bibliography
1939–1958

Dover Publications, Inc.
Mineola, New York

Bibliographical Note

This Dover edition, first published in 1959 and republished in 2004, is an unabridged and revised republication of the first (1939) edition, to which has been added an annotated Copernicus bibliography, 1939–1958.

Library of Congress Cataloging-in-Publication Data

Copernicus, Nicolaus, 1473–1543.
  [Commentariolus. English]
  Three Copernican treatises / translated with introduction and notes by Edward Rosen.
    p. cm.
  Originally published: New York : Columbia University Press, 1939. 2nd ed., rev. with an annotated Copernicus bibliography, 1939–1958.
  Includes bibliographical references and index.
  Contents: The Commentariolus of Copernicus—The letter against Werner—The Narratio prima of Rheticus.
  ISBN 0-486-43605-5 (pbk.)
    1. Astronomy—Early works to 1800. 2. Werner, Johannes, 1468–1528. De motu octavae sphaerae. I. Copernicus, Nicolaus, 1473–1543. Epistola Coppernici contra Wernerum. English. II. Rhäticus, Georg Joachim, 1514–1576. De libris revolutionum Nicolai Copernici narratio prima. English. III. Rosen, Edward, 1906–   IV. Title.

QB41.C84 2004
520—dc22

2003067424

Manufactured in the United States of America
Dover Publications, Inc., 31 East 2nd Street, Mineola, N.Y. 11501

for Sally

PREFACE

IT IS a matter of amazement and regret to many persons interested in the history of civilization that the writings of Copernicus, universally regarded as the founder of modern astronomy, have not yet been made available in the English language. When Professor Frederick Barry suggested that I might attempt to satisfy this need, he pointed out that the *Commentariolus* and the *Narratio prima* are better suited to convey Copernicus's ideas to the general reader than is his classic work *De revolutionibus orbium caelestium*. For these treatises are briefer, and they are relatively free from the extensive calculations necessarily included in the volume that established the heliocentric system. There is, moreover, a historical reason for reproducing the *Commentariolus* and the *Narratio prima;* for it was by these papers from the hands of the rebel cosmic architect and his first disciple that the learned world was first apprised of the revolution in the conceptual structure of the universe.

The *Letter against Werner* possesses intrinsic interest of its own. It throws light on the development of Copernicus's thought. The letter and the *Commentariolus* constitute his minor astronomical works. For these reasons it was included in this book.

I desire to record my gratitude to Professor Austin P. Evans for his editorial guidance. To the many friends and colleagues who have cheerfully given me of their special knowledge I express heartfelt thanks. For the errors that nevertheless appear—and it is rash to hope that a book of this sort can be entirely free from error—full and sole responsibility rests upon the author.

E. R.

College of the City of New York
September 4, 1939

NICHOLAS COPERNICUS

NICHOLAS COPERNICUS was born in 1473 at Thorn on the banks of the Vistula. His father was a prosperous merchant and municipal official in the old Hansa town. But he died when Nicholas was only ten years old; and it was the boy's good fortune to have for maternal uncle Lucas Watzelrode, who became Bishop of Ermland in 1489. The uncle took a fatherly interest in the nephew, guiding his way and smoothing his path. While Copernicus was still a young man, Bishop Lucas designated him a canon of the Cathedral of Frauenburg.[1] He enjoyed the income from this ecclesiastical post until his death (May 24, 1543) at the scriptural age of seventy; and before he was thirty years old, he received in addition an appointment to a sinecure at Breslau.

Copernicus had his elementary schooling in his native city and entered the University of Cracow in 1491. After several years of attendance at the renowned Polish center of learning, he journeyed to Italy in 1496. At Bologna and Padua he studied the liberal arts, medicine, and law, obtaining the doctor's degree in canon law at Ferrara in 1503.

Shortly after his return from Italy his first published work appeared from the press, a translation of an inferior Greek epistolographer into Latin. But it was not only in this concern with classical antiquity that Copernicus showed himself a man of the Renaissance. He also strove to achieve the many-sided accomplishments of that humanistic ideal, the universal man. He was competent in canon law; he practiced medicine; he wrote a tract on coinage; he served his cathedral chapter as an administrator and diplomatic representative; he painted his own portrait; he made many of his own astronomical instruments; and he established the heliocentric system on a firm basis.

[1] Benjamin Ginzburg erred when he referred to Copernicus as a monk (*The Adventure of Science*, New York, 1930, p. 22).

Germans and Poles have bitterly disputed the question of Copernicus's ethnic origin, each national group claiming the distinguished astronomer for its own. Where does the truth lie in this controversy? Politically, Copernicus was a subject of the king of Poland; he remained loyal to the Roman Catholic church; and he wrote chiefly in Latin, but a few of his private letters were composed in German.

GEORGE JOACHIM RHETICUS

George Joachim was born on February 16, 1514, at Feldkirch in the ancient Roman province of Rhaetia. In conformity with the strong classical tradition of his day he assumed the surname "Rheticus." He was apparently reared in comfortable circumstances, for his parents took him in his youth to Italy.

After studying at Zürich, in 1532 he entered the University of Wittenberg, where he obtained his degree. He continued his studies at Nuremberg and Tübingen and then received an appointment as professor of mathematics at Wittenberg. He began his teaching during the academic year 1536–37.

Reports concerning Copernicus's innovations in astronomy had reached the young man, and he was filled with great eagerness to become acquainted with the new system. But how was he to do this? Copernicus had published nothing. Rheticus resolved to seek out the aged scholar at Frauenburg and to master the new astronomy at its source.

Accordingly he set out for Prussia in the spring of 1539. Copernicus received him cordially and was his host for more than two years despite religious difficulties. Rheticus came from the principal stronghold of Protestantism, and there was bitter anti-Lutheran feeling in official Ermland. In this atmosphere of religious animosity the Protestant professor lived with the Catholic canon and studied his system with enthusiasm.

But Rheticus did not confine his studies to astronomy. On the basis of extensive travel during his stay in Prussia, he prepared a map of the region. Though the map has not been preserved, an accompanying essay on the methods of drawing maps is extant.[2] Two other works written during this period have both disappeared. The one was devoted to proving that the new astronomy did not contradict Scripture; the other was a biography of Copernicus. The loss of the latter is particularly unfortunate, for an account written by one so close to the great astronomer would undoubtedly throw valuable light on many obscurities in the life of Copernicus.

Rheticus left Prussia at the end of September, 1541.[3] He returned to Wittenberg, resumed his teaching, and served as dean of the arts faculty in the early months of 1542. He also supervised the separate printing of the trigonometrical portion of Copernicus's *De revolutionibus orbium caelestium*.

He left Wittenberg in 1542 and went to Nuremberg, where the great work was being printed. The early sections were set up under his supervision; but after his departure for Leipzig his place was taken by Andreas Osiander, of whom we shall hear more below.

Rheticus taught at the University of Leipzig from 1542 to 1551. Before he resigned, he published an ephemeris for 1551. After his resignation he devoted himself principally to the calculation of an extensive set of trigonometric tables, for which he has an independent place in the history of mathematics. In this work he received welcome financial assistance from the Emperor Maximilian II and several Hungarian nobles.

It is a curious circumstance that Rheticus was requited for the support and encouragement he brought to the old age of Copernicus. The closing years of his own life were brightened by the interest taken in his project by a young man, Lucius Valentine Otho. The tables on which he worked for a quarter of a century were finally printed in 1596, twenty years after his

---

[2] The essay was published by Franz Hipler in *Zeitschrift für Mathematik und Physik*, XXI (1876), historisch-literarische Abtheilung, 125-50.

[3] He tells us that his stay in Prussia lasted "three years, more or less." This statement appears (fol. a5v) in the preface which Rheticus wrote for John Werner's *De triangulis sphaericis* and *De meteoroscopiis*. While Rheticus's preface was printed separately at Cracow in 1557, the two works by Werner were first published in *Abhandlungen zur Geschichte der mathematischen Wissenschaften*, XXIV, Pt. 1 (1907), Pt. 2 (1913).

death, as the *Opus palatinum de triangulis*, begun by George Joachim Rheticus and completed by L. Valentine Otho.

THE *COMMENTARIOLUS*

Some years before Copernicus consented to the publication of his large work *De revolutionibus orbium caelestium*, he wrote a brief sketch (*commentariolus*) of his astronomical system. The *Commentariolus* was not printed during the life of its author; but a number of handwritten copies circulated for a time among students of the science,[4] and then disappeared from view for three centuries. A copy found in Vienna was published by Maximilian Curtze in 1878.[5] A second copy found in Stockholm was published in 1881.[6] On Curtze's collation[7] of these two manuscripts Leopold Prowe based the text[8] from which the present translation was made. A third manuscript[9] is believed to exist in Leningrad; so far as I know, it has never been published.

The opening section of the *Commentariolus* was translated by Prowe.[10] L. A. Birkenmajer published a partial translation of the work into Polish.[11] The only complete translation previous to the present one was done in German by Adolf Müller.[12]

---

[4] Tycho Brahe states that ". . . a certain little treatise by Copernicus, concerning the hypotheses which he set up, was presented to me in handwritten form some time ago at Ratisbon by that most distinguished man, Thaddeus Hagecius (Hayck), who has long been my friend. Subsequently I sent the treatise to certain other mathematicians in Germany. I mention this fact to enable the persons, into whose hands the manuscript comes, to know its provenience" (*Tychonis Brahe Dani opera omnia*, ed. Dreyer, Copenhagen, 1913-29, II, 428.34-40).

[5] MCV, I(1878), 5-17. The Vienna MS will be referred to as V.

[6] By Arvid Lindhagen in *Bihang till K. Svenska Vet. Akad. Handlingar*, VI(1881), No. 12. The Stockholm MS will be denoted by S.

[7] MCV, IV(1882), 5-9.   [8] PII, 184-202.

[9] Ludwik A. Birkenmajer, *Mikołaj Kopernik Wybór pism* (Cracow, 1926), p. xxviii n.

[10] PI², 288-92.   [11] *Op. cit.*, pp. 3-15.

[12] ZE, XII(1899), 361-82. This translation (with certain revisions) is reprinted from *Osiris*, III(1937), 123-41, by permission of the editor, Dr. George Sarton.

The date of composition of the *Commentariolus* cannot be precisely determined.[13] But an examination of its contents shows conclusively that the *Commentariolus* expounds a heliocentric system which differs in several essential features from the system taught by the mature Copernicus in the *De revolutionibus*. The earlier view may be called "concentrobiepicyclic," and the later "eccentrepicyclic"; the meaning of these terms will be made clear later on. To Ludwik Birkenmajer must be assigned the credit for first pointing out that the two systems are independent, or rather that the *Commentariolus* is a first stage in the development of the heliocentric theory in the mind of Copernicus.

THE *LETTER AGAINST WERNER*

John Werner, a figure of some importance in the history of mathematics, published in 1522 at Nuremberg a collection of papers on mathematics and astronomy.[14] One of these, the

[13] For a summary of the evidence see Aleksander Birkenmajer's "Le Premier Système héliocentrique imaginé par Nicolas Copernic," in *La Pologne au viie Congrès international des sciences historiques* (Warsaw, 1933).

[14] I wish to express my thanks to the Library of the University of Michigan for an opportunity to examine its copy of this rare work. The correspondence of Tycho Brahe shows that even in the sixteenth century it was very difficult to obtain Werner's book. In a letter to Brahe, dated in 1584, Johannes Major wrote: "If I ever find in some old library Werner's books on the motion of the eighth sphere and on the observations of the fixed stars, I shall make you aware that I have been not unmindful of you" (*Tychonis Brahe opera omnia*, ed. Dreyer, VII, 83.31-34).

In 1585 Brahe addressed the following request to Thaddeus Hagecius: "For a long time I have searched far and wide for John Werner's little work on the motion of the eighth sphere, which was printed, I believe, long ago at Nuremberg. But I have not yet obtained it anywhere. Consequently, if you find this book there and make it available to me . . . you will do me a great favor. I shall gladly remit all your expenses, and return the courtesy, whenever possible" (*op. cit.*, VII, 95.10-18). In 1586 Hagecius replied: "Werner's little work on the motion of the eighth sphere is no longer to be found for sale. It was joined with certain other papers of Werner, which are at present unobtainable anywhere" (*op. cit.*, VII, 104.1-3). But in 1588 Hagecius stated: ". . . I am sending you a book which you once requested, Werner on conic elements. It also contains his little treatise on the motion of the eighth sphere, and other papers. I got it from Fabricius in Vienna, who sends you his best wishes" (*op. cit.*, VII, 147.23-26).

De motu octavae sphaerae tractatus primus, was sent to Copernicus by Bernard Wapowski, who had been his fellow student at the University of Cracow and was now a canon at Cracow and secretary to the king of Poland. Wapowski requested Copernicus to pass judgment on Werner's contentions. Copernicus complied, sending to Wapowski under date of June 3, 1524, the *Letter against Werner,* a vigorous attack upon Werner's position. In an age when scientific periodicals had not yet come into existence, such letters served the function now performed by articles and extended book reviews. The *Letter against Werner,* taken together with the *Commentariolus,* may be said to constitute the minor astronomical works of Copernicus; for besides the *De revolutionibus* we have nothing else on astronomy from his pen.

Handwritten copies of the *Letter against Werner* circulated for a time;[15] and from a copy preserved in Berlin the first

Brahe replied in 1589: "For Werner's book on conic elements and the motion of the eighth sphere, which you obtained from Fabricius, of blessed memory (for I hear, to my sorrow, that he is dead), and which you sent me, I thank you most heartily. For a long time I searched vigorously for that book, but I never before could get hold of it. I may say that, insofar as it treats the motion of the stars, the body of the book fails to fulfill the promise of the introduction. I shall demonstrate this elsewhere, when an opportunity arises for dealing with this subject" (*op. cit.,* VII, 213.3-10).

Then in 1590 Giovanni Antonio Magini wrote to Brahe: "Your student, whom I have frequently mentioned, has indicated to me that you want Werner's book on the motion of the eighth sphere. He states that, although you have sought for it all over Germany, you have never been able to find it. I have therefore given him a copy to take to you in my name" (*op cit.,* V, 126.22-25). In the same year Brahe replied: "I have not yet received Werner's little book on the motion of the eighth sphere, which you presented to me. I suppose that on account of the great distance between us and the difficulty of the journey it went astray somewhere. Nevertheless I thank you very much for this not inconsiderable kindness. And I wish to inform you that I obtained the book some time ago from Fabricius, the Imperial Mathematician, through the help of Dr. Thaddeus Hagecius, who is also a remarkably expert mathematician" (*op. cit.,* VII, 295.17-23).

For an examination of Werner's work in physical geography, see Siegmund Günther, *Studien zur Geschichte der mathematischen und physikalischen Geographie* (Halle, 1879), pp. 276-332.

[15] One such copy was in the possession of Tycho Brahe. He states: "A certain letter, which I have in handwritten form, was sent by Copernicus to Bernard Wapowski, cantor and canon at Cracow, under date of June 3, 1524. In this

printed edition was made. It was included in Jan Baranowski's
edition of the *De revolutionibus*.[16] Although the text of this
edition was obviously faulty, it was reproduced by Hipler and
Prowe.[17] Then Maximilian Curtze found a second manuscript
of the *Letter against Werner* in Vienna; he collated both
manuscripts and published a critical text.[18]

The present translation was made from Curtze's text. So
far as I know, there have been two earlier translations, both
into Polish, and both on the basis of the Berlin manuscript
alone.[19]

THE *NARRATIO PRIMA*

It will be recalled that Rheticus left the University of
Wittenberg in the spring of 1539 and set out for Prussia to
study with Copernicus. In the middle of May he reached
Posen, and from there he sent a letter to John Schöner, with
whom he had studied at Nuremberg. In this letter he promised
to inform Schöner as soon as possible whether the achievement
of Copernicus justified his reputation.

Within a short time after his arrival Rheticus became aware
that his host was a genius of the first rank. But Copernicus, for
reasons which will be stated below, was reluctant to publish
his astronomical work. The young professor added his voice to

letter Copernicus analyzes John Werner's little work on *The Motion of the
Eighth Sphere*." Here Brahe quotes a passage from the *Letter against Werner*
and then continues: "This is what Copernicus wrote in the letter which I have
just cited. The copy in my possession was given to me after a second or third
transcription from Copernicus's own draft" (*Tychonis Brahe opera omnia*, ed.
Dreyer, IV, 292.4-20).

[16] *Nicolai Copernici Torunensis De revolutionibus orbium coelestium libri sex*
(Warsaw, 1854), pp. 575-82.

[17] Franz Hipler, *Spicilegium Copernicanum* (Braunsberg, 1873), pp. 172–79;
PII, 145-53.

[18] MCV, I, 23-33; pp. 19-22 describe the two MSS. Prowe gives the history
of a third MS (PII, 171-72) and reproduces Curtze's text (PII, 172-83).

[19] The first appeared in Baranowski's edition of *De rev.* (pp. 575-82), the
second in Ignacy Polkowski, *Kopernikijana czyli materyały do pism i życia
Mikołaja Kopernika* (Gniezno, 1873-75), I, 68-74. I have been unable to make
use of either translation.

the chorus of friends who were urging Copernicus to release his manuscript for publication. In order to test public reaction to the innovations introduced by Copernicus, Rheticus rapidly wrote a survey of the principal features of the new astronomy. He cast it in the shape of a letter to his former teacher Schöner[20] and had it printed at Danzig in 1540.

The response was so favorable that a second edition of the *Narratio prima* was brought out in 1541 at Basel. It is altogether likely that the welcome accorded to the *Narratio prima* was the clinching argument that finally persuaded Copernicus to put his manuscript into the hands of a printer.

The reader of the *Narratio prima* (*First Account*) will notice that Rheticus speaks of his intention to compose a "Second Account" ("Narratio secunda," "Narratio altera"). But Rheticus never wrote the "Second Account." The *Narratio prima* was important, for it was the only book to which astronomers could turn for information about Copernicus's system. But by preparing the way for the publication in 1543 of Copernicus's own work, the *De revolutionibus orbium caelestium*, it made any "Second Account" superfluous.

When the second edition of the *De revolutionibus* appeared in 1566 at Basel, it included the *Narratio prima*. Rheticus's work was printed a fourth and a fifth time as a companion piece to Kepler's *Mysterium cosmographicum* (Tübingen, 1596; Frankfurt, 1621). It received its sixth printing in the

[20] The name is Schöner (*Allgemeine deutsche Biographie*, Leipzig, 1875-1912, XXXII, 295). An incorrect form, Schoner, is frequently used by writers on Copernican astronomy; cf. Joseph Bertrand, *Les Fondateurs de l'astronomie moderne* (Paris, 1865), p. 50; Camille Flammarion, *Vie de Copernic* (Paris, 1872), p. 114; Hipler, *Spicilegium Copernicanum*, p. 208; C. L. Menzzer, *Nicolaus Coppernicus, Über die Kreisbewegungen der Weltkörper* (Thorn, 1879), p. 317; PI², 516; Adolf Müller, *Nikolaus Copernicus, der Altmeister der neuern Astronomie* (Freiburg im Breisgau, 1898), also printed in *Stimmen aus Maria-Laach*, Ergänzungsheft LXXII(1898), 83; Arthur Berry, *A Short History of Astronomy* (London, 1898), p. 98; L. A. Birkenmajer, *Mikołaj Kopernik* (Cracow, 1900), p. 224; J. L. E. Dreyer, *History of the Planetary Systems from Thales to Kepler* (Cambridge, 1906), p. 290; Pierre Duhem, ΣΩΖΕΙΝ ΤΑ ΦΑΙΝΟΜΕΝΑ, *Essai sur la notion de théorie physique de Platon à Galilée* (Paris, 1908), also printed in *Annales de philosophie chrétienne*, 79ᵉ année, t. 156(1908), p. 375; A. Koyré, *Nicolas Copernic, Des révolutions des orbes célestes* (Paris, 1934), p. 10; Angus Armitage, *Copernicus* (London, 1938), p. 58.

Warsaw edition (1854) of the *De revolutionibus* and its seventh in the Thorn edition (1873). Finally Prowe printed it for the eighth time.[21] The present translation was made from Prowe's text.

The Warsaw edition included a translation into Polish, which is, so far as I know, the only one previous to the present.[22]

THE DOCTRINE OF THE SPHERES

The ancient Greek astronomer Eudoxus (about 408–355 B.C.) introduced imaginary spheres into astronomical theory for the purpose of representing the apparent motions of the planets.[23] These spheres were invisible, and the observable planet was regarded as situated, like a spot or point, on the surface of the invisible sphere. The planet was deemed to have no motion of its own, but simply to participate in the motion of the sphere to whose surface it was attached. Now the observed movements of any planet are so complicated that a single sphere, moving at a uniform rate always in the same direction, could not produce the observed phenomena. Hence it became necessary to devise for each planet a set of spheres. These remained an integral part of astronomical theory until Kepler (1571–1630) banished them by demonstrating the ellipticity of the planetary orbits.

Copernicus used these spheres (*orbes*) throughout his work. He avoided taking sides in the controversy over the question whether the spheres were imaginary or real,[24] whether, that is,

[21] PII, 293-377.

[22] Polish translation, pp. 489-544 of the Warsaw edition. The editors of Th stated (p. xxiv) that C. L. Menzzer had translated into German both *De rev.* and the *Narratio prima*. But when Menzzer's book appeared, it contained only *De rev.* I have not been able to find out what became of his *Narratio prima* MS.

[23] For an account of Eudoxus's system see Dreyer, *Planetary Systems*, ch. iv. The basic article is G. Schiaparelli's "Le sfere omocentriche di Eudosso, di Callippo e di Aristotele," reprinted in his *Scritti sulla storia della astronomia antica* (Bologna, 1925-27), II, 3-112.

[24] Kepler understood Copernicus to accept the existence of solid spheres (*orbes solidi*; see *Joannis Kepleri astronomi opera omnia*, ed. Ch. Frisch, Frankfurt am Main, 1858-71, III, 181). But Frisch was undoubtedly right in his comment that Copernicus himself nowhere in his work either explicitly asserts or explicitly denies the reality of the spheres (*op. cit.*, III, 464).

they were simply a mathematical means of representing the planetary motions and a convenient geometrical basis for computing the apparent paths, or whether they really had a physical existence in space and like a piece of machinery produced the observed phenomena.[25] But whether the planets were carried by material balls or hoops or by imaginary spheres or circles through a medium of whatever type, the resultant computation of the actual planetary courses was the same. From Copernicus's language it sometimes appears that he regarded the planet as attached to a three-dimensional sphere; but more often a two-dimensional great circle of the sphere[26] was the geometrical figure to which he affixed the planet. For astronomical, as opposed to cosmological or astrophysical, theory it was a matter of indifference whether a planet was thought to be attached to a sphere or to a great circle thereof.

[25] For the history of this controversy in ancient and medieval astronomy see Pierre Duhem, Le Système du monde (Paris, 1913-17), Vol. II, Pt. I, chs. x-xi; Vol. IV, Pt. II, chs. viii-ix; et passim. Tycho Brahe rejected material spheres: "For it is now quite clear to me that there are no solid spheres in the heavens . . ." (Tychonis Brahe opera omnia, ed. Dreyer, III, 111.17-18). Nevertheless he retained imaginary spheres: "But there really are not any spheres in the heavens . . . and those which have been devised by the authors to save the appearances exist only in the imagination, for the purpose of permitting the mind to conceive the motion which the heavenly bodies trace in their course and, by the aid of geometry, to determine the motion numerically through the use of arithmetic . . . Of course, almost the whole of antiquity and also very many recent philosophers consider as certain and unquestionable the view that the heavens are made of a hard and impenetrable substance, that it is divided into various spheres, and that the heavenly bodies, attached to some of these spheres, revolve on account of the motion of these spheres. But this opinion does not correspond to the truth of the matter . . ." (op. cit., IV, 222.24-35). It will be observed that Kepler's approval of Brahe in this connection is limited to his denial of the existence of material spheres: "But if you remove the solid spheres, as Brahe rightly does . . ." (Kepleri opera omnia, ed. Frisch, III, 182; cf. VI, 312).

[26] Circulus enim bifariam secans sphaeram per centrum est sphaerae et maximus circumscribilium circulus (Th 17.21-22); "For a circle which cuts a sphere in half passes through the center of the sphere, and is the greatest circle that can be described [on the surface of the sphere]." It will be observed that the term for "great circle" is circulus maximus; cf. Definition No. 1 in Werner's De triangulis sphaericis: Circulus maximus in sphaera est, cuius planum sphaeram super ipsius centrum secat aequaliter, "The greatest circle on a sphere [or, A great circle on a sphere] is a circle whose plane cuts the sphere through its center into equal parts" (Abh. zur Gesch. d. math. Wiss., XXIV, 1, p. 1).

This repeated shift from sphere to circle and back again is, I believe, the root of some troublesome difficulties in Copernicus's terminology. Consider first the ambiguity of the word *orbis*, a term of central importance and frequent occurrence in his writings.[27] Now *orbis* may mean either a three-dimensional sphere or a two-dimensional circle;[28] and in fact it is constantly used by Copernicus in both senses, being interchangeable at times with *sphaera* and more frequently with *circulus*.

An example of equivalence between *orbis* and *sphaera* occurs in the first Assumption of the *Commentariolus: Omnium orbium caelestium sive sphaerarum unum centrum non esse;*[29] "There is no one center of all the celestial spheres [*orbium sive sphaerarum*]." For a second instance of interchangeability of *orbis* with *sphaera* we turn to the section of the *Commentariolus* entitled "The Order of the Spheres [*orbium*]": *Orbes caelestes hoc ordine sese complectuntur. Summus est stellarum fixarum immobilis et omnia continens et locans;*[30]

---

[27] The need for a careful investigation of the meanings of *orbis* in astronomical literature was evidently felt by Adolf Müller (ZE, XII, 365, n. 25). Although he made a beginning by setting down some of the necessary distinctions, he overlooked, for example, the use of *orbis* in the sense of universe, *mundus* (Th 5.30, 6.2; cf. Pliny *Natural History* ii.2.5).

[28] This ambiguity was referred to by Ludwik A. Birkenmajer (*Mikołaj Kopernik Wybór pism*, p. 3, n. 1). He observed that *orbis* meant both sphere (*kula* in Polish) and circle (*koło*). However, instead of undertaking a rigorous analysis of each occurrence of *orbis* to determine its precise meaning, Birkenmajer resorted to the dubious device of employing the Polish word *krąg*, which, like *orbis*, means both sphere and circle. The English word "orb" may have both senses (and certain others as well), but fortunately it no longer appears in astronomical literature.

[29] PII, 186.10-11. Prowe rendered (PI², 290) the first Assumption as follows: "Für alle Himmels-Körper und deren Bahnen giebt es nur einen Mittelpunkt" (There is only one center for all the heavenly bodies and their orbits). In addition to mistaking a negative proposition for an affirmative (thereby obscuring the point that the first Assumption is a rejection of the principle of concentricity), this mistranslation shows that Prowe completely failed to grasp the meaning of *orbis*. We need not be surprised that he did not attempt to carry his translation of the *Commentariolus* beyond the opening section (PI², 288-92).

[30] PII, 188.5-6. The corresponding section in *De rev.* furnishes clear proof of the interchangeability of *orbis* with *sphaera*. The title of the section is *De ordine caelestium orbium* (Th 25.12), "The Order of the Celestial Spheres"; but *ordo sphaerarum* is the phrase used in the body of the section (Th 28.30).

"The celestial spheres [*orbes*] are arranged in the following order. The highest is the immovable [sphere] of the fixed stars, which contains and gives position to all things." We may properly hold that for "the sphere of the fixed stars" Copernicus wrote here *stellarum fixarum orbis*,[31] while the expression he regularly employs (in the *De revolutionibus*) is *stellarum fixarum sphaera* or *non errantium stellarum sphaera*.[32] These phrases do not occur anywhere else in the *Commentariolus*; for that paper, devoted almost entirely to planetary theory, seldom refers to the sphere of the fixed stars, and on those occasions it uses *firmamentum*.[33] On the other hand, in the *Letter against Werner*, which is concerned exclusively with the fixed stars, their *sphaera* is mentioned twice.[34] It is of interest to note that, although enormous differences in the distances of stars were demonstrated centuries ago, present-day textbooks of elementary astronomy still retain, for the purposes of preliminary exposition, the concept of an imaginary "sphere" of the fixed stars, a concept adequate enough for ordinary astronomical work.

We have seen that at times *orbis* means a three-dimensional sphere, *sphaera*.[35] We shall next examine some passages in which *orbis* is equivalent to *circulus*, a two-dimensional circle.

Following the ancient and medieval tradition, Copernicus

[31] Cf. Rheticus's expression in the *Narratio prima . . . orbem stellarum, quem octavum vulgo appellamus . . .* (PII, 324.11), ". . . the sphere of the stars, which we commonly call the eighth sphere . . ."

[32] Th 19.23, 114.28, *et passim*.

[33] PII, 186, Assumptions 4 and 5; 189.4; 190.1, 20, 25, 26. *Firmamentum* is equated with *orbis stellarum* by Rheticus: *Primum locum infra firmamentum seu orbem stellarum . . .* (PII, 324.29).

[34] PII, 173.21: *sphaera stellarum fixarum*; 183.1: *non errantium stellarum sphaera*.

[35] An additional decisive example may be adduced from Rheticus: *Nam praeterquam quod nullus in vulgaribus hypothesibus finis effingendarum sphaerarum apparebat, orbes, quorum immensitas nullo sensu aut ratione percipi poterat, tardissimis et velocissimis circumducebantur motibus* (PII, 327.25-28); "For in the common hypotheses there appeared no end to the invention of spheres [*sphaerarum*]; moreover, spheres [*orbes*] of an immensity that could be grasped by neither sense nor reason were revolved with extremely slow and extremely rapid motions."

devises for each of the planets a set of geometrical figures, designed to account as accurately as possible for the observed movements of the planet or, in the familiar phrase, "to save the appearances."[36] These geometrical figures are regularly referred to in the *Commentariolus* as *orbes*; in fact *orbis* occurs there most frequently in two senses: (*a*) *orbis magnus*, the path of the earth's annual revolution about the sun;[37] (*b*) the deferent of the moon, Saturn, Jupiter, Mars, Venus, and Mercury.[38] But in the closing paragraph of the *Commentariolus*, where Copernicus summarizes the plan of the solar system elaborated in that paper, he refers to these same geometrical figures as *circuli*.[39]

A second example of equivalence between *orbis* and *circulus* is a sentence from the *Commentariolus* in which these two terms stand in juxtaposition as synonyms: *Quantitates tamen semidiametrorum orbium in circulorum ipsorum explanatione hic ponentur, e quibus mathematicae artis non ignarus facile percipiet quam optime numeris et observationibus talis circulorum compositio conveniat;*[40] "But in the explanation of the circles [*circulorum*] I shall set down here the lengths of the radii of the circles [*orbium*]; and from these the reader who is not unacquainted with mathematics will readily perceive how

---

[36] The traditional formula was *apparentias salvare*, σώζειν τὰ φαινόμενα (Th 7.4).

[37] *Terra triplici motu circumfertur, uno quidem in orbe magno, quo solem ambiens secundum signorum successionem anno revolvitur, temporibus aequalibus semper aequales arcus describens* (PII, 188.16-189.1); "The earth has three motions. First, it revolves annually in a great circle [*orbe magno*] about the sun in the order of the signs, always describing equal arcs in equal times." The expression "order of the signs" is explained below (p. 40).

[38] PII, 192.16-18, 195.2-6, 198.9-10, 200.20-22. For the term "deferent" see p. 36, below.

[39] *Sicque septem omnino circulis Mercurius currit, Venus quinque, tellus tribus et circa eam luna quattuor, Mars demum, Iuppiter et Saturnus singuli quinque. Sic igitur in universum 34 circuli sufficiunt, quibus tota mundi fabrica totaque siderum chorea explicata sit* (PII, 202.4-8); "Then Mercury runs on seven circles [*circulis*] in all; Venus on five; the earth on three, and round it the moon on four; finally Mars, Jupiter, and Saturn on five each. Altogether, therefore, thirty-four circles [*circuli*] suffice to explain the entire structure of the universe and the entire ballet of the planets." [40] PII, 187.8-11.

closely this arrangement of circles [*circulorum*] agrees with the numerical data and observations."

A striking illustration, drawn also from the *Commentariolus*, may serve as a third example: *Igitur centro terrae in superficie eclipticae semper manente, hoc est in circumferentia circuli magni orbis. . . .*[41] Literal translation is of course impossible here; for those unfamiliar with Latin I give the following rendering, necessarily awkward: "Therefore, while the center of the earth always remains in the plane of the ecliptic, that is, in the circumference of the *orbis magnus*, which is a circle. . . ."[42] We may conclude this discussion of the equivalence between *orbis* and *circulus* with Copernicus's definition of the nodes in his section on "The Superior Planets." The nodes are defined as . . . *sectiones circulorum orbis et eclipticae . . .*,[43] the intersections of two circles: (1) the deferent (*orbis*) of the planet, and (2) the ecliptic.

Although Copernicus wrenched astronomy loose from its geocentric past, his sentences abound in language that presupposes the earth to be in the center of the universe. The revolution in ideas did not at once precipitate a complete transformation of the terminology.[44] Consider, for example, his use of "ecliptic," the geocentric name for the sun's annual circuit of the heavens, a motion demonstrated by Copernicus to be not real but only apparent. In the place of "ecliptic," as we saw just above, Copernicus introduced *orbis magnus* as the heliocentric term for the path of the earth's real annual revolution about the sun.[45] So far as I know, no systematic examina-

[41] PII, 190.15-16.
[42] Although Copernicus uses *orbis magnus* with great frequency, it is significant that he occasionally writes instead *circulus magnus* (Th 380.12).
[43] PII, 197.8.
[44] Copernicus states that "Nobody need be surprised if the rising and setting of the sun and the stars and similar phenomena are still simply so designated by us; but the reader should recognize that we are speaking the familiar language, which can be retained by all" (Th 73.19-22).
[45] The term *orbis magnus* was adopted by Rheticus in the *Narratio prima* (PII, 317.8-15) and became part of the astronomical vocabulary of the sixteenth and seventeenth centuries. It was used, for example, by Digges, Kepler, Galileo, and Newton. Thomas Digges employed it in his *Perfit Description of the Caelestiall Orbes* (London, 1576), reprinted in *Huntington Library Bulletin*, No. 5 (April,

tion has yet been undertaken of the remains of geocentric language in the writings of Copernicus; such a study might well yield fruitful results.

The reader who is familiar with the theory of relativity will note that the distinction made in the preceding paragraph between the real motion of the earth and the apparent motion of the sun is part of the world-view of Copernicus and Newton[46] and is itself an example of terminological lag behind a

1934), p. 88. In his *Epitome astronomiae Copernicanae* Kepler gave the following question and answer: "*In the astronomy of Copernicus what is the* orbis magnus? This is the name applied by Copernicus to the true orbit [*orbita*] of the earth about the sun. This orbit is located in the space between the orbit of Mars outside it and the orbit of Venus within it; and he calls it "great" not on account of its size, since the circular orbits of the superior planets are much larger, but on account of its extraordinary usefulness in saving the apparent motions of not only the sun but also all the primary planets" (*Kepleri opera omnia*, ed. Frisch, VI, 431). In his *Dialogo sopra i due massimi sistemi del mondo* Galileo has Simplicio say on the second day: ". . . more than 2,529 miles an hour; for so great is the distance which the center of the earth in its annual motion traverses in an hour along the circumference of the *orbis magnus*" (*Opere*, ed. nazionale, Florence, 1890-1909, VII, 278.28-30; or *Opere*, ed. Timpanaro, Milan, 1936– , I, 339). In his note on this passage Giulio Dolci erroneously equates "orbe magno" with "equator" (*Galileo Galilei, I dialoghi su i massimi sistemi*, Milan, 1925, p. 126, n. 4). Newton also employed *orbis magnus*: *Assumendo radium orbis magni 1,000 et eccentricitatem terrae* $16^7/_8$ *. . . Sed eccentricitas terrae paulo maior esse videtur . . . Et sit DF ad DC ut dupla eccentricitas orbis magni ad distantiam mediocrem solis a terra . . . id est, ut* $33^7/_8$ *ad 1,000 . . .;* "Assume that the radius of the *orbis magnus* is 1,000, and the earth's eccentricity 16⅞ . . . But the earth's eccentricity seems to be somewhat greater . . . And let *DF* be to *DC* as twice the eccentricity of the *orbis magnus* to the sun's mean distance from the earth . . . that is, as 33⅞ to 1,000 . . ." (*Philosophiae naturalis principia mathematica*, Book III, Scholium to Prop. 35; ed. Horsley, London, 1779-85, Vol. III, 99.36-100.1; 100.3; 102.24-26, 28). With wider acceptance of the two ideas that the earth is one of the planets, and that the planetary paths are elliptical, the term *orbis magnus* tended to become obsolete; for in whatever sense *orbis* was used, it implied circularity.

[46] "Therefore the planets Saturn, Jupiter, Mars, Venus, and Mercury are not really retarded in their perigees, nor do they become really stationary, nor retrograde with a slow motion. All these phenomena are merely apparent; and the absolute motions, by which the planets continue to revolve in their orbits, are always direct and nearly uniform. These motions, as we have proved, are performed about the sun; and therefore the sun, as the center of the absolute motions, is at rest; i r the proposition that the earth is at rest must be completely denied . . ." (Newton, *De mundi systemate*, paragraph 27; ed. Horsley, III, 197.17-24).

revolutionary change in fundamental ideas. For according to the relativist, the only meaningful assertion that can be made in the present context is that one of these bodies, earth and sun, is moving with respect to the other. Which of them shall be declared at rest and which in motion? In other words, which motion shall be regarded as real and which as apparent? Such a question is decided for the relativist by considerations of mathematical simplicity and ease of calculation. To avoid confusion with regard to the historical development of ideas and to simplify the presentation of the subject matter, it is well to separate the Copernican-Newtonian statement of the case from the relativistic statement.

Let us return to the doctrine of the spheres. We have seen that Copernicus accepted them; and that at times he regarded them as three-dimensional figures (*orbis* or *sphaera*), but more frequently as two-dimensional circles (*orbis* or *circulus*). His indifference to the distinction even led him to write *concentricos circulos*[47] in his only reference to the system of concentric or homocentric spheres worked out by Eudoxus and Callippus. We may be sure that the presence of *circulos* in this passage was due to neither a slip of the pen on the part of the author nor an error on the part of the copyist. For in the *De revolutionibus* Copernicus writes: *Quorum causa alii nonam sphaeram, alii decimam excogitaverunt, quibus illa sic fieri arbitrati sunt, nec tamen poterant praestare quod pollicebantur. Iam quoque undecima sphaera in lucem prodire coeperat, quem circulorum numerum uti superfluum facile refutabimus in motu terrae;*[48] "Therefore

---

[47] PII, 185.6. Although Adolf Müller translated these words correctly by "conzentrische Sphären," his note on the passage shows that he had not completely mastered Copernicus's terminology. For he says that *circulos* is here used in opposition to *orbes*, ". . . *circulos*, welches hier im Gegensatz zu *orbes* gebraucht ist" (ZE, XII, 361, n. 8); the truth is that *circulos* is here used interchangeably with *orbes*. L. A. Birkenmajer was also confused. In his translation he rendered *concentricos circulos* by *kół wspólśrodkowych* (concentric circles); but in his note on the passage he referred three times to the concentric spheres, each time using *kręgów*, which means both sphere and circle (*Mikołaj Kopernik Wybór pism*, p. 4, n. 3).

[48] Th 158.13-17.

some writers devised a ninth sphere [*sphaeram*], and others a tenth, by which they thought to account for these phenomena; but they were unable to fulfill their promises. Then an eleventh sphere [*sphaera*] began to appear. This number of circles [*circulorum*] is excessive, as we shall easily demonstrate by using the motion of the earth." Again, in the Preface to the *De revolutionibus* he refers to the homocentric spheres as *circulis homocentris.*[49]

In the main, when Copernicus is discussing the details of planetary theory, he reserves *orbis* for the deferent. But when he is speaking more generally about the structure of the universe or the principles of astronomy, *orbis* regularly means sphere. In this connection it should be noted that the title of his major work is sometimes misunderstood: *De revolutionibus orbium caelestium* means "Concerning the Revolutions of the Heavenly Spheres," not "Planets."[50] In the Preface addressed to Pope Paul III Copernicus speaks of . . . *hisce meis libris quos de revolutionibus sphaerarum mundi scripsi,*[51] "the book which I have written about the revolutions of the heavenly spheres [*sphaerarum*]." The spheres may be denoted by either *sphaera* or *orbis* (or even *circulus*); but *orbis* does not mean "planet."

The distinction between *orbis* and *planeta*[52] may be seen

---

[49] Th 5.9.

[50] This error appears throughout in Menzzer, *Coppernicus, Über die Kreisbewegungen der Weltkörper*, the only complete translation since Baranowski's Polish rendering. Menzzer's otherwise excellent book is marred by a misunderstanding of the doctrine of the spheres. Henry Osborn Taylor, *Thought and Expression in the Sixteenth Century* (New York, 1920) renders the title "The Revolutions of the Celestial Bodies" (II, 335). Henry Hallam, *Introduction to the Literature of Europe in the Fifteenth, Sixteenth, and Seventeenth Centuries* (London, 1837-39) mentions ". . . the treatise of Copernicus on the revolutions of the heavenly bodies . . ." (I, 634). In his *Vie de Copernic* Flammarion gives the title as "Les Révolutions des corps célestes" (p. 238).

[51] Th 3.6-7.

[52] *Et cum sol suo semper et directo itinere proficiscatur, illi variis modis errant, modo in austrum modo in septemtrionem evagantes, unde planetae dicti sunt* (Th 14.31-15.2); "And while the sun always moves on its own direct path, they wander in various ways, turning sometimes to the south and sometimes to the north; for this reason they are called planets [*planetae*]."

very clearly in a passage of the *Narratio prima* where Rheticus is speaking of the sphere of the fixed stars, the so-called "eighth sphere": *ideo Deum tot eum orbem, nostra quippe causa, insignivisse globulis stellantibus, ut penes eos, loco nimirum fixos, aliorum orbium et planetarum contentorum animadverteremus positus ac motus;*[53] "Hence this sphere [*orbem*] was studded by God for our sake with a large number of twinkling stars, in order that by comparison with them, surely fixed in place, we might observe the positions and motions of the other enclosed spheres [*orbium*] and planets [*planetarum*]." As Rheticus continues, the distinction becomes utterly certain, beyond possibility of confusion or quibble. He writes: . . . *quo orbium quinque planetarum centra circa solem reperirentur;*[54] ". . . so that the centers of the spheres [*orbium*] of the five planets [*planetarum*] are located in the neighborhood of the sun."

For "planet" Copernicus employs, in the *Commentariolus*, usually *sidus*[55] and occasionally *corpus*.[56] An occurrence of the latter term in the *Narratio prima* may serve to illustrate how the spheres were conceived in relation to the planets: *Sed generalibus his praelibatis accedamus sane ad lationum circularium, quae competunt singulis orbibus et sibi adhaerentibus ac incumbentibus corporibus, enumerationem;*[57] "But now that we have touched on these general considerations, let us proceed to an exposition of the circular motions which are appropriate to the several spheres [*orbibus*] and to the bodies [*corporibus*] that cleave to and rest upon them."

In the *De revolutionibus* Copernicus uses, in addition to *planeta*, *sidus*, and *corpus*, other expressions for "planet," a few of which may be indicated here: *sidus vagans, sidus errans, stella, stella soluta, stella errans, errans, globus.*[58] Does *orbis*

---

[53] PII, 324.19-22.   [54] PII, 325.2-3.   [55] PII, 184.5; 185.8, 9; *et passim.*
[56] PII, 185.2, 194.1.   [57] PII, 329.19-21.
[58] A single instance of each will suffice: Th 109.5-6, 403.15, 415.3, 107.20, 307.5-6, 412.19, 108.30. Note also *erraticis* in the *Commentariolus* (PII, 187.2).

in his works ever mean "planet"? No certain example of such use has come to my attention.[59] If on occasion he employed the term in this sense, we should remember that he regarded the planets, together with all the other heavenly bodies, as spherical: *Principio advertendum nobis est globosum esse mundum, sive quod ipsa forma perfectissima sit omnium . . . sive etiam quod absolutae quaeque mundi partes, solem dico, lunam et stellas, tali forma conspiciantur . . . quo minus talem formam divinis corporibus attributam quisquam dubitaverit;*[60] "The first point for us to notice is that the universe is spherical, either because the sphere is the most perfect of all figures . . . or else because all the finite parts of the universe, I mean the sun, moon, and stars,[61] are seen to be of this shape . . . so that no one will question the attribution of this form to the divine bodies."

Let us now recapitulate the results of our inquiry. As we have seen, Copernicus accepted the doctrine of the spheres, ignoring the question whether they were imaginary or real. In referring to them he used the terms *sphaera, orbis,* and even *circulus,* for at times he regarded them as three-dimensional bodies, but more frequently as two-dimensional circles. When he dealt with planetary theory, he used *orbis* to mean the "great circle" in the case of the earth, and the deferent in the cases of the other planets. Seldom or never did he employ *orbis* in the sense of "planet"; his words for "planet" were chiefly *sidus, sidus errans, planeta, stella errans, errans,* and *corpus.*

[59] The meaning of *orbis* in Assumptions 3 and 6 of the *Commentariolus* is unclear; in the former case *orbis* may mean either "sphere" or "planet"; in the latter, either "(great) circle" or "globe."

[60] Th 11.8-16.

[61] I take it that *stella* here includes both the planets and the fixed stars. It is perhaps possible that only the planets are meant; if so, this use of *stella* resembles Rheticus's *ad stellarum imitationem,* "by analogy with the planets" (PII, 330.3-4). On one occasion, when there was no need to distinguish between the planets and the fixed stars, in a statement intended to cover both, Copernicus used *sidus* (Th 22.24).

THE TITLE OF THE *COMMENTARIOLUS* AND THE VIEWS
OF COPERNICUS CONCERNING THE NATURE
OF ASTRONOMICAL HYPOTHESES

The title of the *Commentariolus* is given in both our manuscripts as *Nicolai Copernici de hypothesibus motuum caelestium a se constitutis commentariolus*, "Nicholas Copernicus, Sketch of his Hypotheses for the Heavenly Motions." Maximilian Curtze, who first published the *Commentariolus*, called special attention to the title, remarking that in it Copernicus himself spoke of the hypotheses set up in his astronomical system.[62] This statement was attacked by Prowe, who denied the authenticity of the title on the ground that Copernicus would not have designated his work as a mere hypothesis.[63] While Prowe's contention has been widely echoed in the subsequent literature of the subject, it is unsound, for it rests upon a misapprehension of Copernicus's views concerning the nature of astronomical hypotheses.

This topic was discussed in an exchange of letters between Copernicus and Osiander, a prominent Lutheran preacher and theologian who was interested in the mathematical sciences. We are told that Copernicus wrote to Osiander on July 1, 1540, but unfortunately the text of his letter has been lost.[64] Osiander replied on April 20, 1541, urging Copernicus to present his astronomical system, not as a true picture of the universe, but rather as a device, true or false, for saving the phenomena. From inner conviction as well as for reasons of expediency Osiander wrote in part as follows: "I have always felt about hypotheses that they are not articles of faith but the basis of computation; so that even if they are false it does not matter, provided that they reproduce exactly the phenomena of the motions. For if we follow the hypotheses of Ptolemy, who will inform us whether the unequal motion of the sun occurs on

[62] MCV, I, 5 n.

[63] PI², 288 n; PII, 185 n.

[64] The source of our information about this correspondence is Kepler's incomplete essay, *Apologia Tychonis contra Ursum*; see *Kepleri opera omnia*, ed. Frisch, I, 245-46.

account of an epicycle or on account of the eccentricity, since either arrangement can explain the phenomena? It would therefore appear to be desirable for you to touch upon this matter somewhat in your Introduction. For in this way you would mollify the peripatetics and theologians, whose opposition you fear."[65]

On the same day Osiander addressed a letter to Rheticus, then living in Frauenburg with Copernicus. This second letter continued along the lines laid down in the first: "The peripatetics and theologians will be readily placated if they hear that there can be different hypotheses for the same apparent motion; that the present hypotheses are brought forward, not because they are in reality true, but because they regulate the computation of the apparent and combined motion as conveniently as may be; that it is possible for someone else to devise different hypotheses; that one man may conceive a suitable system, and another a more suitable, while both systems produce the same phenomena of motion; that each and every man is at liberty to devise more convenient hypotheses; and that if he succeeds, he is to be congratulated. In this way they will be diverted from stern defense and attracted by the charm of inquiry; first their antagonism will disappear, then they will seek the truth in vain by their own devices, and go over to the opinion of the author."[66]

Although we do not have Copernicus's letter, it is evident that he did not share his correspondent's opinions. As Kepler put it, "Strengthened by a stoical firmness of mind, Copernicus believed that he should publish his convictions openly, even though the science should be damaged."[67] Despite this disagreement Osiander's views appeared in the Preface to the first edition of the De revolutionibus (1543). For a long time it was not known that Osiander, entrusted with the supervision of the printing of the great work, suppressed the Introduction

[65] The Latin text of the extant portion of this letter may be found in Kepler's Opera, ed. Frisch, I, 246; German translation in PI², 521-22.
[66] Text in Kepler's Opera, ed. Frisch, I, 246; German translation in PI², 523.
[67] Opera, ed. Frisch, I, 246.

written by Copernicus and replaced it by an unsigned Preface of his own composition.[68] The views expressed therein concerning the nature of astronomical hypotheses conflicted with the fundamental principles of the book.

Osiander's Preface, printed on the verso of the title page, read as follows:[69]

TO THE READER

CONCERNING THE HYPOTHESES OF THIS WORK

Since the novelty of the hypotheses of this work has already been widely reported, I have no doubt that some learned men have taken serious offense because the book declares that the earth moves, and that the sun is at rest in the center of the universe; these men undoubtedly believe that the liberal arts, established long ago upon a correct basis, should not be thrown into confusion. But if they are willing to examine the matter closely, they will find that the author of this work has done nothing blameworthy. For it is the duty of an astronomer to compose the history of the celestial motions through careful and skillful observation. Then turning to the causes of these motions or hypotheses about them, he must conceive and devise, since he cannot in any way attain to the true causes, such hypotheses as, being assumed, enable the motions to be calculated correctly from the principles of geometry, for the future as well as for the past. The present author has performed both

---

[68] It was Kepler who identified the anonymous author: "It is a most absurd fiction, I admit, that the phenomena of nature can be demonstrated by false causes. But this fiction is not in Copernicus. He thought that his hypotheses were true, no less than did those ancient astronomers of whom you speak. And he did not merely think so, but he proves that they are true. As evidence, I offer this work.

Do you wish to know the author of this fiction, which stirs you to such great wrath? Andreas Osiander is named in my copy, in the handwriting of Jerome Schreiber, of Nuremberg. Andreas, who supervised the printing of Copernicus's work, regarded the Preface, which you declare to be most absurd, as most prudent (as can be inferred from his letter to Copernicus) and placed it on the title page of the book when Copernicus was either already dead or certainly unaware [of what Osiander was doing]. Therefore Copernicus is not composing a myth but is giving earnest expression to paradoxes, that is, is philosophizing, which is what you required in an astronomer" (Kepler's *Opera*, ed. Frisch, III, 136). For an account of the incident see PI², 520-39, and Kepler, *Neue Astronomie*, tr. by Max Caspar (Munich, Berlin, 1929), p. 399.

[69] Th 1.1-2.9; German translation in PI², 526-28; a condensed paraphrase in Dreyer's *Planetary Systems*, pp. 319-20. For the position of Osiander's Preface in the first edition see Hipler's *Spicilegium Copernicanum*, p. 108; PI², 542.

these duties excellently. For these hypotheses need not be true nor even probable; if they provide a calculus consistent with the observations, that alone is sufficient. Perhaps there is someone who is so ignorant of geometry and optics that he regards the epicycle of Venus as probable, or thinks that it is the reason why Venus sometimes precedes and sometimes follows the sun by forty degrees and even more. Is there anyone who is not aware that from this assumption it necessarily follows that the diameter of the planet in the perigee should appear more than four times, and the body of the planet more than sixteen times, as great as in the apogee, a result contradicted by the experience of every age? In this study there are other no less important absurdities, which there is no need to set forth at the moment. For it is quite clear that the causes of the apparent unequal motions are completely and simply unknown to this art. And if any causes are devised by the imagination, as indeed very many are, they are not put forward to convince anyone that they are true, but merely to provide a correct basis for calculation. Now when from time to time there are offered for one and the same motion different hypotheses (as eccentricity and an epicycle for the sun's motion), the astronomer will accept above all others the one which is the easiest to grasp. The philosopher will perhaps rather seek the semblance of the truth. But neither of them will understand or state anything certain, unless it has been divinely revealed to him. Let us therefore permit these new hypotheses to become known together with the ancient hypotheses, which are no more probable; let us do so especially because the new hypotheses are admirable and also simple, and bring with them a huge treasure of very skillful observations. So far as hypotheses are concerned, let no one expect anything certain from astronomy, which cannot furnish it, lest he accept as the truth ideas conceived for another purpose, and depart from this study a greater fool than when he entered it. Farewell.

For the purposes of our present inquiry Osiander's contentions may be restated as follows: Since divine revelation is the only source of truth, astronomical hypotheses are not concerned therewith, and serve only as a basis of calculations.

Copernicus dissented from this view, as we have seen. For him the doctrine of the earth's motion, which he properly regarded as the principal innovation of his system, was true; the motion of the earth was a physical reality. In the *De revolutionibus* he stated:

If therefore any motion is attributed to the earth, there will appear in all the bodies outside the earth a motion of equal velocity, but in the opposite direction, as though these bodies were moving past the earth. Among such motions the daily rotation is of first importance. For it appears to affect the entire universe except the earth and the elements near it. But if it is granted that the firmament has no part in this motion and that the earth rotates from west to east, upon earnest examination it will be found that this is indeed the case, as far as the apparent rising and setting of the sun, moon, and stars are concerned.[70]

And further on he asked:

Why then do we still hesitate to assign to the earth the motion consistent by nature with its figure, in preference to accepting a rotation of the entire universe, the bounds of which are unknown and unknowable; and why do we not declare that the appearance of the daily rotation is in the firmament, but the reality in the earth?[71]

Copernicus buttressed this position by refuting at some length the traditional objections to the earth's motion.[72]

But with regard to the relation between science and religion he proceeded with the utmost caution. He believed that human reason can attain to truth and that the philosopher must seek the truth. However, he was careful not to contradict explicitly those who, like Osiander, asserted that divine revelation was the only source of truth; and he emphasized the limitations of human reason and the necessity of divine assistance: "The philosopher endeavors in all matters to seek the truth, to the extent permitted to human reason by God";[73] "With the favor of God, without whom we can accomplish nothing, I shall attempt to press further the inquiry into these questions."[74]

To fortify his scientific work against religious attacks, Copernicus, a cleric and doubtless a devout Catholic, gave the first place in the De revolutionibus to a laudatory letter from Cardinal Nicholas Schönberg.[75] Next came the Preface, dedicating

---

[70] Th 16.10-18.
[71] Th 22.3-6. It is repeatedly affirmed in the *Commentariolus* that the real motion of the earth produces apparent motions in the sun, moon, planets, and fixed stars (PII, 186-87, Assumptions 5, 6, 7; 187.12-188.3; 189.3-190.28; 196.10-197.3; 197.27-31; 198.22-199.8).   [72] *De rev.* Bk. I, chs. vii-viii.
[73] Th 3.11-12.   [74] Th 10.32-11.1.   [75] Th 2.10-32.

the volume to Pope Paul III.[76] Here Copernicus explained that it was only the importunities of his friends, among others Cardinal Schönberg and Tiedemann Giese, bishop of Kulm, that overcame his reluctance to publish the book.[77] Among these friends he avoided naming Rheticus, who was a Lutheran; and yet we know how highly Giese estimated Rheticus's share in helping to bring out the *De revolutionibus*.[78] The combination of bishop, cardinal, and pope was intended to provide a stout bulwark against Roman Catholic assaults.

But Copernicus did not rely for defense merely on the approval of prelates. He indicated that his doctrines did not contradict Scripture;[79] and Rheticus wrote a pamphlet, no

---

[76] Preface, Th 3.1-8.7; dedication, 7.8-15.

[77] "For a long time I debated with myself whether I should publish the book which I had written to prove the motion of the earth ... While then I pondered on these matters, the scorn which I had reason to fear on account of the novelty and absurdity of my opinion, had almost persuaded me to abandon completely the work which I had begun. But while I long hesitated and even resisted, my friends drew me back. Among them the foremost was Cardinal Nicholas Schönberg, of Capua, distinguished in every field of learning. Next to him was Tiedemann Giese, bishop of Kulm, a man who loved me dearly, a close student of sacred literature, as he is of all good literature. He frequently urged upon me and, sometimes adding reproaches, demanded that I should publish and finally permit to appear the book which, hidden by me, had lain concealed not merely nine years but already four times that period. The same conduct was recommended to me by not a few other very eminent and learned men. They exhorted me no longer to refuse, on account of the fear which I had conceived, to apply my work to the general advantage of students of mathematics. To the degree, they declared, that my doctrine of the earth's motion now appeared absurd to most men, to the same degree would it gain admiration and thanks after men saw, through the publication of my writings, the fog of absurdity dispelled by most lucid proofs. Influenced therefore by the persuasion of these men and by this hope, I finally permitted my friends, who had addressed the request to me for a long time, to bring out an edition of the book" (Th 3.17-18, 4.4-24).

[78] In a letter to Rheticus, dated July 26, 1543, Giese wrote: "For I know of no one more suitable or more eager to lay this matter before the council of Nuremberg than you, who played a prominent role while the drama was enacted, so that now the author's concern is evidently no greater than your own in restoring what has been dropped from the authentic version" (Latin text in PII, 420.9-12; German translation in PI², 539).

[79] "If perhaps there are babblers who, although completely ignorant of mathematics, nevertheless take it upon themselves to pass judgment on mathematical questions and, improperly distorting some passage of Scripture to their purpose,

longer extant, reconciling the motion of the earth with Holy Writ.[80] Even so distinguished a Father of the Church as Lactantius, Copernicus pointed out, could go sadly astray in matters scientific.[81] The immobility of the earth was an idea emanating from antiquity, to be sure, but not from Biblical antiquity;[82] and there was ample ancient support for the contrary belief in the earth's motion.[83] Why should he not enjoy the freedom of inquiry permitted to his predecessors?[84] It was, after all, not any desire to refute Scripture or revered church doctrine or tradition or common sense,[85] but the sheer perplexities of the professional astronomer that impelled Copernicus to try the earth's motion.[86] He tried it; it worked; and he believed it to be real.

Then Copernicus would hold as against Osiander that astronomy deals with reality. If its fundamental propositions are verified by the empirical evidence—the observations and resultant calculations—they are true; if they conflict with the empirical evidence, they are false. Some such position is implicit in Copernicus's criticism of the procedure adopted by his predecessors: "Therefore in the process of demonstration, which is called 'method,' they are found either to have omitted something essential, or to have admitted something extraneous and wholly irrelevant. This would not have happened to them, had they followed sure principles [*principia*]. For if the hypotheses [*hypotheses*] assumed by them were not false, everything which follows from their hypotheses would be verified beyond any doubt."[87]

---

dare to find fault with my system and censure it, I disregard them even to the extent of despising their judgment as uninformed" (Th 7.16-20).

[80] In his letter to Rheticus, Giese expressed a desire to see this pamphlet printed with *De rev.*: "I should like bound with it your little work, in which you have very properly defended the earth's motion from the charge of being in conflict with the Sacred Scriptures" (PII, 420.25-27).

[81] "For it is not unknown that Lactantius, otherwise an illustrious writer, but no mathematician, speaks quite childishly about the earth's form, when he mocks those who have stated that the earth has the form of a sphere" (Th 7.21-23).

[82] It was a "received opinion of the mathematicians" (Th 4.29-30), "confirmed by the judgment of many centuries" (Th 3.15).

[83] Th 6.2-14, 16.20-24, 17.6-10.

[84] Th 6.15-21.   [85] Th 4.30.   [86] Th 4.31-6.2.   [87] Th 5.22-27.

Such fundamental propositions are termed by Copernicus *principium, assumptio,* and *hypothesis* without any distinction: "Furthermore astronomy, that divine rather than human science, which inquires into the loftiest things, is not free from difficulties. Especially with regard to its principles [*principia*] and assumptions [*assumptiones*], which the Greeks call 'hypotheses' [*hypotheses*] . . ."[88]

Before these principles, assumptions, or hypotheses can be accepted as true, they must meet two requirements. First, they must save the appearances (*apparentias salvare*): the results deduced from them must agree with the observed phenomena within satisfactory limits of error.[89] Secondly, they must be consistent with certain preconceptions, called "axioms of physics," such as that every celestial motion is circular, every celestial motion is uniform, and so forth.[90] Disagreement with the observations is no more grave a defect than departure from the axiom of uniform motion: *apparentias salvare* and *aequalitatem tueri*[91] are equally essential. Thus in the *Commentariolus* Copernicus concedes that the Ptolemaic system was consistent with the numerical data; but he rejects it on the ground that it violated the axiom of uniformity.[92]

[88] Th 10.13-16. For additional examples of the equivalence of these three terms, the student should examine Th 5.7, 25-26; 36. lines 9-14 of the notes; 109.4-5.

[89] This first requirement is expressed in several ways: *apparentiis consentire* (Th 275.5-8); *apparentiis congruere* (Th 281.29-30); *apparentiis sufficere* (Th 327.18-19). In the *Narratio prima* Rheticus also employs *apparentiis satisfacere* (PII, 316.25-26).

[90] PII, 186.1-4; Th 14.12-14

[91] In *De rev.* "to observe the axiom of uniform motion" is expressed by *aequalitatem tueri* (Th 233.30). In the *Commentariolus* Copernicus uses the variant *aequalitatem motuum servare* (PII, 187.5-6).

[92] PII, 185.13-20; see also Th 5.13-16. His objection is not that Ptolemy was unable "to attribute uniform velocity to the planetary motions," as E. A. Burtt stated in *The Metaphysical Foundations of Modern Physical Science* (revised edition; London, New York, 1932), p. 27; the motions were all uniform, but some of them were arranged by Ptolemy to be uniform with reference to some point other than their own center. What Copernicus desired was not merely a simpler system, as Burtt thought, but a more reasonable one (*rationabilior*, PII, 186.2; Ptolemy's system is adjudged "not sufficiently absolute nor sufficiently pleasing to the mind," PII, 185.18-20). Of highest importance in this connection is the passage in *De rev.* in which Copernicus sets aside the lunar theory of the

The doctrine of the earth's motion was, according to Copernicus, a fundamental proposition that satisfied the double requirement; consequently he called his most important innovation a hypothesis:

For this reason some thinkers believed that the sphere of the fixed stars also moves, and hence they adopted a surmounting ninth sphere. This ninth sphere having proved insufficient, the moderns now add a tenth, and even so do not attain their end. By the motion of the earth we hope to attain the goal. For we shall use the earth's motion as a principle [*principio*] and hypothesis [*hypothesi*] in demonstrating the other motions.[93]

The statement that Copernicus never called the earth's motion a hypothesis must be set aside as false.[94]

He likewise referred to the other principal tenets of his astronomy as hypotheses. Thus he regarded the immobility of the fixed stars as a hypothesis: ". . . among our principles and hypotheses we have assumed that the sphere of the fixed stars is altogether immovable."[95] He described as a hypothesis his explanation of the precession of the equinoxes, and his account of the change in the obliquity of the ecliptic.[96] He used

ancients (Th 233.7-234.12); see also in the *Narratio prima* Rheticus's fourth reason for abandoning the hypotheses of the ancient astronomers (PII, 318.27-31).

[93] Th 34.14-19. Lest this be considered an isolated example, I add other passages of *De rev.* in which Copernicus called the earth's motion a hypothesis: *Cum igitur mobilitati terrenae tot tantaque errantium siderum consentiant testimonia, iam ipsum motum in summa exponemus, quatenus apparentia per ipsum tamquam hypothesim demonstrentur* (Th 31.2-4); . . . *ex hypothesi motus terrae* . . . (Th 163.2); . . . *nostrae hypothesi mobilitatis terrae* . . . (Th 345.20-21); . . . *per hanc hypothesim mobilitatis terrae* . . . (Th 357.12); . . . *nostrae hypothesi mobilitatis terrenae* . . . (Th 365.5-6). Similarly in the *Narratio prima* Rheticus spoke *de hypothesibus motuum terreni globi* (PII, 329.22).

[94] This error was made by Dreyer, who said: ". . . to Copernicus the motion of the earth was a physical reality and not a mere working hypothesis. Not to speak of the fact that he nowhere in his work calls it a hypothesis . . ." (*Planetary Systems*, p. 320).

[95] . . . *inter principia et hypotheses assumpserimus non errantium stellarum sphaeram omnino immobilem esse* . . . (Th 109.4-5). Menzzer (p. 90) grotesquely mistranslated *inter principia et hypotheses* as "in the struggle of principles against hypotheses," (in dem Streite der Prinzipien gegen die Hypothesen).

[96] Th 162.28-29.

the term hypothesis to denote his lunar theory and also his account of the motions of the planets in longitude and in latitude.[97]

Similarly in the *Letter against Werner* Copernicus designated as a hypothesis Werner's reconstruction of the entire movement of the fixed stars from the earliest observations to his own time;[98] and also Werner's contention that the fixed stars had a perfectly uniform motion during the four centuries before Ptolemy.[99]

Rheticus's usage in the *Narratio prima*, written during his stay with Copernicus, confirms our conclusion that the master termed his basic ideas "hypotheses." What is called nowadays "the Copernican revolution"—essentially the shift from geocentrism to heliocentrism, with the corollary that the earth is one of the planets—was regularly referred to by Rheticus as a revision of the hypotheses, *renovare hypotheses*;[100] the Copernican system as the new hypotheses, *novae hypotheses*;[101] the ancient astronomical systems as the hypotheses of the ancient astronomers, *veterum astronomorum hypotheses*;[102] and the accepted astronomy of his own age as the common hypotheses, *communes hypotheses*[103] or *vulgares hypotheses*.[104]

We have seen that Copernicus used the term "hypotheses" for the basic ideas of his system, and that these hypotheses must meet the twofold requirement of agreement with the observational data and consistency with the presuppositions or "axioms." Hence we may confidently reject Prowe's argument against the authenticity of the title of the *Commentariolus*, as given in both our manuscripts; for the argument rests solely on the mistaken assertion that Copernicus would not have presented his system as a "mere hypothesis."[105]

[97] Th 275.6-7, 327.18-19, 415.1.

[98] This is the meaning, I take it, of *hypothesis* in PII, 178.21 and 179.8-9.

[99] . . . *in ipsa eius hypothesi in qua existimat CCCC annis ante Ptolemaeum aequali tantummodo motu non errantia sidera mutata fuisse* (PII, 175.26-28).

[100] PII, 321.27.   [101] PII, 321.24.   [102] PII, 317.19-20.

[103] PII, 314.8.   [104] PII, 333.12.

[105] "The title of the *Commentariolus*, as formulated in the two MSS known at present, cannot have originated with Copernicus himself. He would not have designated his conception of the system of the universe as a mere hypothesis"

Unfortunately Prowe's error has infected the later literature of the subject. Thus in the introduction to his translation of the *Commentariolus* into German Adolf Müller stated:

Nothing lay further from Copernicus's intention than to recommend his system as a mere computing hypothesis. He even avoids with evident care the term *hypothesis*. Hence we may well conclude that the title of the *Commentariolus*, in the form which we have placed at the head of this article, does not come from his pen.[106]

Ludwik A. Birkenmajer, we are told by his son,[107] likewise regarded the title as not authentic. We noted just above Dreyer's assertion that Copernicus nowhere called the earth's motion a hypothesis. In his *History of Modern Culture* Preserved Smith wrote: "Copernicus never spoke of his hypotheses."[108] A refinement was introduced by Aleksander Birkenmajer, who contended that in the *Commentariolus* Copernicus attributed to his cosmological ideas only the value of a working hypothesis, but later acquired a firm conviction of their truth.[109] Needless to say, there is no distinction in this respect between the *Commentariolus* and the *De revolutionibus*, for the conviction prevails equally in both works that the fundamental ideas advanced in them are true. If there is any difference at all on this head between the earlier and later writings of Copernicus, it is that the term "hypothesis" does not occur in the body

(PII, 185 n). Prowe's blunder is áll the more surprising, because he himself quoted the following statement by Kepler about Copernicus: "He thought that his hypotheses were true . . . And he did not merely think so, but he proves that they are true" (PI², 532 n; see above, p. 24, n. 68). Note also *vera hypothesi* in Rheticus (PII, 373.15).

[106] ZE, XII, 360-61. Müller took the same position in his *Nikolaus Copernicus, der Altmeister der neuern Astronomie* (*Stimmen aus Maria-Laach*, Ergänzungsheft LXXII, 80, n. 1).

[107] Aleksander Birkenmajer, in the article "Le Premier Système héliocentrique imaginé par Nicolas Copernic," p. 3.

[108] I, 22 n (New York, 1930).

[109] "Le Premier Système héliocentrique imaginé par Nicolas Copernic," pp. 3-4. In the writings of Copernicus *hypothesis* did not connote an uncertain supposition, as it does in popular language; it retained the traditional meaning of a fundamental proposition, basic to a theory. His term for a tentative suggestion was *coniectura* (PII, 178.6-7).

of the *Commentariolus,* but, as we have seen, is frequently used in the *De revolutionibus.*

On the other hand Pierre Duhem recognized that Copernicus regarded his fundamental propositions as hypotheses; but he blamed Copernicus for believing in the truth of his hypotheses, for being, in the learned Frenchman's terminology, a realist.[110] According to Duhem, only those ideas about the physical universe are true which can be shown to admit, now and for all future time, no possible alternative.[111] Of course proof of this sort is unobtainable. Hence it follows, according to Duhem's view, in explicit agreement with Osiander's, that truth is unattainable in the natural sciences, and that the hypotheses of physics are only mathematical devices intended to save the phenomena.[112] Does it also follow, it may be asked, that since the way to truth is closed to science, the way to Truth is open to some other human activity? While this is not the place to discuss the role of hypotheses in science, it should be indicated that Duhem's view is not without alternative. We are not limited to the choice offered by Duhem between realism and fictionalism:[113] any proposition or hypothesis is either the Ultimate Truth or a mere fiction. We may properly accept a hypothesis as the best statement at the moment and be ready to revise or to reject it when fresh empirical data require a modification of it, or a rival and superior hypothesis emerges to replace it.

[110] "Essai sur la notion de théorie physique" (*Annales de philosophie chrétienne,* 79ᵉ année, t. 156, pp. 373-75).

[111] "To prove that an astronomical hypothesis conforms to the nature of things, it is necessary to prove not only that the hypothesis is sufficient to save the phenomena, but also that these phenomena could not be saved if the hypothesis were abandoned or modified" (*ibid.,* pp. 374-75). "If the hypotheses of Copernicus succeed in saving all the known appearances, the conclusion will be that these hypotheses may be true, not that they are certainly true. To make the latter conclusion valid it would be necessary first to prove that no other combination of hypotheses could be devised which permitted the appearances to be saved equally well; and this demonstration has never been given" (*ibid.,* pp. 584-85).

[112] *Ibid.,* pp. 588, 591-92.

[113] The rival views are named "realism" and "formalism" by Augustin Sesmat in his *Systèmes de référence et mouvements (physique classique), II: L'Ancienne Astronomie d'Eudoxe à Descartes* (Paris, 1937), 105-30 (II, 29-54).

DEFERENT AND EPICYCLE, ECCENTRIC AND EQUANT

In the present section I shall attempt to set forth just so much of the elements of Copernicus's astronomy as is essential to an understanding of our three treatises.

The system of concentric or homocentric spheres,[114] devised by Eudoxus, improved by Callippus, and incorporated by Aristotle into his philosophy, was unable to answer the fatal objection that it represented the planets as moving at immutable distances from the earth, whereas observation indisputably revealed variations in the distances.[115] Consequently the principle of concentricity was abandoned, and recourse was had to the eccentric circle (*eccentricus, eccentrus*).

Draw a circle with center at C (Fig. 1). Let any point E, within the circle but outside the center, represent the earth.[116]

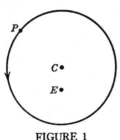

FIGURE 1

Let any point P on the circumference of the circle represent the planet. Finally let the circle revolve uniformly about its center C.

It is obvious that on this arrangement the distance between the planet P and the earth E varies. This distance is greatest when PCE form a straight line, with C between P and E (Fig. 2); in this juncture the planet was said to be in its apogee (*apogium, apogaeon, apogaeum*) and opposite the earth. When CEP form a straight line, with E and P on the same side of C (Fig. 3), the distance between earth and planet is least, and the planet was said to be in its perigee (*perigium, perigaeon, perigaeum*).[117] The apogee was also called the higher apse

---

[114] They received this name because they shared a common center; "concentric" is derived from the Latin, "homocentric" from the Greek, and both mean "having the same center." The word "homocentric" has nothing whatever to do with the Latin word for man, *homo*, as Benjamin Ginzburg thought (*The Adventure of Science*, p. 53).

[115] PII, 185.6-11; Th 5.10-13.

[116] The distance CE (Fig. 1) was called the eccentricity, *eccentricitas* (PII, 305.6) or *eccentrotes* (Th 219.29).

[117] In MSS and early printed books *aux* was frequently written for "apogee," and *oppositum augis* for "perigee."

(*summa absis*), and the perigee, the lower apse (*infima absis*, *ima absis*).[118] The line connecting the apogee with the perigee was called the line of apsides or apse-line (*linea absidum*).[119]

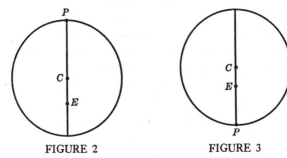

FIGURE 2   FIGURE 3

This arrangement of an eccentric circle revolving about a fixed center was adequate for certain purposes. But to accommodate some portions of astronomical theory to the observed patterns of celestial motion, it was necessary to provide the revolving eccentric with a moving center.

Let there be an eccentric circle revolving uniformly about its center C, while C revolves on the circumference of a smaller circle ABC (Fig. 4). Then the point S will not describe a circle, but will trace a nearly circular path[120] determined by (*a*) the relative velocities of the outer and inner circles; (*b*) the relative lengths of the radii of the two circles; and (*c*) the direction of their motions, for they

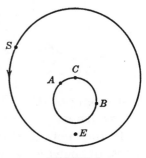

FIGURE 4

may turn in the same direction or in opposite directions. If ABC is itself an eccentric circle, we have the case of an eccentric

---

[118] "Let the apogee, which is called *summa absis* in Latin, be at A, the point most remote from the center of the universe; and let the perigee, which is called *infima absis*, be at D, the point nearest to the center of the universe" (Th 204.10-13); cf. PII, 306.28-33; for an example of *ima absis* see PII, 333.25. When *absis* was used alone without any qualifying adjective, it designated the apogee.

[119] PII, 349.33.

[120] . . . *non describit circulum perfectum, sed quasi* . . . (Th 326.31).

on an eccentric (*eccentricus eccentrici, eccentreccentricus, eccentri eccentrus*).[121]

When Copernicus wrote the *Commentariolus,* he accepted the Ptolemaic doctrine of the fixity of the solar apogee.[122] Hence in that earlier work he conceived the apparent annual revolution of the sun (or real annual revolution of the earth) as an eccentric revolving about a fixed center.[123] Having subsequently learned that the sun's apogee is not fixed but moves,[124] in the *De revolutionibus* he employed for the annual revolution an eccentric on an eccentric;[125] and this arrangement accordingly appeared in Rheticus's *Narratio prima.*[126]

An alternative device for producing a nearly circular orbit was the deferent (*deferens*) and epicycle (*epicyclus*). Let a

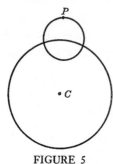

FIGURE 5

circle revolve uniformly about its center C (Fig. 5). Choose any point on the circumference, and about it describe a smaller circle. Let any point P on the circumference of the smaller circle represent the planet. As the larger circle, the deferent, revolves about C, let the smaller circle, the epicycle,[127] revolve uniformly about its own moving center. Here again in this combination of an epicycle with a homocentric deferent, it was possible to adjust the relative lengths of the radii of the two circles, the relative rates of their rotation, and also the direction of their motion.

In the effort to obtain closer agreement with the observations, the deferent was sometimes made eccentric; or a second

---

[121] PII, 317.8, 319.17; Th 218.3. The outer circle was also called "second eccentric" and "movable eccentric": . . . *qui hic eccentricus eccentrici, eccentricus secundus et mobilis vocabitur* . . . (PII, 349.31-32).

[122] PII, 189.12-190.2.

[123] PII, 188.16-189.3.

[124] Th 209.26-210.1, 216.3-5, 217.1-11.

[125] Th 217.15-218.4.

[126] PII, 304.12-16.

[127] The diminutive *epicyclium* was used if it was desired to emphasize the smallness of an epicycle as compared with its deferent (Th 325.20).

epicycle was introduced.[128] In the latter case, the celestial body was located on the circumference of the second epicycle, which revolved uniformly about a moving center (Fig. 6). This center was situated on the circumference of the first epicycle, which in turn revolved uniformly about a moving center. This center was a point on the circumference of the deferent, which revolved uniformly on its own center. An example of the use of two epicycles with a homocentric deferent may be seen in Copernicus's lunar theory.[129] He employed the same device in the *Commentariolus* for planetary theory; [130] but in the *De revolutionibus* he dropped this scheme in favor of a single epicycle moving on an eccentric deferent.[131]

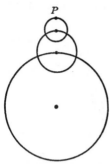

FIGURE 6

It had been established in Greek antiquity that the eccentric and epicyclic arrangements were equivalent and yielded the same results;[132] and this equivalence was perfectly familiar to Copernicus.[133] To take the simplest case, let there be a concentric deferent with radius CQ, and an epicycle with radius PQ (Fig. 7). Now draw a circle about C with radius AC equal and parallel to PQ; and describe an eccentric circle with center A and radius AP=CQ. Then if A revolves on the circumference of the circle AC, the eccentric circle AP has a movable center in A. In Figure 7 the system of concentric deferent with epicycle has been represented by unbroken lines, and the sys-

---

[128] One term for an epicycle moving on an eccentric deferent was *eccentrepicyclus* (Th 219.11); and for a second epicycle moving on a first epicycle, *epicyclepicyclus* (Th 218.29).

[129] PII, 192.15-194.10, 315.29-316.23; Th 235.14-236.8.

[130] PII, 195.6-8, 198.7-13, 200.20-21.

[131] Th 325.16-326.31; the reason for the substitution is given in Th 327.13-16.

[132] See for example HI, 216.18-217.6.

[133] "From all these considerations it is clear that the same inequality of the appearances is always produced, whether by an epicycle on a homocentric deferent or by an eccentric circle equal to the homocentric; and that these devices do not differ from each other at all, provided that the distance between the centers [AC in Fig. 7] is equal to the radius of the epicycle" (Th 207.2-8).

tem of eccentric circle with movable center by dotted lines.
Let corresponding circles turn with equal velocities in the
same sense; that is, let the motion of P on the epicycle equal
in velocity and direction that
of the moving center A, and
let the motion of the deferent
equal that of the eccentric.
Identical results will follow
from either system; that is,
whether we adopt the eccentric or the epicyclic device, the
point P will occupy the same
position at any given moment.

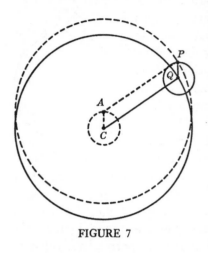

FIGURE 7

In the use of eccentrics
there was one traditional practice[134] which Copernicus regarded as improper, on the
ground that it violated the
principle of uniformity. Let P describe an eccentric circle with
fixed center C (Fig. 8). Let E represent the earth. On the
apse-line measure CQ=EC. As P revolves on the circumference of the eccentric, let ∠ PQR in-
crease uniformly, that is, increase equally
in equal intervals of time. Then, although P will always remain at an equal
distance from the center C, the point
with reference to which its motion is
uniform will be the equant Q, not the
center C.

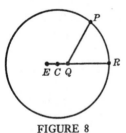

FIGURE 8

Despite its usefulness in representing
the observed phenomena, Copernicus rejected the equant; for
in his view the "rule of absolute motion"[135] required every
circle to move not only uniformly but also uniformly with respect to its own center. As Rheticus put it, it was "an essential
property of circular motion that all the circles in the universe

[134] See for an example HI, 298.11-299.23.
[135] *ratio absoluti motus* (PII, 186.4).

should revolve uniformly and regularly about their own centers, and not about other centers."[136] Copernicus objected to the use of the equant in the lunar theory of the ancients: "For when they assert that the motion of the center of the epicycle is uniform with respect to the center of the earth, they must also admit that the motion is not uniform on the eccentric circle which it describes . . . But if you say that the motion uniform with respect to the center of the earth satisfies the rule of uniformity, what sort of uniformity will this be which holds true for a circle on which the motion does not occur, since it occurs on the eccentric?"[137] Similarly in criticizing the planetary theory of the ancients, he said: "Therefore they admit, in the present case also, that the motion of a circle can be uniform with respect to some center other than its own."[138]

Having completed this brief exposition of deferent and epicycle, eccentric and equant, let us turn to other aspects of Copernicus's astronomy. He regarded the universe as a huge sphere; imbedded in its surface were the fixed stars, all at equal distances from the center of the universe. Did Copernicus believe the universe to be finite or infinite?[139] He discussed the question[140] and, aware that either position involved logical difficulties, decided that the subject was one proper not to astronomy but rather to speculative philosophy: "Let us therefore leave the question whether the universe is finite or infinite to the discussion of the natural philosophers."[141] Nevertheless Copernicus is an important figure in the modern development of the idea of an infinite universe. For he was compelled by

[136] PII, 318.28-31.

[137] Th 233.11-13, 233.29-234.1.

[138] Th 322.26-27.

[139] In an article entitled "Nicolas Copernicus and an Infinite Universe" (*Popular Astronomy*, XLIV [1936], 525-33), Grant McColley contended that Copernicus's universe was infinite. McColley's misconception was set aright by Francis R. Johnson's *Astronomical Thought in Renaissance England* (Baltimore, 1937), p. 107, n. 15; see also Arthur O. Lovejoy, *The Great Chain of Being* (Cambridge, Mass., 1936), pp. 103-4, and Armitage, *Copernicus*, p. 81.

[140] Th 21.13-22.6.

[141] Th 21.30-22.1; *physiologi* were not physicists, as McColley thought (*op. cit.*, p. 527), the term for them being *physici* (PII, 322.6).

the lack of evidence of stellar parallax[142] to make the fixed stars enormously remote. He declared that "the heavens are immense by comparison with the earth and appear to be of infinite size";[143] and that "the universe is an immense sphere, similar to the infinite."[144] "But it is not at all clear how far this immensity extends";[145] "the bounds of the universe are unknown and unknowable."[146] In short, Copernicus expanded the traditional finite universe to dimensions approaching the infinite, but an infinite universe was not part of his world-system.[147]

The portion of this enormous spherical surface within which the apparent motions of the sun, moon, and planets are confined was called the zodiac. It was divided into twelve equal parts of 30° each, the signs (*signum, dodecatemorion*) of the zodiac: Aries, Taurus, Gemini, Cancer, Leo, Virgo, Libra or Chelae, Scorpio or Scorpius,[148] Sagittarius, Capricornus, Aquarius, and Pisces. Distances in the zodiac were reckoned eastward from the beginning or first point of Aries; thus a star 212° east of the first point of Aries was said to be in 2° of Scorpio.[149] Similarly the direction of a motion in the zodiac was indicated by reference to the signs. Thus, eastward motion was said to take place in the order of the signs or in consequence,[150] and westward motion in the reverse order of the signs or in precedence.[151]

[142] This topic is treated below (pp. 51-52).

[143] Th 18.16-17.

[144] Th 36. lines 10-11 of the notes; cf. pp. 143-45, below.

[145] Th 19.16-17.   [146] Th 22.4-5.

[147] The views of Copernicus on this question were fully formulated when he wrote the *Commentariolus*, as the fourth Assumption proves; hence McColley's suggestion (*op. cit.*, p. 532, n. 4) of a subsequent development must be rejected.

[148] Current usage prefers Scorpio; but when the genitive is required, it employs Scorpii.

[149] PII, 181.24-182.1.

[150] *Secundum ordinem signorum* (PII, 190.8); *ad ordinem signorum* (PII, 194.24); *secundum signorum successionem* (PII, 188.17); *secundum signorum consequentiam* (PII, 300.7-8); *in signorum consequentiam* (PII, 309.11-12); *in consequentia* (PII, 309.8).

[151] *Contra signorum ordinem* (PII, 194.16); *secundum signorum antecedentiam* (PII, 310.23-24); *in antecedentia(m) signorum* (PII, 304.7, 311.20); *in antecedentia* (PII, 316.11-12); *in praecedentia* (PII, 300.6). The westward motion of the equinoxes is still called "precession."

"To the east of" was expressed by *post,* and "to the west of" by *ante.*[152] After a motion had been described as occurring in consequence or in precedence, if it was desired to indicate that a second motion took place in the opposite direction, the term used was *reflecti contra motum* or *obviare motum.*[153]

The fixed stars were grouped, as they still are, into constellations. To designate individual stars Copernicus used the cumbersome descriptions employed by the ancient Greek astronomers; for the modern compact method had not yet been introduced.[154] Thus for the star now called β Scorpii Copernicus wrote *quae ex tribus in fronte Scorpii borealior est,* "the star which is the most northerly of the three in the brow of Scorpio."[155] Certain of the fixed stars, distinguished by unusual brilliance or color, bore in addition special names, which have remained fairly stable during the ages.[156]

Just as the sun appears each day to rise in the east and travel across the sky until it sets in the west, so the stars at night seem to have a westward motion, or a motion in precedence as Copernicus would have said. This apparent daily motion from east to west, involving the entire universe except the earth, was called the first motion (*primus motus*);[157] and it was accepted as real in the geocentric system, which placed the immovable earth at the center of the universe. The first motion was attributed not to the fixed stars themselves, but to the firmament or eighth sphere in which they were imbedded; and the entire universe was thought to rotate daily about an axis joining the earth with the polestar. But Copernicus regarded the diurnal revolution of the heavens as an appearance, explained by the real daily rotation of the earth, the second motion of the earth in the *Commentariolus.*[158]

[152] PII, 197.11-12.

[153] PII, 192.21, 195.9-10, 200.24; Th 31.24.

[154] In his *Uranometria* (Augsburg, 1603) John Bayer initiated the practice of identifying the component stars of each constellation by assigning Greek and Roman letters to them.

[155] PII, 181.17-18.

[156] Examples are Spica Virginis (PII, 192.8) and Vergiliae (PII, 197.13).

[157] PII, 296.11.

[158] PII, 186, fifth Assumption; 190.7-10.

In addition to its participation in the first motion—the apparent diurnal rotation of the heavens—the sun seems to have another motion, a continual movement eastward or in consequence among the fixed stars. The time required by the sun to complete its circuit from any given star to the same star is a year (*annus*). The path of the annual circuit as projected on the firmament is called the ecliptic (*ecliptica, signifer*). Copernicus regarded the sun's motion on the ecliptic as an appearance, explained by the real annual revolution of the earth on the great circle about the immovable sun; this is the first motion of the earth in the *Commentariolus.*[159]

Let us now imagine that we are looking at the spherical universe from the outside (Fig. 9). On the surface of the

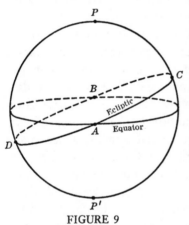

FIGURE 9

sphere draw a great circle halfway between the poles P, P′; this circle is the celestial equator (*aequinoctialis*), the projection of the earth's equator upon the firmament. The ecliptic cuts the equator obliquely at an angle called the obliquity of the ecliptic (*obliquitas eclipticae, obliquitas signiferi*). The points of intersection (A, B) are named the equinoxes or equinoctial points (*puncta aequinoctialia*). The point A where the sun crosses from the south side of the equator to the north is the vernal equinox ♈ (*aequinoctium vernum, aequinoctium vernale, punctum aequinoctii vernalis*); the other point of intersection (B) is the autumnal equinox (*aequinoctium autumni*). The points C, D on the ecliptic, which are midway between the two equinoxes and 90° from each, are the solstices (*puncta solstitialia, conversiones*). The great circle of the celestial sphere, which is drawn from pole to pole through the equinoxes, is the equinoctial colure (*colurus aequinoctiorum, colurus dis-*

tinguens aequinoctia); and through the solstices, the solstitial colure (*colurus solstitiorum, colurus distinguens solstitia*).

Any great circle on the sphere, similarly drawn from pole to pole, is perpendicular to the equator and is known as an hour-circle. The hour-circle which passes through a given star S measures its right ascension (*ascensio recta*), the arc ♈A intercepted on the equator between the vernal equinox ♈ and the point A, where the star's hour-circle cuts the equator (Fig. 10). Right ascension is reckoned eastward or in consequence from the vernal equinox ♈. If through a star on the celestial sphere a circle is drawn

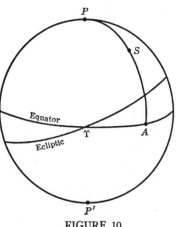

FIGURE 10

parallel to the equator, it intersects the hour-circle of the star. The arc AS of the hour-circle, which is thus intercepted between this point of intersection and the equator, is called the star's declination (*declinatio*), reckoned in degrees and minutes north or south of the equator. Declination and right ascension, taken together as a pair of co-ordinates, completely define the position of a body on the celestial sphere; for example, in Figure 10 the location of the star S on the sphere is fixed by its right ascension ♈A and its declination AS. The primary circle of reference for these co-ordinates is the celestial equator, so that declination and right ascension correspond, in terrestrial geography, to latitude and longitude on the surface of the earth.

Just as the equator has its poles (celestial poles, poles of the daily rotation), so the points 90° distant from the ecliptic are called the poles (P, P′) of the ecliptic (*eclipticae poli*). Through these poles a great circle may be drawn so as to pass through any body X on the celestial sphere (Fig. 11). Such a circle is perpendicular to the ecliptic; and the distance LX on this circle between the body and the ecliptic is the body's celestial latitude (*latitudo*). If we measure eastward along the ecliptic from the vernal equinox ♈ to the point L where the great circle intersects the ecliptic, we have found the body's

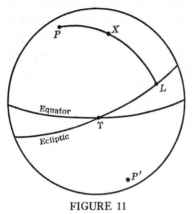

celestial longitude (*longitudo*) ♈L. Celestial longitude and latitude, taken together, constitute a second pair of co-ordinates that completely define the position of a body on the celestial sphere; for example, in Figure 11 the location of the body X on the sphere is fixed by its celestial latitude LX and its celestial longitude ♈L. It must be particularly noticed that the primary circle to which

FIGURE 11

these co-ordinates are referred is the ecliptic. Hence they do not correspond to terrestrial longitude and latitude, for which the circle of reference is the terrestrial equator.

While the latitudes of the fixed stars have remained fairly constant during the past two thousand years, their longitudes have changed considerably. The phenomenon may be represented as follows (Fig. 12). Let ♈R denote the equator, and ♈T the ecliptic. Let S mark the position of a fixed star, and through S draw a great circle perpendicular to ♈T. Then TS is the latitude, and ♈T the longitude. Now let S′ mark the position of the same star as observed at a later time; through S′ draw a great circle perpendicular

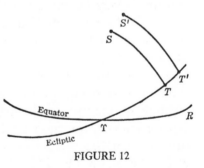

FIGURE 12

to ♈T and intersecting it at T′. Then the new latitude T′S′ is equal to the former latitude TS; but the new longitude ♈T′ is greater than the former longitude ♈T.

Astronomers were accordingly confronted with the problem

of explaining a general increase in the longitude of the fixed stars unaccompanied by any change of latitude. The solution generally accepted in Copernicus's time supposed that all the fixed stars moved slowly eastward: the eighth sphere rotated slowly in consequence. For this motion of the eighth sphere Copernicus substituted a rotation of the earth's axis, the third motion of the earth, which he called the motion in declination (*motus declinationis*).[160]

Since the earth's axis is perpendicular to the plane of the terrestrial equator, and hence of the celestial equator also, the rotation of the axis causes a corresponding alteration in the plane of the earth's equator, and hence of the celestial equator. This shift of the celestial equator produces a slow westward motion of its points of intersection with the ecliptic—the precession of the equinoxes. The general increase in longitude indicates, then, not an eastward motion of the fixed stars, which were conceived by Copernicus to be completely immovable, but a westward motion of the vernal equinox ♈, the point from which longitude is measured. The phenomenon may be represented as follows (Fig. 13). During the interval between two observations of the same fixed star S, the celestial equator ♈R has shifted to the new position ♈'R'. Hence the new longitude ♈'T is greater than the former longitude ♈T, but the latitude TS has remained unchanged.

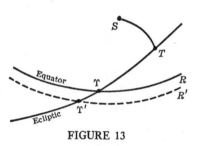

FIGURE 13

While Copernicus's explanation of precession as arising from a motion of the earth's axis has been adopted by modern astronomy, a part of his treatment of the topic has been rejected. For after his death it was established that the annual rate of precession is a constant, as is likewise the related phenomenon, the yearly diminution of the obliquity of the ecliptic. But Copernicus accepted the erroneous idea (trepidation, *trep-*

[160] PII, 190.12-33.

idatio) that these values varied and were not constant. When he wrote the *Commentariolus*, he had not yet worked out the elaborate account of trepidation[161] which he included in the *De revolutionibus*, and which accordingly appeared in the *Narratio prima*.

We have already defined the year as the period of the sun's apparent circuit of the heavens. Now if we compute the time required by the sun to move from a given fixed star eastward on the ecliptic back again to the same star, we have measured the sidereal year (*annus sidereus*). But the time required by the sun to move from, say, a vernal equinox round the ecliptic back again to the next vernal equinox is somewhat shorter than the sidereal year; for while the sun is moving eastward along the ecliptic, the equinox is shifting slowly westward. The period that elapses between successive occurrences of the equinox is known as the tropical year (*annus vertens*).[162]

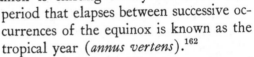

Let the circle in Figure 14 represent the ecliptic. Let the sun S start from the equinoctial point ♈ and traverse the ecliptic in an eastward motion. Meanwhile the equinoctial point is moving slowly westward along the ecliptic. When the sun reaches the equinox at ♈′, the tropical year is completed; but the sidereal year is not completed until the sun describes the arc ♈′♈, thereby returning to its original position among the fixed stars.

FIGURE 14

[161] Observe his hesitation in PII, 190.27-33.

[162] Other terms used to designate the tropical year are *annus naturalis, annus temporalis, annus temporarius, annus ab aequinoctiis numeratus, annus ab aequinoctiis, annus ab aequinoctio*. Dorothy Stimson misunderstood *annus vertens* to be a period of 15,000 years (*The Gradual Acceptance of the Copernican Theory of the Universe*, Hanover, N. H., 1917, p. 111). But Copernicus says plainly enough that the ancient mathematicians before Hipparchus "failed to distinguish the tropical year [*annum vertentem sive naturalem*], which is measured by the equinox or solstice, from the year which is determined by one of the fixed stars" (Th 157.8-10); and again "We must differentiate the tropical year from the sidereal year, and define them. We call that year 'natural' which regulates the four seasons of the year for us, but that year 'sidereal' which returns to one of the fixed stars" (Th 191.20-24).

The moon, as was indicated above, shares in the first motion. But in addition it revolves about the earth. The observer who watches this revolution from the earth sees the moon move eastward among the stars at a rate far more rapid than that of the sun's annual motion. The phases of the moon depend upon its apparent position with respect to the sun (Fig. 15). Thus at new moon the moon is between the earth E and the sun S; in this position (M_1) it is entirely invisible, and it is said to be in conjunction (*conjunctio*). At full moon the earth is between the moon and the sun, and the moon (M_3) is said to be in opposition (*oppositio, obiectio*). At half-moon the moon (M_2, M_4) is midway between conjunction and opposition, and it is said to be in quadrature (*quadratura*).[163] It should

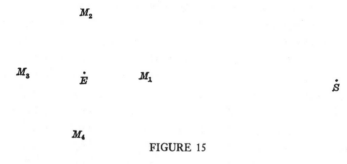

FIGURE 15

also be noted that the moon's apparent path on the celestial sphere is inclined to the ecliptic at a small angle, the two points of intersection being named the nodes (*nodi*). The point where the moon crosses from the south side of the ecliptic to the north is called the ascending node; the other point of intersection is termed the descending node.

Whereas the sun and the moon both appear always to move steadily eastward across the background of the fixed stars, the planets follow a more complicated pattern of motion. Most of the time they seem to move eastward or in consequence; they are then said to be direct (*directus*).[164] The rate of this

[163] The term *quadratura* was also occasionally used to denote the quadrant of a circle (PII, 201.4, 19), but this idea was more commonly expressed by *quadrans*.

[164] Direct motion was called *progressus* (PII, 187.2), *progressio* (Th 308.5), or *directio* (PII, 354.6).

direct motion is not uniform, but increases until it reaches its maximum and then decreases until the planet seems hardly to move at all (*stare, subsistere*). At this time the planet is said to be stationary (*stationarius*) and to have reached its station or stationary point (*statio*). Thereafter the planet moves with gradually increasing speed westward; it is then said to be retrograde (*retrogradus*) or to retrograde (*retrocedere, regredi*).[165] After a time this regression becomes slower and then ceases, the planet having reached its second stationary point (*statio secunda*). Thereafter it again reverses the direction of its motion, proceeding eastward once more at first slowly and then faster. The planet has now returned to the position with which we began our account of its apparent motion on the celestial sphere; it has closed the cycle and begins to repeat it.

This movement of the planets now forward, now backward at a varying, unequal rate was accepted as real in the geocentric system. But Copernicus held that these phenomena were only appearances, explained by the real orbital revolution of the earth on the great circle.[166] For he believed that a celestial body could move only with uniform velocity always in the same direction. Hence the sole real motion of a planet took place at a uniform rate without change of direction. However, we observe this motion not from a stationary position, but from the moving earth. The motion of the terrestrial observer produces, according to Copernicus, the appearances of stations, retrogradations, and variations in the velocity of the planetary movements; to the irregularities thus imposed upon the real motion of the planet he gave the name "motion in commutation" (*motus commutationis*).[167]

In the *Commentariolus* he did not use the term "motion in commutation," but spoke instead of the "second inequality." Any motion that departed from a perfectly circular path, or from a perfectly uniform rate, was called an unequal motion (*motus diversus, motus inaequalis*) or an inequality (*diversitas,*

[165] Retrograde motion (retrogradation, regression) was called *retrocessio* (PII, 187.2), *repedatio* (PII, 353.9), *regressus* (Th 308.5), or *regressio* (Th 407.25).
[166] PII, 187, seventh Assumption; 196.10-13.
[167] PII, 354.12-17; Th 308.2-20.

inaequalitas, anomalia). As we saw above, the epicyclic and eccentric arrangements were contrived to produce a nearly circular motion, the first inequality in the case of the planets.[168] Although it had long been known that the planetary orbits were not perfectly circular, Copernicus adhered to the traditional precept that the motion of the celestial bodies was either circular or composed of circular movements.[169] The ancient axiom of circularity was not shattered until Kepler demonstrated the ellipticity of the planetary orbits. But it should be remembered that the ellipses which are the true planetary orbits have a very small eccentricity and actually differ but little from perfect circles.

As a planet moves across the sky, its position with respect to the sun changes. Let the sun be fixed at S (Fig. 16). Assume that the earth in the course of its annual revolution about the sun has reached the position E on the great circle. The three superior planets (Saturn, Jupiter, and Mars) describe orbits which enclose the great circle. When a superior planet is at P₁, so that the line of sight from the earth to the planet passes through the apparent place of the sun, the planet is said to be in conjunction (*conjunctio*). When the planet is at P₂, so

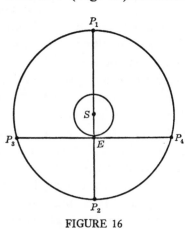

FIGURE 16

that the earth lies on a straight line between the planet and the sun, the planet is said to be in opposition (*oppositio*).[170] When the line drawn from the earth to the planet (P₃, P₄) is perpendicular to the line drawn from the earth to the sun, the planet is said to be in quadrature (*quadratura*).

---

[168] PII, 196.8-9.

[169] Th 14.10-11.

[170] "To be in opposition" was expressed by *opponi* (PII, 199.5); "to be in conjunction" by *coniungi* (PII, 355.25).

The inferior planets (Venus and Mercury) describe orbits enclosed within the great circle. Let the sun be at S and the earth at E (Fig. 17). Venus or Mercury at P_1 or P_2 is in conjunction; but an inferior planet cannot come to either opposition or quadrature.

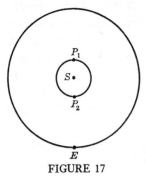

FIGURE 17

In modern astronomy the term "elongation" is used to denote the angle between the lines drawn from the terrestrial observer to the planet and to the sun; for example, when a superior planet is in quadrature, its elongation is 90°. In Copernicus's time the word *elongatio* had not yet acquired this specialized meaning and was simply a synonym for "distance" (*distantia*).[171]

In like manner, *elongari* and *sese elongare* meant "to be at a distance" or "to increase the distance."[172] Since there was no equivalent of the modern term "elongation," rather cumbersome expressions were employed. Thus for an inferior planet the elongation reaches a certain maximum value, now named the "greatest elongation," whereas Rheticus called it "the outermost point in the curvature of the deferent" of the planet,[173] and Copernicus spoke of the planet's "greatest distance from the place of the sun."[174]

So far we have treated the planetary motions as though they took place in the plane of the earth's orbit, and were simply forward and backward movements along the ecliptic (motion in longitude). But in fact the orbits of the planets are inclined at small angles to the ecliptic, so that in addition to their motions in longitude the planets have also motions in latitude. As in the case of the moon, the point where the planet crosses

[171] PII, 342.19-23.

[172] PII, 351.31, 200.5, 309.19. Just as *elongatio* indicated distance from a point of reference, so *accessio* and *accessus* meant approach toward it (PII, 195.15-16, 341.20); and *accedere*, to approach (Th 360.7-11).

[173] PII, 355.25-26.

[174] Th 364.25-26, 365.10-11.

from the south side of the ecliptic to the north is called the ascending node, and the other point of intersection is termed the descending node.

Before we conclude our discussion of the elements of astronomy, we must deal with two or three other matters. Suppose that the moon M is observed from a point A on the earth's surface (Fig. 18). Since the moon is nearer to us than are the fixed stars, the line of sight from the observer's eye to the moon may be prolonged to meet the firmament in M_1. Now draw a straight line from the center of the earth C to the moon, and station a second observer at B. As seen by him,

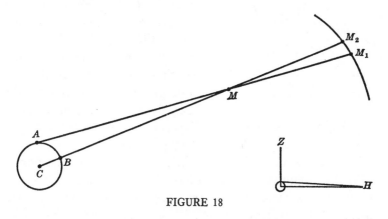

FIGURE 18

the moon's direction will be M_2. The difference between the direction of the moon as seen from A and as seen from the center of the earth C (or B) is called the parallax (*diversitas aspectus, parallaxis, commutatio*), and it is measured by $\angle$ AMC. It is obvious that the lunar parallax varies, being greatest when the moon is on the horizon H, and zero when it is in the zenith Z.

In an analogous manner the fixed stars are said to have an annual parallax. Let EE′ represent the great circle, the path of the earth's yearly revolution about the sun (Fig. 19). If the direction of one of the nearer stars S is observed when the earth is at E, and then six months later when it is at E′, there will be a slight difference between the two directions. This displacement in the apparent position of S is due to the orbital motion of the earth, and resembles the shift in the apparent place of the moon caused by a change in the observer's station on the surface of the earth. But stellar parallax was not detected until almost three centuries after the death of Coper-

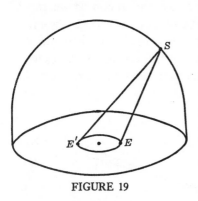

nicus. Hence he was compelled to assume that all the fixed stars were enormously remote from the earth; for then the apparent annual motion conferred upon them by the earth's real motion would be so minute as to be imperceptible. But since the planets are very much nearer to the earth than are the fixed stars, the effect upon them of the earth's orbital motion is corre-

FIGURE 19

spondingly greater; and what Copernicus called the "motion in commutation" is in effect planetary parallax.[175]

We have been speaking of "line of sight"; for this phrase Copernicus's equivalent was *radius visualis* or simply *radius*.[176] He employed *radius* also in the sense of "ray," writing *radii solis* for "the rays of the sun."[177] In short, with him *radius* retained its ancient meanings and had not yet come to stand for "the semidiameter of a circle or sphere." In the *Commentariolus* such a line was called by Copernicus *semidiameter* or *semidimetiens*;[178] but in the *De revolutionibus* he also used *quae ex centro*, a literal translation of the Greek ἡ ἐκ τοῦ κέντρου.[179]

[175] PII, 354.12-17.
[176] PII, 196.14, 197.2.
[177] PII, 200.17.
[178] PII, 186, fourth Assumption; 189.3.
[179] See for example HI, 35.6; Thomas L. Heath, *The Works of Archimedes* (Cambridge, 1897), p. clxii; and Heath's *The Thirteen Books of Euclid's Elements* (2d ed.; Cambridge, 1926), I, 199. In the *Narratio prima* Rheticus used both *quae ex centro* and *semidiameter*; but when he was thinking of the radius as a length, he wrote *extensio*=διάστημα (PII, 335.4; 349.12, 29).

Besides the customary division of time into days, hours, minutes, and seconds, an alternative system was employed which did not include the hour. In this system the day was divided into sixty minutes, so that each minute (*minutum diei*) equaled twenty-four of the ordinary minutes. The *minutum diei* was divided into sixty seconds, so that similarly each of these seconds equaled twenty-four of the ordinary seconds.[180]

[180] Cf. HI, 208.8-12; Th 195.31-196.2; and p. 116, n. 33, below.

THE *COMMENTARIOLUS* OF COPERNICUS

NICHOLAS COPERNICUS
SKETCH OF HIS HYPOTHESES FOR THE
HEAVENLY MOTIONS

OUR ANCESTORS assumed, I observe, a large number of celestial spheres for this reason especially, to explain the apparent motion of the planets by the principle of regularity. For they thought it altogether absurd that a heavenly body, which is a perfect sphere, should not always move uniformly.[1] They saw that by connecting and combining regular motions in various ways they could make any body appear to move to any position.

Callippus and Eudoxus, who endeavored to solve the problem by the use of concentric spheres, were unable to account for all the planetary movements; they had to explain not merely the apparent revolutions of the planets but also the fact that these bodies appear to us sometimes to mount higher in the heavens, sometimes to descend; and this fact is incompatible with the principle of concentricity. Therefore it seemed better to employ eccentrics and epicycles, a system which most scholars finally accepted.

Yet the planetary theories of Ptolemy and most other astronomers, although consistent with the numerical data, seemed likewise to present no small difficulty. For these theories were not adequate unless certain equants were also conceived; it then appeared that a planet moved with uniform velocity neither on its deferent nor about the center of its epicycle. Hence a system of this sort seemed neither sufficiently absolute nor sufficiently pleasing to the mind.

Having become aware of these defects, I often considered whether there could perhaps be found a more reasonable arrangement of circles, from which every apparent inequality would be derived and in which everything would move uni-

[1] Perhaps this sentence should be translated: For they thought it altogether absurd that a heavenly body should not always move with uniform velocity in a perfect circle.

formly about its proper center, as the rule of absolute motion requires. After I had addressed myself to this very difficult and almost insoluble problem, the suggestion at length came to me how it could be solved with fewer and much simpler constructions than were formerly used, if some assumptions (which are called axioms) were granted me. They follow in this order.

Assumptions[2]

1. There is no one center of all the celestial circles or spheres.

2. The center of the earth is not the center of the universe, but only of gravity[3] and of the lunar sphere.

3. All the spheres revolve about the sun as their mid-point, and therefore the sun is the center of the universe.

4. The ratio of the earth's distance from the sun to the height of the firmament is so much smaller than the ratio of the earth's radius to its distance from the sun that the distance from the earth to the sun is imperceptible in comparison with the height of the firmament.

5. Whatever motion appears in the firmament arises not from any motion of the firmament, but from the earth's motion. The earth together with its circumjacent elements[4] performs a complete rotation on its fixed poles in a daily motion, while the firmament and highest heaven abide unchanged.

6. What appear to us as motions of the sun arise not from

---

[2] In his description of the *Commentariolus* Dreyer incorrectly states the number of assumptions as six (*Planetary Systems*, p. 317). The source of his mistake is probably the oversight in PI², 291, which Prowe himself calls attention to and corrects (PII, 187 n).

[3] "Now the element of earth is the heaviest; and all heavy objects are borne to the earth, tending toward its inmost center. In accordance with their nature, heavy objects are borne from all directions at right angles to the surface of the earth; and since the earth is spherical, they would come together at its center, were they not checked at its surface. For a straight line which is at right angles to the tangential plane at the point of tangency leads to the center" (Th 19.28-20.3).

[4] These are (*a*) the atmosphere and (*b*) the waters that lie upon the surface of the earth. See p. 63, below and in *De rev.*: ". . . not only does the earth so move together with the watery element that is joined with it, but also no small part of the air and whatever else is related in the same way to the earth" (Th 22.15-17).

its motion but from the motion of the earth and our sphere, with which we revolve about the sun like any other planet. The earth has, then, more than one motion.

7. The apparent retrograde and direct motion of the planets arises not from their motion but from the earth's. The motion of the earth alone, therefore, suffices to explain so many apparent inequalities in the heavens.

Having set forth these assumptions, I shall endeavor briefly to show how uniformity of the motions can be saved in a systematic way. However, I have thought it well, for the sake of brevity, to omit from this sketch mathematical demonstrations, reserving these for my larger work.[5] But in the explanation of the circles I shall set down here the lengths of the radii; and from these the reader who is not unacquainted with mathematics will readily perceive how closely this arrangement of circles agrees with the numerical data and observations.

Accordingly, let no one suppose that I have gratuitously asserted, with the Pythagoreans, the motion of the earth; strong proof will be found in my exposition of the circles. For the principal arguments by which the natural philosophers attempt to establish the immobility of the earth rest for the most part on the appearances; it is particularly such arguments that collapse here, since I treat the earth's immobility as due to an appearance.

The Order of the Spheres

The celestial spheres are arranged in the following order. The highest is the immovable sphere of the fixed stars, which contains and gives position to all things. Beneath it is Saturn, which Jupiter follows, then Mars.[6] Below Mars is the sphere on which we revolve; then Venus; last is Mercury. The lunar sphere revolves about the center of the earth and moves with

[5] From this reservation we may infer that when Copernicus wrote the *Commentariolus* he had already planned *De rev.* or was at work upon it.

[6] S, V: *sub eo Saturnus; hunc sequitur Martius.* In S, after *Saturnus*, the words *quem sequitur Iovius* have been inserted above the line by a second hand. These readings provide a clue to the relationship between S and V; see the following note.

the earth like an epicycle. In the same order also, one planet surpasses another in speed of revolution, according as they trace greater or smaller circles. Thus Saturn completes its revolution in thirty years, Jupiter in twelve, Mars in two and one-half,[7] and the earth in one year; Venus in nine months, Mercury in three.

[7] S: *Sic quidem Saturnus anno 30, Jupiter 12, Mars, tellus annua revolutione restituuntur;* V: *Sic quidem Saturnus anno trigesimo, Iuppiter duodecimo, Mars, tellus annua revolutione restituitur.* The number for Mars has dropped out of both S and V, but it may be restored from a later section in the *Commentariolus*, where the sidereal period of Mars is given as twenty-nine months (p. 74, below). In *De rev.* Copernicus reduced the period to two years, bringing it closer to the true value of 687 days, or one year and ten and one-half months (Th 29.6: *Deinde Mars, qui biennio circuit;* the explanatory figure likewise has *Martis bima revolutio;* confirmation from Rheticus on p. 146, below).

In his edition of V, Curtze filled the lacuna by inserting, without any supporting argument, [*tertio*] after *Mars* (MCV, I, 7.27). This erroneous reading was accepted by Prowe, who unwisely dropped the square brackets (PII, 188.13). Adolf Müller evidently accepted Prowe's text unquestioningly (ZE, XII, 360); hence his translation of the *Commentariolus* assigned the grossly inaccurate value of three years to Mars' sidereal period (ZE, XII, 364). With no more warrant L. Birkenmajer brought Copernicus into close agreement with modern astronomy; for his translation runs: "Mars revolves in not quite two years" (*Mikołaj Kopernik Wybór pism,* p. 9). It will be observed that none of these scholars noted the later passage in the *Commentariolus* from which the lacuna may be filled without hesitation.

The omission of the number of years or months in Mars' sidereal period furnishes a clue to the relation between S and V. Since they share this omission, they are both derived from a copy of the *Commentariolus* in which the error had already appeared. Now it is most unlikely that the *Commentariolus* came from the hands of Copernicus with so glaring a defect in it. Let us assume C as the text issued by Copernicus. Between C, as the original text, and S and V, as later copies, there intervenes X, one or more copies in which the omission occurred. The stemma here proposed may be represented by an inverted Y. The foregoing analysis is supported by the readings cited in the preceding note, where both S and V omit Jupiter from the list of the celestial spheres. Certainly no two independent scribes, copying from accurate texts, would both of them have omitted Jupiter and dropped the number for Mars. A study of the other variants makes it equally unlikely that S was copied from V, or V from S.

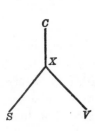

In the preface to his edition of S, Lindhagen reports the opinion of paleographers that S was written in Switzerland or northern Italy during the late sixteenth or early seventeenth century. Curtze thought that V was written in the late sixteenth century (MCV, I, 2).

The Apparent Motions of the Sun

The earth has three motions. First, it revolves annually in a great circle[8] about the sun in the order of the signs, always describing equal arcs in equal times; the distance from the center of the circle to the center of the sun is ½₅ of the radius of the circle.[9] The radius is assumed to have a length imperceptible in comparison with the height of the firmament;[10] consequently the sun appears to revolve with this motion, as if the earth lay in the center of the universe. However, this appearance is caused by the motion not of the sun but of the earth, so that, for example, when the earth is in the sign of Capricornus, the sun is seen diametrically opposite in Cancer, and so on. On account of the previously mentioned distance of the sun from the center of the circle, this apparent motion of the sun is not uniform, the maximum inequality being 2⅙°.[11]

[8] This great circle is the *orbis magnus* discussed above (p. 16).

[9] Here Copernicus accepts Ptolemy's view that the eccentricity was fixed (HI, 233.11-16). However, Ptolemy had put the eccentricity at ½₄ (HI, 236.19-21). Hence we may say that in the *Commentariolus* Copernicus retains a fixed eccentricity, but offers an improved determination of it. On the other hand, in *De rev.* he finds that the eccentricity is ⅓₁ (Th 211.23-25; cf. p. 160, below). Consequently he there abandons the idea of a fixed eccentricity (Th 209.27-210.1), and holds that it varies between a maximum of ½₄ and a minimum of ⅓₁ (Th 219.31-220.6, 209.11-13, 211.18).

[10] See Assumption 4, above.

[11] Let the apparent motion of the sun (or real motion of the earth) take place on the great circle (*orbis magnus*) AEP (Fig. 20). Let the motion be uniform

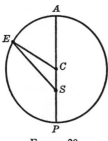

FIGURE 20

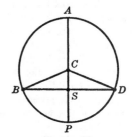

FIGURE 21

with respect to the center at C. Let the sun be at S. Let the apogee be at A, and the perigee at P. Assume that the earth starts from A and has reached any point E on the circumference. Then the line of sight ES will give the observed place of the sun, and ∠ ASE will measure the observed motion. But ∠ ACE will measure

The line drawn from the sun through the center of the circle is invariably directed toward a point of the firmament about 10° west of the more brilliant of the two bright stars in the head of Gemini;[12] therefore when the earth is opposite this point, and the center of the circle lies between them, the sun is seen at its greatest distance from the earth.[13] In this circle, then,

---

the uniform or mean motion. Now the inequality to which Copernicus refers is the difference between the uniform and the observed motions; and it is measured by $\angle$ CES. It is evident that when the earth (or the observed place of the sun) is at A or P, the inequality is zero.

If we draw BD $\perp$ ACSP at S (Fig. 21), the inequality attains its maximum at B and D (Th 207.15-208.7; cf. HI, 220.12-16, 221.9-223.3). It is obvious that the smaller the eccentricity CS is, the smaller the maximum inequality will be. Now Ptolemy had put the maximum inequality at 2° 23′ (HI, 238.22-239.1), corresponding to an eccentricity of ¹⁄₂₄ (CS:AC=1:24). Since in the *Commentariolus* Copernicus reduces the eccentricity to ¹⁄₂₅, he diminishes the maximum inequality to 2° 10′. And in *De rev.*, where he further reduces the eccentricity to ¹⁄₃₁, the maximum inequality is 1° 51′ (Th 212.14-16).

When he states in the *Commentariolus* that the maximum inequality, corresponding to an eccentricity of ¹⁄₂₅, is 2⅙° (*duobus gradibus et sextante unius*), he is evidently writing a convenient fraction. For ¹⁄₂₅ × 100,000 = 4,000; and by his Table of Chords, 4,000 subtends 2° 17½′ (Th 44.20-21). For the equivalence of Copernicus's Table of Chords with a modern table of sines see Armitage, *Copernicus*, pp. 171-73.

[12] With what fixed star are we to identify "the more brilliant of the two bright stars in the head of Gemini" (*stella lucida quae est in capite Gemelli splendidior*)? Both Gemini 1 (Castor, α Geminorum) and Gemini 2 (Pollux, β Geminorum) were described as being in the head of Gemini. They were usually differentiated as western and eastern (HII, 20.17-18, 21; 92.3-4; Th 132.29-32); thus in a later section of the *Commentariolus*, where Copernicus refers to "the star which is described as being in the head of the eastern of the two Gemini" (*stellam quae in capite Geminorum orientalis dicitur*), he is speaking of Pollux (p. 78). But in the present passage he does not employ the customary designation, and he relies on *splendidior* to indicate whether he is referring to Castor or to Pollux. Now, in the catalogues these stars were both listed as of the the second magnitude; and it therefore seems impossible to decide the question by appealing to a difference in brilliancy. However, Pollux is distinguished by its color; and it is perhaps possible that Copernicus is using *splendidior* as a color term. On this uncertain basis let us tentatively identify the star of our text with Pollux.

[13] If the preceding note is correct, then Copernicus is here locating the solar apogee about 10° west of Pollux. Now Ptolemy had put the longitude of the solar apogee at 65° 30′ (HI, 237.9-11) and the longitude of Pollux at 86° 40′ (HII, 93.4); hence the apogee was 21° 10′ west of Pollux. He held that the

the earth revolves together with whatever else is included within the lunar sphere.

The second motion, which is peculiar to the earth, is the daily rotation on the poles in the order of the signs, that is, from west to east. On account of this rotation the entire universe appears to revolve with enormous speed. Thus does the earth rotate together with its circumjacent waters and encircling atmosphere.[14]

The third is the motion in declination. For the axis of the daily rotation is not parallel to the axis of the great circle, but is inclined to it at an angle that intercepts a portion of a circum-

apogee was fixed in relation to the vernal equinox (HI, 232.18-233.16), that the equinoctial and solstitial points were constant (HI, 192.12-22), and that the fixed stars moved eastward 1° in 100 years (HII, 15.15-17). Had Ptolemy been right, the apogee should have been found, at the time of Copernicus, about 35° west of Pollux.

Copernicus reverses Ptolemy's explanation of precession; for he regards the fixed stars as constant (Assumption 5) and attributes the precessional motion to the equinoctial and solstitial points (p. 67, below). Hence, when in the present passage he asserts that the solar apogee is fixed, he means that it is fixed in relation to the fixed stars; but its distance from the vernal equinox increases, because the equinoctial points move steadily westward.

Furthermore, so long as the vernal equinox was regarded as constant, it had served as the point from which celestial longitude was measured (HII, 36.16-17). Hence if Copernicus is to utilize without error the work of the ancient astronomers, he must first reconstruct the entire history of precession. In the next section of the *Commentariolus* he lays down two of the main propositions. But it is evident that he has not yet completely formulated the theory which is outlined briefly in the *Letter against Werner* (pp. 99-101, below) and expounded fully in *De rev.* (Th 157-173; cf. pp. 111-17, below). Moreover, since celestial longitude can no longer be reckoned from the vernal equinox, some fixed star must be selected in its stead. But Copernicus has apparently not yet made a choice; throughout the *Commentariolus*, when he states a celestial position, he gives it in terms of neighboring stars, never in terms of longitude reckoned from a fixed origin. Because he does not give us sufficient data to make the correction for precession, we cannot say with precision what he then believed the longitude of the solar apogee to be.

In *De rev.* he chooses a fixed star from which to measure longitude (Th 114. 22-33, 130.6-7); he determines the longitude of the solar apogee as 96° 40′ (Th 211.20-21, 25-26); and he no longer regards the doctrine of a fixed apogee as tenable: "There now emerges the more difficult problem of the motion of the solar apse . . . which Ptolemy thought was fixed . . ." (Th 216.3-4).

[14] Cf. above, p. 58, n. 4.

ference, in our time about 23½°.[15] Therefore, while the center of the earth always remains in the plane of the ecliptic, that is, in the circumference of the great circle, the poles of the earth rotate, both of them describing small circles about centers equidistant from the axis of the great circle.[16] The period of this motion is not quite a year and is nearly equal to the annual revolution on the great circle. But the axis of the great circle is invariably directed toward the points of the firmament which are called the poles of the ecliptic. In like manner the motion in declination, combined with the annual motion in their joint effect upon the poles of the daily rotation, would keep these poles constantly fixed at the same points of the heavens, if the periods of both motions were exactly equal.[17] Now with the long passage of time it has become clear that this inclination of the earth to the firmament changes. Hence it is the common opinion that the firmament has several motions in conformity with a law not yet sufficiently understood. But the motion of

[15] In De rev. Copernicus states that he and certain of his contemporaries have found this angle (which is equal to the obliquity of the ecliptic) to be not greater than 23° 29' (Th 76.29-77.1); and again that "in our times it is found to be not greater than 23° 28½'" (Th 162.24-25). Newcomb's determination of the obliquity for 1900 was 23° 27' 8".26; on the basis of an annual diminution of 0".4684, the value for 1540 would be 23° 29' 57" (American Ephemeris and Nautical Almanac for 1940, Washington, D. C., 1938, p. xx).

[16] Müller's version is faulty. He translated: "beschreiben die beiden Pole der Erdachse bei stets gleichbleibendem Abstand kleine Kreise um die Pole der Ekliptik" (the two poles of the earth's axis, always maintaining an equal distance, describe small circles about the poles of the ecliptic; ZE, XII, 367). But Copernicus says plainly enough that it is the centers of the small circles that are equidistant from the poles of the ecliptic: circulos utrobique parvos describentes in centris ab axe orbis magni aequidistantibus; hence the poles of the earth are not, as Müller thought, equidistant from the poles of the ecliptic. This blunder led Müller into another error, as we shall see below (p. 73, n. 45).

[17] This obviously requires the direction of the motion in declination to be opposite to the direction of the annual motion. The explicit statement appears in De rev. (where the annual revolution of the earth about the sun is termed "the annual motion of the center" or more briefly "the motion of the center"): "Then there follows the third motion of the earth, the motion in declination, which is also an annual revolution but which takes place in precedence, that is, in the direction opposite to that of the motion of the center. Since the two motions are nearly equal in period and opposite in direction. . ." (Th 31.22-25; cf. p. 148, below).

the earth can explain all these changes in a less surprising way. I am not concerned to state what the path of the poles is. I am aware that, in lesser matters, a magnetized iron needle always points in the same direction. It has nevertheless seemed a better view to ascribe the changes to a sphere, whose motion governs the movements of the poles. This sphere must doubtless be sublunar.

Equal Motion Should Be Measured Not by the Equinoxes but by the Fixed Stars

Since the equinoxes and the other cardinal points of the universe shift considerably, whoever attempts to derive from them the equal length of the annual revolution necessarily falls into error.[18] Different determinations of this length were made in different ages on the basis of many observations. Hipparchus computed it as $365\frac{1}{4}$ days, and Albategnius the Chaldean as $365^d\ 5^h\ 46^m$,[19] that is, $13\frac{3}{5}^m$ or $13\frac{1}{3}^m$ less than Ptolemy.[20] Hispalensis increased Albategnius's estimate by the

---

[18] This assertion is directed against the Ptolemaic doctrine that the length of the year must be measured by the solstices and equinoxes (HI, 192.12-22; cf. Th 309.4-9).

[19] In *De rev.* Copernicus cites Albategnius's estimate more fully as $365^d5^h46^m24^s$ (Th 193.7-8). It is to this value that he adds $13\frac{3}{5}^m$ (= 13^m36^s), in order to obtain the sum $365\frac{1}{4}^d$ (= 365^d6^h). For Albategnius's determination see C. A. Nallino, *Al-Battānī sive Albatenii opus astronomicum* (*Pubblicazioni del Reale osservatorio di Brera in Milano*, No. 40, 1899-1907), Pt. I, 42.17. It seems clear that Copernicus did not draw from a single source the historical statements made in this section. But it is altogether likely that they were in large part based upon the *Epitome in Almagestum Ptolemaei* (Venice, 1496), begun by George Peurbach and completed by Regiomontanus (for Rheticus's use of this work, see below, p. 117, n. 35). For the *Epitome* (Bk. III, Prop. 2) gave Albategnius's determination as $365^d5^h46^m24^s$ or $13\frac{3}{5}^m$ less than $365\frac{1}{4}^d$.

[20] When Copernicus wrote the *Commentariolus*, he was misinformed about the value accepted by Hipparchus and Ptolemy, for he put it at $365\frac{1}{4}$ days. But in *De rev.* he correctly states that they found the year less than $365\frac{1}{4}$ days by $\frac{1}{300}$th of a day, or $365^d5^h55^m12^s$ (Th 191.31-192.3, 192.21-23, 237.13-15; HI, 207.24-208.14). The *Epitome* (*loc. cit.*) cited Hipparchus's determination as $365\frac{1}{4}^d$, but quoted Ptolemy's value correctly. It should be noted that a work contemporary with the *Commentariolus* states: "Hipparchus thought that the year consisted of $365\frac{1}{4}$ days. Although he says that it was a fraction less than the complete quarter, he ignored the fraction, since he judged it to be imperceptible" (Augustinus Ricius, *De motu octavae sphaerae*, Trino, 1513, fol. e6r; Paris, 1521, p. 40 r).

20th part of an hour, since he determined the tropical year as $365^d\ 5^h\ 49^m$.[21]

[21] Prowe (PII, 191 n) and Müller (ZE, XII, 368, n. 41; the reference to *De rev.* should be III, xiii, not III, liii) followed Curtze (MCV, I, 10 n) in supposing that Hispalensis, i.e., from Hispalis = Seville, here means Isidore of Seville. In Copernicus's view precession attained its greatest rapidity in the time of Albategnius; thereafter diminution set in: "From these computations it is clear that in the 400 years before Ptolemy the precession of the equinoxes was less rapid than in the period from Ptolemy to Albategnius, and that in this same period it was more rapid than in the interval from Albategnius to our times" (Th 162.14-17; cf. p. 113, below). Therefore the shortest length of the tropical year fell in the time of Albategnius; and the increase noted by Hispalensis must be associated with a later astronomer. This chronological consideration rules out Isidore immediately. Moreover, an examination of the astronomical portions of his extant works (J. P. Migne, *Patrologia Latina*, Vols. LXXXI–LXXXIV) shows that he gives 365 days as the length of both the tropical and sidereal years.

Who, then, is Hispalensis? Jābir ibn Aflaḥ? In 1534 Peter Apian's *Instrumentum primi mobilis* was published together with Gebri filii Affla Hispalensis . . . *Libri IX de astronomia*. A copy was given by Rheticus to Copernicus (MCV, I, 36), and hence it did not get into his hands before 1539 (PII, 377.11-12). But all our evidence points to 1533 as the very latest year in which the *Commentariolus* could have been written. Moreover, Jābir (*op. cit.*, pp. 38-39) simply repeats the Hipparchus-Ptolemy estimate of the length of the tropical year. Clearly he is not the Hispalensis to whom Copernicus refers.

In his *Stromata Copernicana* (Cracow, 1924), p. 353, Birkenmajer correctly identified Copernicus' "Hispalensis" with Alfonso de Cordoba Hispalensis. The latter, who usually called himself *Alfonsus artium et medicinae doctor*, corrected Abraham Zacuto's *Almanach perpetuum exactissime nuper emendatum omnium celi motuum cum additionibus in eo factis tenens complementum* (Venice, 1502). On fol. a2r a letter is addressed to him as *Alfonso hispalensi de corduba artium et medicinae doctori*. His correction of Zacuto's *Almanach perpetuum* was published by Peter Liechtenstein at Venice on July 15, 1502, while Copernicus was a student at the nearby University of Padua. Alfonso Hispalensis' statement concerning the length of the year occurs on fol. a1v, where he corrects a computation of Zacuto and says: . . . *dividas per numerum dierum anni .365. et quartam minus undecim minutis hore* . . . (divide by the number of days in a year, 365¼ minus eleven minutes = $365^d5^h49^m$). This direct statement was overlooked by Birkenmajer, who thought he found nearly the same length of the tropical year by implication in the tables (which, however, were due to Zacuto and not to Alfonso Hispalensis). Birkenmajer also misread the second word in the volume's title, where "perpetuu3" = "perpetuum," not "perpetuum et" (Adriano Cappelli, *Lexicon abbreviaturarum*, 5th ed., Milan, 1954, p. XXXII). The *Almanach perpetuum* belonging to the library of the Ermland cathedral chapter (ZE, V, 375) may or may not have been a copy of the Venice, 1502 edition. The copy of that edition in the library of Upsala University (Pehr Fabian Aurivillius, *Catalogus librorum impressorum bibliothecae r. academiae Upsaliensis*, Upsala, 1814, p. 1002) lacks the page on which the entry *Liber capit. Varm.* would have appeared, had the volume

Lest these differences should seem to have arisen from errors of observation, let me say that if anyone will study the details carefully, he will find that the discrepancy has always corresponded to the motion of the equinoxes. For when the cardinal points moved 1° in 100 years, as they were found to be moving in the age of Ptolemy,[22] the length of the year was then what Ptolemy stated it to be. When however in the following centuries they moved with greater rapidity, being opposed to lesser motions, the year became shorter; and this decrease corresponded to the increase in precession. For the annual motion was completed in a shorter time on account of the more rapid recurrence of the equinoxes. Therefore the derivation of the equal length of the year from the fixed stars is more accurate. I used Spica Virginis[23] and found that the year has always been 365 days, 6 hours, and about 10 minutes,[24] which is also the estimate of the ancient Egyptians.[25] The same method must be employed also with the other motions of the

---

once belonged to the library of the Ermland chapter (Birkenmajer, *Stromata*, p. 300).

Since "Hispalensis" in the *Commentariolus* means the *Almanach perpetuum* of 1502, it follows that Copernicus wrote the *Commentariolus* after July 15 of that year. If the entry . . . *sexternus Theorice asserentis Terram moveri, Solem vero quiescere* . . . (a manuscript of six leaves expounding the theory of an author who asserts that the earth moves while the sun stands still) in the catalogue of his books drawn up on May 1, 1514, by Matthew of Miechow (1457–1523), professor at the university of Cracow, refers to the *Commentariolus*, then its date of composition is narrowed down to the dozen years between July 15, 1502 and May 1, 1514.

[22] HII, 15.6–16.2.   [23] Virgo 14 (HII, 102.16; Th 136.10), α Virginis.

[24] Copernicus's estimate of the length of the sidereal year is stated more exactly in *De rev.* as $365^d6^h9^m40^s$ (Th 195.29–196.2); Curtze misquotes the estimate as $365^d6^h8^m40^s$ (MCV, I, 10 n), and Prowe repeats the misstatement (PII, 191 n). Newcomb's determination (1900) is $365^d.25636042 = 365^d6^h9^m9^s.54$ (*American Ephemeris for 1940*, p. xx).

[25] Copernicus apparently derived this information from the *Epitome*. It stated (*loc. cit.*) that the value found by the ancient Egyptians was $365\frac{1}{4}^d + \frac{1}{130}^d$ ($= 365^d6^h11^m$). The Latin translation of Albategnius, which was printed at Nuremberg in 1537, likewise ascribed to certain ancient Egyptian and Babylonian astronomers a year consisting of $365\frac{1}{4}^d + \frac{1}{131}^d = 365^d6^h11^m$ (Nallino, *Al-Battānī*, I, 40.28–29, 204–9; cf. below, p. 117, n. 34). So far as I am aware, no determination of the length of the year more precise than $365\frac{1}{4}^d$ has been discovered among the papyri or other documents surviving from ancient Egypt.

planets, as is shown by their apsides, by the fixed laws of their motion in the firmament, and by heaven itself with true testimony.

The Moon

The moon seems to me to have four motions in addition to the annual revolution which has been mentioned. For it revolves once a month on its deferent circle about the center of the earth in the order of the signs.[26] The deferent carries the epicycle which is commonly called the epicycle of the first inequality or argument, but which I call the first or greater epicycle.[27] In the upper portion of its circumference this greater epicycle revolves in the direction opposite to that of the deferent,[28] and its period is a little more than a month. Attached

[26] The loss of a leaf from V creates a lacuna which begins at this point and ends near the close of the present section. For the intervening text we must rely on S alone.

[27] The meaning of *anni* is not clear to me, and I have omitted it from the translation. Müller rendered the passage as follows: "wir nennen ihn einfach den ersten, den Haupt- oder Jahres-Epicykel" (but which I call the first, the chief, or annual epicycle; ZE, XII, 370). There are three objections to Müller's version of *anni*. It is syntactically unsound; in Copernicus's system the first lunar epicycle has no connection with the year; Copernicus regularly employs in his lunar theory the terms "first epicycle" and "greater epicycle," but never "annual epicycle" or "epicycle of the year" (cf. Th 235.14-15, 257.7-8, 262.26, 277.22, 288.23).

[28] When the motion of a circle, in the upper portion of its circumference, is in precedence, i.e., from east to west, in the lower portion it is in consequence, from west to east; and vice versa. "Now let *abc* (Fig. 22) be the epicycle . . .

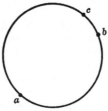

FIGURE 22

and let the motion of the epicycle be understood to be from *c* to *b* and from *b* to *a*, that is, in precedence in the upper portion and in consequence in the lower portion" (Th 251.26-252.1; cf. also Th 323.26-28, 325.21-23; PII, 349.14-16). When the direction of a motion is stated without reference to the portion of the circumference, it is the upper circumference that is understood.

to it is a second epicycle. The moon, finally, moving with this second epicycle, completes two revolutions a month in the direction opposite to that of the greater epicycle, so that whenever the center of the greater epicycle crosses the line drawn from the center of the great circle through the center of the earth (I call this line the diameter of the great circle), the moon is nearest to the center of the greater epicycle. This occurs at new and full moon; but contrariwise at the quadratures, midway between new and full moon,[29] the moon is most remote from the center of the greater epicycle. The length of the radius[30] of the greater epicycle is to the radius of the def-

---

Müller was evidently unfamiliar with this usage, for he detached *in superiore quidem portione* from *contra motum orbis reflexus*. He translated: "dabei führt er auf seiner Aussenseite einen ferneren Epicykel mit sich" (as the first epicycle revolves, it carries with it on its surface another epicycle; ZE, XII, 370). But Rheticus explicitly states: "As the first epicycle revolves uniformly about its own center, in its upper circumference it carries the center of the small second epicycle in precedence, in its lower circumference, in consequence" (p. 134, below).

[29] Here, too, Müller blundered. For he translated *in quadraturis mediantibus iisdem* by: "zur Zeit der mittleren Quadraturen" (at the time of the mean quadratures; ZE, XII, 371). This version ignores *iisdem* and mistakes *mediare* (to halve) for *medius* (the technical astronomical term for "mean"). But Copernicus has not yet begun to discuss the lunar inequalities; all that he is stating here is the elementary fact (see p. 47, above) that the quadratures are midway between new and full moon (*iisdem*).

[30] Although *diametri*, the reading of S, cannot be checked on account of the lacuna in V, it is certainly wrong and must be changed to *semidiametri*. Computational support for this emendation is adduced in n. 32. Additional support comes from a calculation jotted down by Copernicus in his copy of the Tables of Regiomontanus (see Curtze in *Zeitschrift für Mathematik und Physik*, XIX(1874), 454-56). The note reads: *Semidiametrus orbis lunae ad epicyclium a* $\frac{10}{1\frac{1}{18}}$; *epicyclus a ad b* $\frac{19}{4}$ (PII, 211); "Radius of deferent of moon to first epicycle 10:1⅛; first epicycle to second epicycle 19:4." Throughout this series of calculations Copernicus is comparing radius with radius, never diameter with radius.

While the note was properly used by Curtze to emend another false reading (*parte* for *quarta*) in this same sentence of S, he overlooked *diametri*. Curiously enough, in citing Curtze's work Prowe speaks of the note as containing "values calculated by Copernicus for the radii of the planetary epicycles" (PII, 193 n); yet he too failed to notice the discrepancy. Had Müller compared his computations (ZE, XII, 372, n. 51) for the *Commentariolus* with the lunar numerical ratios in *De rev.*, he would surely have caught the copyist's error. It should be observed that Rheticus compares diameter with diameter when he gives the ratio of the lunar epicycles (p. 134, below).

erent as 1⅛8:10;[31] and to the radius of the smaller epicycle as 4¾:1.[32]

By reason of these arrangements the moon appears, at times rapidly, at times slowly, to descend and ascend; and to this first inequality the motion of the smaller epicycle adds two irregularities.[33] For it withdraws the moon from uniform motion on the circumference of the greater epicycle, the maximum inequality being 12¼° of a circumference of corresponding size or diameter;[34] and it brings the center of the greater epicycle

[31] I have adopted this form for the sake of clarity and compactness. What Copernicus actually wrote may be literally translated as follows: "The length of the radius of the greater epicycle contains a tenth part of the radius of the deferent plus one-eighteenth of such tenth part." This ratio may be numerically represented by the expression ⅒ + ⅛8 · ⅒ : 1 or 1 ⅛8 : 10.

[32] Literally: "(The length of the radius of the greater epicycle) contains the radius of the smaller epicycle five times minus one-fourth of the smaller radius." While Copernicus incorporated in *De rev.* the lunar theory sketched in this section, he altered the numerical components slightly (Th 258.10-11). The ratio of first epicycle to deferent is given here as 1⅛8 : 10, which may be written 1055:10,000; in *De rev.* it has been changed to 1097:10,000, which may be written 1⅒ : 10. The ratio of first epicycle to second epicycle appears there as 1097:237, which may be written 4.63:1; it is given above as 4.75:1.

[33] Although the meaning of the passage is clear, the text is faulty and simply does not parse. We might have expected *et primae quidem diversitati dupliciter variationem motus epicycli minoris ingerit* (cf. Th 257.20-21). The distance from the moon to the center of the earth varies, because the moon's orbit around the earth is really an ellipse; and the rate of the moon's apparent motion varies for the same reason. Copernicus uses the term "first inequality" to denote the variation in the moon's distance from the center of the earth and employs the first epicycle to account for it. Both the term and the geometrical device were traditional (cf. HI, 300.16-301.1).

[34] The inequality is measured by an arc of the greater epicycle, or of a circle of equal dimensions. Let AB be the greater epicycle with center at C (Fig. 23). Choose any point E on the circumference, and with E as center describe the second epicycle. Draw CM and CL tangent to the second epicycle. When the moon is at M or L, the inequality attains its maximum. Now in the *Commentariolus* CE:EM = 4.75:1 = 100,000:

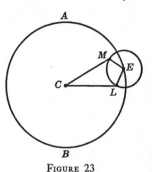

FIGURE 23

21,053. Then by the Table of Chords ∠ECM, which measures the maximum inequality, = 12° 9′ (Th 45.19-20). Hence the reading of S, *17 gradus et quadrantem*, is certainly wrong, and must be corrected to *12 gradus et quadrantem*. As in the case of the solar inequality (see above, p. 62, n. 11), Copernicus is writing a convenient fraction.

at times nearer the moon, at times further from it, within the limits of the radius of the smaller epicycle.[35] Therefore, since the moon describes unequal circles about the center of the greater epicycle, the first inequality varies considerably. In conjunctions and oppositions to the sun its greatest value does not exceed 4° 56', but in the quadratures it increases to 6° 36'.[36] Those who employ an eccentric circle to account for this variation[37] improperly treat the motion on the eccentric as unequal,[38]

While the reading of S cannot be checked on account of the lacuna in V, the proposed emendation is confirmed by a comparison with *De rev.* We saw above (n. 32) that in the later work Copernicus diminished the ratio CE:EM, making it 1097:237 = 4.63:1 = 100,000:21,604. It is obvious that since CE has been shortened in relation to EM, $\angle$ECM must increase; by the Table of Chords, it is 12° 28' (Th 45.21-22, 258.32–259.4, 264.31). Hence any such value as 17¼° for the maximum inequality in the *Commentariolus* must be rejected as a copyist's error.

[35] Reading *cum* for *eum* (PII, 193.11).

[36] The difference between the maximum in the quadratures and the maximum in conjunctions and oppositions is 6° 36' − 4° 56' = 1° 40'. According to Ptolemy, the difference was 2° 40' (HI, 362.1-6). In *De rev.* it is 2° 44' (Th 262.23-32, 265.10-11). Hence I suggest that the figure in our text should be changed from 6° 36' to 7° 36'. Again the reading of S cannot be checked on account of the lacuna in V.

From the following table it can be seen how closely Copernicus adhered to the Ptolemaic determination of the lunar inequalities. The second column contains the maximum inequality in conjunctions and oppositions; the third column shows the greatest additional inequality in the quadratures; and the fourth column sums the second and third.

Ptolemy	5° 0'	2° 40'	7° 40'
Commentariolus	4° 56'	2° 40'	7° 36'
De revolutionibus	4° 56'	2° 44'	7° 40'

Although Ptolemy's table for the first lunar inequality gives 5° 1' as the maximum (HI, 337.21, 390.24), he generally uses the round number 5° in his calculations (HI, 338.22–339.3, 362.1-6, 363.10-12, 364.20-22; cf. Th 257.26-29).

[37] Ptolemy is credited with having discovered the second inequality (HI, 294.9-14, 354.18–355.20); to account for it, he represented the center of the lunar epicycle as revolving on a circle eccentric to the earth (HI, 355.20-22; cf. Th 232.1-3).

[38] This charge that the representation employed by Ptolemy and his successors violates the axiom of uniform motion is amplified in *De rev.*: "For when they assert that the motion of the center of the epicycle is uniform with respect to the center of the earth, they must also admit that the motion is not uniform on the circle which it describes, namely, the eccentric" (Th 233.11-13). Müller was apparently puzzled by the words *praeter ineptam in ipso circulo motus inaequalitatem* and omitted them from his translation (ZE, XII, 373).

and, in addition, fall into two manifest errors. For the consequence by mathematical analysis is that when the moon is in quadrature, and at the same time in the lowest part of the epicycle, it should appear nearly four times greater (if the entire disk were luminous) than when new and full, unless its magnitude increases and diminishes in no reasonable way.[39] So too, because the size of the earth is sensible in comparison with its distance from the moon, the lunar parallax[40] should increase very greatly at the quadratures. But if anyone investigates these matters carefully, he will find that in both respects the quadratures differ very little from new and full moon, and accordingly will readily admit that my explanation is the truer.

With these three motions in longitude, then, the moon passes through the points of its motion in latitude.[41] The axes of the epicycles are parallel to the axis of the deferent, and therefore the moon does not move out of the plane of the deferent. But the axis of the deferent is inclined to the axis of the great circle

[39] Müller translated the last clause: "es sei denn, man behauptete thörichterweise ein wirkliches Wachsen und Abnehmen der Mondkugel" (unless they absurdly maintained that there is a real increase and decrease in the size of the moon; ZE, XII, 373). This version misses the point. The apparent size of the moon (as measured by its apparent diameter) varies, because the distance of the moon from the earth is not constant (see n. 33). The first of the "two manifest errors" produced by the eccentric is, not that it causes the apparent size of the moon to vary, but that it grossly exaggerates the variation (cf. Th 234.31–235.8).

[40] Müller failed to recognize *diversitas aspectus* as the technical term for parallax (see p. 51, above). Hence he was unable to distinguish the second of the "two manifest errors," and his translation (ZE, XII, 373) speaks only of the apparent variation in the size of the moon, "der scheinbare Unterschied in der Grösse." Consequently, in the next sentence, where Copernicus refers to both (*utrumque*) disagreements with the observational data which are produced by the eccentric (1. exaggeration of the variation in the apparent size of the moon; 2. exaggeration of the variation in the lunar parallax), Müller does not know how to render *utrumque*, and falls back on "Grössenunterschied" (variation in size). The explicit statement of Rheticus puts the matter beyond all question: "But experience has shown my teacher that the parallax and size of the moon, at any distance from the sun, differ little or not at all from those which occur at conjunction and opposition, so that clearly the traditional eccentric cannot be assigned to the moon" (p. 134, below).

[41] Here the lacuna in V ends.

or ecliptic;[42] hence the moon moves out of the plane of the ecliptic. Its inclination is determined by the size of an angle which subtends 5° of the circumference of a circle.[43] The poles of the deferent revolve at an equal distance from the axis of the ecliptic,[44] in nearly the same manner as was explained regarding declination.[45] But in the present case they move in the reverse order of the signs and much more slowly, the period of the revolution being nineteen years.[46] It is the com-

[42] Müller rendered *axi magni orbis sive eclipticae* by: "die Achse des grössten Kreises der Ekliptik" (the axis of the great circle of the ecliptic; ZE, XII, 373). This faulty translation shows that Müller did not quite grasp the meaning of *orbis magnus*, which he interpreted (ZE, XII, 365, n. 25) as meaning "great circle" in the geometrical sense, i.e., a circle drawn on the surface of a sphere with its center in the center of the sphere. However, Copernicus's term for "great circle" in the geometrical sense (see above, p. 12, n. 26) is not *orbis magnus* but *circulus maximus* (Th 57-66, *passim*). In the present passage *orbis magnus* bears its usual sense of the real annual revolution of the earth about the sun (see p. 16, above). The *orbis magnus* and the ecliptic lie in the same plane and have a common axis: "But the axis of the great circle is invariably directed toward the points of the firmament which are called the poles of the ecliptic" (p. 64, above).

[43] This estimate of 5° for the maximum latitude of the moon was derived from Ptolemy (HI, 388.11–389.7, 391.52; cf. Th 272.13-15, 274.8-9) and, subject to the correction mentioned in the following note, is retained in modern astronomy.

[44] Therefore the inclination of the moon's orbit to the ecliptic would be constant. That this inclination in fact varies was discovered by Tycho Brahe; see *Tychonis Brahe opera omnia*, ed. Dreyer, II, 121-30, 413.13-21; IV, 42.27-43.22; VI, 170.1–171.8; VII, 151.28–154.35; XI, 162-63; XII, 399-400; Dreyer's remarks on p. liv of the Introduction to Vol. I; and his *Tycho Brahe* (Edinburgh, 1890), pp. 342-44.

[45] See above, p. 64, n. 16. Müller missed the force of *propemodum sicut*, which he translated (ZE, XII, 373) by "ähnlich" (like); whereas "almost like," or something of the sort, is required. In the case of the moon, the poles of the deferent revolve at an equal distance from the axis of the ecliptic, *in aequidistantia axis eclipticae*; but in the case of the motion in declination, the poles of the earth revolve on circles having centers equidistant from the axis of the ecliptic.

[46] This estimate of nineteen years for the period during which the lunar nodes perform their regression was also derived from Ptolemy. He measured the rate of regression by subtracting the moon's mean motion in longitude from the mean motion in latitude (HI, 301.18-23, 356.4-9), the difference being about 3' a day (HI, 356.25-357.6, 358.6-11). By reference to his tables for the moon's motion (HI, 282-293) we can determine the period required for the completion of the circuit as 18 years, 7 months, and 16 days. The discovery that the regression of the nodes is not uniform was made by Tycho Brahe (see the references cited in n. 44, above).

mon opinion that the motion takes place in a higher sphere, to which the poles are attached as they revolve in the manner described. Such a fabric of motions, then, does the moon seem to have.

The Three Superior Planets
Saturn—Jupiter—Mars

Saturn, Jupiter, and Mars have a similar system of motions, since their deferents completely enclose the great circle and revolve in the order of the signs about its center as their common center. Saturn's deferent revolves in 30 years, Jupiter's in 12 years, and that of Mars in 29 months;[47] it is as though the size of the circles delayed the revolutions. For if the radius of the great circle is divided into 25 units, the radius of Mars' deferent will be 38[48] units, Jupiter's 130 5/12, and Saturn's 230 1/6.[49] By "radius of the deferent" I mean the distance from the center of the deferent to the center of the first epicycle. Each deferent has two epicycles,[50] one of which carries the

[47] See above, p. 60, n. 7.

[48] Although both S and V read 30, I propose to substitute 38, for the reasons stated in n. 50, below.

[49] S: *230 et sextantem unius*; V: *236 et sextantem unius*. Prowe accepted V, but S is to be preferred, for the reasons given in n. 50, below.

[50] In the *Commentariolus* Copernicus employs for the planets what we have called the concentrobiepicyclic arrangement (see pp. 7, 37, above), consisting of two epicycles upon a deferent which is concentric with the great circle. In *De rev.* this device is replaced, for the three superior planets, by an eccentrepicyclic arrangement, i.e., by a single epicycle upon an eccentric deferent (Th 325.16-21); after indicating the geometric equivalence of the two devices (Th 325.11-16, 327.6-13), Copernicus points to the variation in the eccentricity of the great circle as the reason for his choice of the eccentrepicyclic arrangement (Th 327.13-16). When he wrote the *Commentariolus*, he regarded this eccentricity as constant (see above, p. 61, n. 9).

Now if the two arrangements are to produce identical results, then, as Copernicus points out, the radius (R) of the concentric deferent (*Commentariolus*) must be equal to the radius (R) of the eccentric deferent (*De rev.*). Let r denote the radius of the great circle. By a comparison of the ratio R:r, as given here, with the values in *De rev.*, we may discover whether in shifting from the concentrobiepicyclic to the eccentrepicyclic arrangement Copernicus altered the relative sizes of the deferent and great circle. In *De rev.*, for Saturn r = 1090 (Th 341.29), for Jupiter r = 1916 (Th 353.15-16), and for Mars r = 6580 (Th 364.8-9), when in each case R = 10,000.

other, in much the same way as was explained in the case of the moon,[51] but with a different arrangement. For the first epicycle revolves in the direction opposite to that of the deferent, the periods of both being equal. The second epicycle, carrying the planet, revolves in the direction opposite to that of the first with twice the velocity. The result is that whenever the second epicycle is at its greatest or least distance from the center of the deferent, the planet is nearest to the center of the first epicycle; and when the second epicycle is at the midpoints, a quadrant's distance from the two points just men-

	$R:r$	
	De revolutionibus	*Commentariolus*
Saturn	$10,000:1090 = 229\frac{1}{3}:25$	$230\frac{1}{6}:25$
Jupiter	$10,000:1916 = 130\frac{1}{2}:25$	$130\frac{5}{12}:25$
Mars	$10,000:6580 = 38:25$	$38:25$

The table enables us to deal with a variant reading in this passage. For R in the case of Saturn, S has $230\frac{1}{6}$, while V gives $236\frac{1}{6}$ (Curtze's collation [MCV, IV, 7] inaccurately assigns to Jupiter the reading of S for Saturn). Prowe accepted the reading of V, but S is clearly preferable, as the following analysis will show.

I have already referred (see p. 69, n. 30) to the series of notes made by Copernicus in his copy of the Tables of Regiomontanus. Curtze correctly pointed out that the ratios contained in these notes are identical with those adopted in the *Commentariolus* (MCV, IV, 7 n.); and he used the statement about the moon to emend a false reading in our text. However, he failed to make any further use of these entries. Now for the radius (not diameter, as MCV, I, 12 n. and PII, 195 n. have it; cf. PII, 211) of Saturn's deferent, they give $230\frac{5}{6}$ (not $230\frac{5}{8}$, as PII, 195 n.). Hence we are justified in preferring the reading of S to that of V. This judgment is confirmed by the fact that Tycho Brahe's reference to the *Commentariolus* agrees with S (see his *Opera omnia*, ed. Dreyer, II, 428.40–429.2).

Moreover, these notes of Copernicus show that S and V agree on a false reading for R in the case of Mars. The statement in the Tables of Regiomontanus gives the radius of Mars' deferent as approximately 38 (*Martis semidiametrus orbis 38 fere*). Now a value of 30, which is the reading of both our MSS, would make the ratio R:r for Mars $30:25 = 10,000:8333$, at wide variance from the corresponding ratio in *De rev.* But reference to the table will show that the agreement between *De rev.* and the *Commentariolus* for both Saturn and Jupiter is quite close. Hence I have adopted 38, the number written by Copernicus in his Tables of Regiomontanus, in place of 30. Writing o for 8 is not an uncommon error of copyists (cf. below, p. 82, n. 74).

[51] See the opening paragraph of the section on "The Moon."

tioned,[52] the planet is most remote from the center of the first epicycle. Through the combination of these motions of the deferent and epicycles, and by reason of the equality of their revolutions, the aforesaid withdrawals and approaches occupy absolutely fixed places in the firmament, and everywhere exhibit unchanging patterns of motion. Consequently the apsides are invariable;[53] for Saturn, near the star which is said to be on the elbow of Sagittarius;[54] for Jupiter, 8° east of the star which is called the end of the tail of Leo;[55] and for Mars, 6½° west of the heart of Leo.[56]

[52] Müller rendered *in quadrantibus autem mediantibus* by: "zur Zeit der mittleren Quadraturen" (at the time of the mean quadratures; ZE, XII, 374). With regard to *mediantibus*, this version repeats the blunder pointed out above in n. 29 on p. 69; and, in addition, it mistakes *quadrans* (quadrant, the fourth part of a circumference) for *quadratura* (quadrature; cf. above, p. 47, n. 163).

[53] This was Ptolemy's view. He held that the planetary apsides were fixed in relation to the sphere of the fixed stars, since, as measured by the equinoxes and solstices, both the apsides and the fixed stars moved in the same direction at the same slow rate (HII, 251.24–252.7, 252.11-18, 257.3-12, 269.3-11; cf. Th 308.20-24).

[54] The star is here described as *quae super cubitum esse dicitur Sagittatoris*. It is unquestionably to be identified with Sagittarius 19 in Ptolemy's catalogue (HII, 114.10), for that star was described in the first printed translation of the *Syntaxis* into Latin (Venice, 1515, p. 84r) as *quae est super cubitum dextrum*. In *De rev.* Copernicus uses instead the name *In dextro cubito* (Th 139.14).

[55] This star is Leo 27 in Ptolemy's catalogue and in *De rev.* (HII, 100.7; Th 135.12). Its Bayer name is β Leonis.

[56] This star is Leo 8 (HII, 98.6; Th 134.23-24). It was called Basiliscus or Regulus, and its Bayer name is α Leonis.

Ptolemy had put the apogee of Saturn at 23° of Scorpio; of Jupiter, at 11° of Virgo; and of Mars, at 25° 30′ of Cancer (HII, 412.12-17, 380.22-381.4, 345.12-20). In his catalogue of the fixed stars these places are, respectively, 31° 50′ west of Sagittarius 19, 16° 30′ east of Leo 27, and 7° west of Leo 8. From them Copernicus's determinations differ, respectively, by 31° 50′ eastward, 8° 30′ westward, and ½° eastward. Hence we may say that although in the *Commentariolus* Copernicus accepted Ptolemy's doctrine of the fixity of the planetary apsides, he intended to put forward improved determinations of them.

In *De rev.* the places are again altered. But now they are all east of Ptolemy's determinations; for Saturn's apogee is 17° 49′ west of Sagittarius 19; Jupiter's, 21° 10′ east of Leo 27; and Mars', 3° 50′ east of Leo 8 (Th 338.15-18, 350.15-16, 360.3-5). Hence Copernicus abandons the idea of the fixed apogee and enunciates the discovery that the longitude of the planetary apogees increases: "Moreover, the position of the higher apse of [Saturn's] eccentric has in the meantime advanced 13° 58′ in the sphere of the fixed stars. Ptolemy believed that this posi-

The radius of the great circle was divided above into 25 units. Measured by these units, the sizes of the epicycles are as follows. In Saturn the radius of the first epicycle consists of 19 units, 41 minutes; the radius of the second epicycle, 6 units, 34 minutes. In Jupiter the first epicycle has a radius of 10 units, 6 minutes; the second, 3 units, 22 minutes. In Mars the first epicycle, 5 units, 34 minutes; the second, 1 unit, 51 minutes.[57] Thus the radius of the first epicycle in each case is three times as great as that of the second.[58]

The inequality which the motion of the epicycles imposes upon the motion of the deferent is called the first inequality; it follows, as I have said, unchanging paths everywhere in the firmament. There is a second inequality, on account of which the planet seems from time to time to retrograde, and often to become stationary. This happens by reason of the motion, not of the planet, but of the earth changing its position in the great circle. For since the earth moves more rapidly than the planet, the line of sight directed toward the firmament regresses, and the earth more than neutralizes the motion of the planet. This regression is most notable when the earth is nearest to the planet, that is, when it comes between the sun and the planet at the evening rising of the planet. On the other hand, when

---

tion, like the others, was fixed; but it is now clear that it moves about 1° in 100 years" (Th 339.7-11; cf. Th 351.2-5, 359.33-360.7; and Dreyer, *Planetary Systems*, p. 338).

[57] I resume the comparison instituted above in n. 50 on p. 74. As Copernicus points out (Th 327.7-8), the radius (E) of the first epicycle (*Commentariolus*) must be equal to the eccentricity (E) of the eccentric (*De rev.*). Now in *De rev.* for Saturn E = 854 (Th 330.18), for Jupiter E = 687 (Th 343.23-28), and for Mars E = 1460 (Th 358.28); we already have the values of r.

	r:E *De revolutionibus*	*Commentariolus*
Saturn	1090: 854 = 25:19^{p}35^m	25:19^{p}41^m
Jupiter	1916: 687 = 25: 8^{p}58^m	25:10^p 6^m
Mars	6580:1460 = 25: 5^{p}33^m	25: 5^{p}34^m

[58] Hence the radius of the second epicycle in the *Commentariolus* is equal to the radius of the single epicycle in *De rev.*, since both = ⅓ E (Th 325.19-20). An exception will be noted in the case of Mars, where Copernicus reduces the eccentricity from 1,500 (Th 354.29-355.2) to 1,460, but leaves the radius of the epicycle at 500 (Th 358.24-31, 360.7-11, 362.26-28).

the planet is setting in the evening or rising in the morning, the earth makes the observed motion greater than the actual. But when the line of sight is moving in the direction opposite to that of the planets and at an equal rate, the planets appear to be stationary, since the opposed motions neutralize each other; this commonly occurs when the angle at the earth between the sun and the planet is 120°.[59] In all these cases, the lower the deferent on which the planet moves, the greater is the inequality. Hence it is smaller in Saturn than in Jupiter, and again greatest in Mars, in accordance with the ratio of the radius of the great circle to the radii of the deferents. The inequality attains its maximum for each planet when the line of sight to the planet is tangent to the circumference of the great circle. In this manner do these three planets move.

In latitude they have a twofold deviation. While the circumferences of the epicycles remain in a single plane with their deferent, they are inclined to the ecliptic. This inclination is governed by the inclination of their axes, which do not revolve, as in the case of the moon,[60] but are directed always toward the same region of the heavens. Therefore the intersections of the deferent and ecliptic (these points of intersection are called the nodes) occupy eternal places in the firmament.[61] Thus the node where the planet begins its ascent toward the north is, for Saturn, 8½° east of the star which is described as being in the head of the eastern of the two Gemini;[62] for Jupiter, 4° west of

[59] Cf. Pliny Natural History ii.15(12).59: "In the trine aspect, that is, at 120° from the sun, the three superior planets have their morning stations, which are called the first stations . . . and again at 120°, approaching from the other direction, they have their evening stations, which are called the second stations"; cf. also ii.16(13).69-71. It has been shown that Copernicus read carefully a copy of the Rome, 1473 edition of Pliny's Natural History (L. A. Birkenmajer, Stromata Copernicana, Cracow, 1924, pp. 327-34); and also a copy of the Venice, 1487 edition (MCV, I, 40-41).

[60] See the closing paragraph of the section on "The Moon."

[61] Copernicus derived from Ptolemy the view that the nodes, like the apsides, are fixed (HII, 530.8-11; cf. Karl Manitius, Des Claudius Ptolemäus Handbuch der Astronomie, Leipzig, 1912-13, II, 426). But in De rev., having discovered the motion of the apsides, Copernicus holds that this motion is shared by the nodes (Th 413.7-15, 415.20-25).

[62] Gemini 2 (HII, 92.4; Th 132.31-32), Pollux, β Geminorum; cf. above, p. 62, n. 12.

the same star; and for Mars, 6½° west of Vergiliae.[63] When the planet is at this point and its diametric opposite, it has no latitude. But the greatest latitude, which occurs at a quadrant's distance from the nodes,[64] is subject to a large inequality. For the inclined axes and circles seem to rest upon the nodes, as though swinging from them. The inclination becomes greatest when the earth is nearest to the planet, that is, at the evening rising of the planet; at that time the inclination of the axis is, for Saturn 2⅔°, Jupiter 1⅔°, and Mars 1⅝°.[65] On the other hand, near the time of the evening setting and morning rising, when the earth is at its greatest distance from the planet, the inclination is smaller,[66] for Saturn and Jupiter by ⁵⁄₁₂°, and for

[63] Taurus 30 (HII, 90.2; Th 132.5-6). Authorities differ about the identification of Taurus 30; see Christian H. F. Peters and Edward B. Knobel, *Ptolemy's Catalogue of Stars* (Carnegie Institution of Washington, Publication No. 86, 1915), p. 115.

[64] Ptolemy had put the points of greatest northern latitude for Saturn and Jupiter at 0° of Libra, and for Mars at 30° of Cancer (HII, 526.6-11; cf. Th 413.7-11). If we compare these places with his determinations of the apogees (see above, p. 76, n. 56), we find that for Saturn the point of greatest northern latitude is 53° west of the apogee; for Jupiter, 19° east; and for Mars, 4° 30' east. Ptolemy states these differences of position in round numbers as 50° west, 20° east, and 0° (HII, 587.5-9; cf. Manitius, *Ptolemäus Handbuch*, II, 425, n. 21).

In the present passage Copernicus gives the places of the ascending nodes. By adding 90° to these places, we obtain the points of greatest northern latitude. They turn out to be, for Saturn, 79° 40' west of the apogee; for Jupiter, 20° 10' east; and for Mars, 0° 20' west. In the *Commentariolus*, then, Copernicus not only adheres to the Ptolemaic ideas of the fixed apogee and the fixed node, but he also retains Ptolemy's distance between apogee and node for Jupiter and Mars, although increasing the distance by 30° for Saturn.

In *De rev.*, although the apogee moves, the distance between apogee and node remains constant, since the node shares the motion of the apogee. Copernicus finds the points of greatest northern latitude, for Saturn at 7° of Scorpio; for Jupiter at 27° of Libra; and for Mars at 27° of Leo (Th 413.11-13). If we compare these places with his determinations of the apogees (see above, p. 76, n. 56), we find that for Saturn the point of greatest northern latitude is 23° 21' west of the apogee; for Jupiter, 48° east; and for Mars, 27° 20' east.

[65] S: *dextante*; V: *sextante*. Prowe, followed by Müller (ZE, XII, 377), adopted the reading of V; but *sextante* is clearly impossible, for the following sentence of the text states that the inclination diminishes in the case of Mars by 1⅔°.

[66] The inclination is greatest when the planet is in opposition, smallest when the planet is in conjunction; and the greatest difference between maximum and

Mars by $1\frac{2}{3}°$. Thus this inequality is most notable in the greatest latitudes, and it becomes smaller as the planet approaches the node, so that it increases and decreases equally with the latitude.

The motion of the earth in the great circle also causes the observed latitudes to change, its nearness or distance increasing or diminishing the angle of the observed latitude, as mathematical analysis demands. This motion in libration occurs along a straight line, but a motion of this sort can be derived from two circles. These are concentric, and one of them, as it revolves, carries with it the inclined poles of the other. The lower circle revolves in the direction opposite to that of the upper, and with twice the velocity. As it revolves, it carries with it the poles of the circle which serves as deferent to the epicycles. The poles of the deferent are inclined to the poles of the circle

---

minimum occurs at the points of greatest latitude (Th 415.9-14). The following table compares the maximum and minimum angles of inclination as given here with those in *De rev*. (Th 421.22-25; 421.31–422.1; 422.7-8, 10-11).

	Angles of Inclination	Commentariolus	De revolutionibus
Saturn	Greatest	2° 40'	2° 44'
	Least	2° 15'	2° 16'
Jupiter	Greatest	1° 40'	1° 42'
	Least	1° 15'	1° 18'
Mars	Greatest	1° 50'	1° 51'
	Least	0° 10'	0° 9'

From the table we see that the main inclinations and their limits of variation are as follows:

	Commentariolus	De revolutionibus
Saturn	2° 27½' ± 12½'	2° 30' ± 14'
Jupiter	1° 27½' ± 12½'	1° 30' ± 12'
Mars	1° ± 50'	1° ± 51'

In Ptolemy's treatment of the latitudes, for the three superior planets the angle at which the eccentric deferent was inclined to the ecliptic was constant (HII, 529.3-9). His values were: for Saturn 2° 30', for Jupiter 1° 30', and for Mars 1° (HII, 540.13-14, 542.5-9). But the epicycle was inclined to the eccentric at a varying angle (HII, 529.12-530.8). It will be observed that in Copernicus's theory the epicycles and deferent are coplanar; hence the angle at which the deferent is inclined to the ecliptic cannot be fixed, but must vary (Th 413.1-3, 29-31).

halfway[67] above at an angle equal to the inclination of these poles to the poles of the highest circle.[68] So much for Saturn, Jupiter, and Mars and the spheres which enclose the earth.

Venus

There remain for consideration the motions which are included within the great circle, that is, the motions of Venus and Mercury. Venus has a system of circles like the system of the superior planets,[69] but the arrangement of the motions is different. The deferent revolves in nine months, as was said above,[70] and the greater epicycle also revolves in nine months. By their composite motion the smaller epicycle is everywhere brought back to the same path in the firmament, and the higher apse is at the point where I said the sun reverses its course.[71] The period of the smaller epicycle is not equal to that of the deferent and greater epicycle,[72] but has a constant relation to

[67] S: mediate; V: mediale. Before S was known, Curtze emended V to immediate, which Prowe prints. But S is undoubtedly correct.

[68] Since motion in a straight line would violate the principle of circularity, Copernicus is at pains to prove that a rectilinear motion may be produced by a combination of two circular ones. A less concise account of this geometric device, employed in connection with the theory of precession, as well as an explanation of the term "libration," will be found in the Narratio prima (pp. 153-54, below; cf. Th 165.18-169.22).

[69] In De rev. Copernicus replaces the concentrobiepicyclic arrangement for Venus by an eccentreccentric arrangement, i.e., by two eccentrics (Th 368.23-29). The larger, outer eccentric which carries the planet has for its center a point which revolves on the smaller eccentric (Th 368.30-369.6).

[70] Page 60.

[71] In placing the apogee of Venus at the solar apogee Copernicus retains the Ptolemaic idea of the fixed apse, but he offers an improved determination. For Ptolemy had put the apogee of Venus at 25° of Taurus (HII, 300.15-16; cf. Th 365.20-25; 366.3-7, 17-20), and the solar apogee at 5° 30' of Gemini (see above, p. 62, n. 13). Hence for him the apogee of Venus was 10° 30' west of the solar apogee. Now we have already seen that in the Commentariolus Copernicus advances the solar apogee 11° 10', as measured by the fixed stars, over Ptolemy's determination. Hence he advances the apogee of Venus 21° 40', again as measured by the fixed stars, over Ptolemy's determination.

[72] This is the difference between the arrangement of the motions, on the one hand, for the three superior planets, and on the other hand, for Venus. In the former case the period of the smaller epicycle is one-half the period of the deferent and greater epicycle (see the opening paragraph of the section on "The

the motion of the great circle. For one revolution of the latter the smaller epicycle completes two. The result is that whenever the earth is in the diameter drawn through the apse, the planet is nearest to the center of the greater epicycle; and it is most remote, when the earth, being in the diameter perpendicular to the diameter through the apse, is at a quadrant's distance from the positions just mentioned. The smaller epicycle of the moon moves in very much the same way with relation to the sun.[73] The ratio of the radius of the great circle to the radius of the deferent of Venus is 25:18;[74] the greater epicycle has a value of ¾ of a unit, and the smaller ¼.[75]

Three Superior Planets"). Müller completely missed the distinction. His translation runs: "Die Umlaufszeit dieses kleineren Epicykels ist verschieden von der der oben genannten Kreise; so entsteht längst der Ekliptik eine ungleichförmige Bewegung. Vollführen jene einen Umlauf, so führt der kleinere einen doppelten aus" (The period of this smaller epicycle is different from that of the above-mentioned circles [i.e., deferent and greater epicycle]; thus there appears along the ecliptic an unequal motion. While those circles [i.e., deferent and greater epicycle] complete one revolution, the smaller epicycle completes two; ZE, XII, 378). The source of Müller's difficulty seems to have been the unusual expression *Minor autem epicyclus impares cum illis revolutiones habens, motui orbis magni imparitatem reservavit.* This may be literally translated as follows: "The smaller epicycle, having revolutions unequal with those of the deferent and greater epicycle, has reserved the inequality for the motion of the great circle." The next sentence in the text makes Copernicus's meaning clear beyond dispute. The revolution of the smaller epicycle takes half the time required by the motion on the great circle.

[73] See the opening paragraph of the section on "The Moon."

[74] S has the false reading 10, instead of 18 (Lindhagen reproduces this page of the MS). I call attention to the copyist's error of writing o for 8, in connection with the emendation proposed in the last paragraph of n. 50 (p. 75, above).

[75] To discover whether in shifting from the concentrobiepicyclic arrangement in the *Commentariolus* to the eccentreccentric arrangement in *De rev.* Copernicus altered the relative sizes of the circles, we may make the following comparisons. The radius (R) of the concentric deferent (*Commentariolus*) corresponds to the radius (R) of the outer eccentric (*De rev.*). Similarly, the radius (E) of the first epicycle (*Commentariolus*) corresponds to the eccentricity (E) of the outer eccentric (*De rev.*); and since the eccentricity varies, we take its mean value. Let r denote the radius of the great circle. Now in *De rev.* R = 7193, r = 10,000, and E = 312 (Th 367.13-14, 368.12-22, 371.11). Then in *De rev.* r:R = 10,000:7193 = 25:17.98, while in the *Commentariolus* r:R = 25:18; in *De rev.* r:E = 10,000:312 = 25:0.78, while in the *Commentariolus* r:E =

Venus seems at times to retrograde, particularly when it is nearest to the earth, like the superior planets, but for the opposite reason. For the regression of the superior planets happens because the motion of the earth is more rapid than theirs, but with Venus, because it is slower; and because the superior planets enclose the great circle, whereas Venus is enclosed within it. Hence Venus is never in opposition to the sun, since the earth cannot come between them, but it moves within fixed distances on either side of the sun. These distances are determined by tangents to the circumference drawn from the center of the earth, and never exceed $48°$ in our observations.[76] Here ends the treatment of Venus' motion in longitude.

Its latitude also changes for a twofold reason. For the axis of the deferent is inclined at an angle of $2\frac{1}{2}°$,[77] the node whence the planet turns north being in the apse. However, the deviation which arises from this inclination, although in itself it is one and the same, appears twofold to us.[78] For when the earth is on the line drawn through the nodes of Venus, the deviations on the one side are seen above, and on the opposite

---

25:0.75. The radius of the second epicycle $= \frac{1}{3}$ E, a ratio which is applied in the *Commentariolus* to all the planets. In *De rev.* the radius of the smaller eccentric, being one-third of the mean eccentricity of the outer eccentric (Th 368.18-22), also $= \frac{1}{3}$ E. Hence the second epicycle (*Commentariolus*) corresponds to the smaller eccentric (*De rev.*).

Despite *dodrantem* in the text, Müller's translation makes E $= \frac{2}{3}$ (ZE, XII, 378). He was evidently confused by a misprint in Prowe's footnote (PII, 198). Yet in that same footnote, five lines below the misprint, the correct value of $\frac{3}{4}$ appears (cf. PII, 211 and MCV, I, 14-15 n).

[76] This value of $48°$ for the greatest elongation of Venus was derived from Ptolemy (HII, 522.14), and is accepted by modern astronomy.

[77] Müller wrote $2°$ (ZE, XII, 379). He was evidently unfamiliar with *s.* as the abbreviation of *semissis*, "one-half" (cf. Th 71.23, 167.4, 425.25). For in his note on the matter he misinterpreted *s.* as the abbreviation of *scrupula*, "minutes" (this word was not assigned to the masculine gender, as Müller thought). Had he consulted Curtze's collation of S and V, his difficulty would have been obviated. For Curtze, confronted by a variant reading (MCV, IV, 8), showed that $2\frac{1}{2}°$ is supported by *De rev.* (Th 424.23-24). Moreover, in Ptolemy's treatment of the latitude of Venus, there are two inclinations of the epicycle, and each is given as $2\frac{1}{2}°$ (HII, 535.15-18, 536.8-11).

[78] S, V: *duplex non ostenditur*. Müller correctly emended to *duplex nobis ostenditur* (ZE, XII, 379, n. 72).

side below; these are called the reflexions.[79] When the earth
is at a quadrant's distance[80] from the nodes, the same natural
inclinations of the deferent appear, but they are called the
declinations. In all the other positions of the earth, both latitudes mingle and are combined, each in turn exceeding the
other; by their likeness and difference they are mutually increased and eliminated.

The inclination of the axis is affected by a motion in libration
that swings, not on the nodes as in the case of the superior
planets,[81] but on certain other movable points. These points
perform annual revolutions with reference to the planet.
Whenever the earth is opposite the apse of Venus, at that time
the amount of the libration attains its maximum for this planet,
no matter where the planet may then be on the deferent. As a
consequence, if the planet is then in the apse or diametrically
opposite to it, it will not completely lack latitude, even though
it is then in the nodes. From this point the amount of the
libration decreases, until the earth has moved through a quadrant of a circle from the aforesaid position, and, by reason of
the likeness of their motions, the point of maximum deviation[82]
has moved an equal distance from the planet. Here no trace
of the deviation is found.[83] Thereafter the descent of the deviation continues.[84] The initial point drops from north to south,

[79] An alternative name was obliquation: "They call this deviation of the planet
the obliquation, but some call it the reflexion" (Th 418.22-23). In *De rev.*
Copernicus generally uses obliquation, but in the *Narratio prima* Rheticus favors
reflexion.

[80] Müller's translation: "in den Quadraturen" (in the quadratures; ZE, XII,
379) again confuses *quadrantibus* with *quadraturis* (cf. above, p. 76, n. 52).
An inferior planet cannot come to quadrature (see above, p. 50); Copernicus has
just stated that the greatest elongation of Venus is 48°.

[81] See the penultimate paragraph of the section on "The Three Superior
Planets."

[82] Müller correctly emended *maxime* (S, V) to *maximae* (ZE, XII, 380, n. 75).

[83] Since the deviation vanishes when the earth is 90° from the apse-line of the
planet, the deviation has no effect upon the declinations, but only upon the reflexions. Copernicus employs the deviation "because the angle of inclination . . .
is found to be greater in the obliquation [reflexion] than in the declination"
(Th 418.27-29).

[84] S, V: *continuato*. Prowe's *continuatio* is a misprint (PII, 200.3).

constantly increasing its distance from the planet in accordance with the distance of the earth from the apse. Thereby the planet is brought to the part of the circumference which previously was south. Now, however, by the law of opposition, it becomes north and remains so until the limit of the libration is again reached upon the completion of the circle. Here the deviation becomes equal to the initial deviation and once more attains its maximum. Thus the second semicircle is traversed in the same way as the first. Consequently this latitude, which is usually called the deviation, never becomes a south latitude. In the present instance, also, it seems reasonable that these phenomena should be produced by two concentric circles with oblique axes, as I explained in the case of the superior planets.[85]

Mercury

Of all the orbits in the heavens the most remarkable is that of Mercury, which traverses almost untraceable paths, so that it cannot be easily studied. A further difficulty is the fact that the planet, following a course generally invisible in the rays of the sun, can be observed for a very few days only. Yet Mercury too will be understood, if the problem is attacked with more than ordinary ability.

Mercury, like Venus, has two epicycles which revolve on the deferent.[86] The periods of the greater epicycle and deferent are equal, as in the case of Venus. The apse is located 14½° east of Spica Virginis.[87] The smaller epicycle revolves with twice the velocity of the earth. But by contrast with Venus, whenever the earth is above the apse or diametrically opposite

[85] For a fuller account of Copernicus's theory for the latitudes of Venus see pp. 180-85, below.

[86] In *De rev.* Copernicus replaces the concentrobiepicyclic arrangement for Mercury by an eccentreccentric arrangement (Th 377.2-3).

[87] Since Ptolemy had put the apogee of Mercury at 10° of Libra (HII, 264.12-14, 271.2-4; cf. Th 380.6-7), and Spica at 26° 40' of Virgo (HII, 103.16), the apse was 13° 20' east of Spica Virginis. Hence in the *Commentariolus* Copernicus retains the idea of the fixed apse and modifies its position slightly. But in *De rev.* he puts the apse 41° 30' east of Spica (Th 136.10, 389.5-6, 393.5-8), and extends to Mercury the principle that the longitude of the planetary apogees increases (Th 393.16-19, 27-29; cf. n. 56 on pp. 76-77, above).

to it, the planet is most remote from the center of the greater epicycle; and it is nearest, whenever the earth is at a quadrant's distance[88] from the points just mentioned. I have said[89] that the deferent of Mercury revolves in three months, that is, in 88 days. Of the 25 units into which I have divided the radius of the great circle, the radius of the deferent of Mercury contains 9⅖. The first epicycle contains 1 unit, 41 minutes; the second epicycle is ⅓ as great, that is, about 34 minutes.[90]

But in the present case this combination of circles is not sufficient, though it is for the other planets. For when the earth passes through the above-mentioned positions with respect to the apse the planet appears to move in a much smaller path[91] than is required by the system of circles described above; and in a much greater path,[91] when the earth is at a quadrant's distance[92] from the positions just mentioned. Since no other inequality in longitude is observed to result from this, it may be reasonably explained by a certain approach of the planet to and withdrawal from the center of the deferent[93] along a

[88] Again Müller erroneously translates by "in the quadratures" (ZE, XII, 381). Mercury, like Venus, cannot come to quadrature (cf. above, p. 84, n. 80).

[89] Page 60, above.

[90] The analysis made above (p. 82, n. 75) for Venus is equally applicable here. In *De rev.* R (mean value) = 3,763, r = 10,000, and E (mean value) = 736 (Th 382.9-10, 382.27-383.2). Then in *De rev.* r:R = 10,000:3763 = 25: 9.41, while in the *Commentariolus* r:R = 25:9.40; in *De rev.* r:E = 10,000:736 = 25:1.84, while in the *Commentariolus* r:E = 25:1.68. The radius of the second epicycle (*Commentariolus*) = ⅓ E. But the radius of the smaller eccentric (*De rev.*) = ⅓ E, only where E denotes the eccentricity of the outer eccentric (Th 377.11-15), as set down in conformity with the general planetary theory used in *De rev.* As in the case of Mars (see above, p. 77, n. 58), Copernicus modifies the ratio; the radius of the smaller eccentric = 212 (Th 382.8-9), or 2/7 E, where E denotes the mean eccentricity of the outer eccentric.

[91] Müller translates *longe minori apparet ambitu sidus moveri* by: "so scheint der Planet sich viel langsamer zu bewegen" (the planet appears to move much more slowly); and *longe etiam maiore* by: "viel schneller" (much more swiftly; ZE, XII, 381). However, Copernicus is concerned here with the variations, not in Mercury's velocity, but in its distance from the center of the great circle.

[92] Failing to recognize that *quadratura* is used here and again near the close of this paragraph in the sense of "quadrant" (see above, p. 47, n. 163), Müller inaccurately translates by "in the quadratures" (ZE, XII, 381, 382).

[93] S: *a centro orbis*; V: *centri orbis*. Prowe accepted V, although S is certainly correct.

straight line. This motion must be produced by two small circles stationed about the center of the greater epicycle, their axes being parallel to the axis of the deferent. The center of the greater epicycle, or of the whole epicyclic structure, lies on the circumference of the small circle that is situated between this center and the outer small circle. The distance from this center to the center of the inner circle is exactly[94] equal to the distance from the latter center to the center of the outer circle.[95] This distance has been found to be 14½ minutes[96] of one unit of the

[94] S, V: *asse.* For this sound reading Curtze incorrectly substituted *axe* (MCV, I, 17.5), which Prowe accepted (PII, 201.10). By ignoring the rules of syntax Müller contrived to incorporate *axe* in his translation.

[95] Let the dotted circumference (Fig. 24) represent the inner small circle with

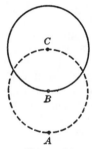

FIGURE 24

its center at B; and the unbroken circumference, the outer small circle with its center at C. The center of the greater epicycle is at A; and AB = BC.

[96] S: *minut. 14 et medio;* V: *minutibus 24 et medio.* Prowe accepted V, although S is certainly correct, as the closing words of this paragraph show. Again Copernicus's notations in his copy of the Tables of Regiomontanus aid us. For the entry concerning Mercury gives values for the deferent and epicycles that agree with those in our text. Then it adds that the inequality of the diameter is 29 minutes (*diversitas diametri 0.29*). Now Curtze, Prowe, and Müller quoted the entry in their notes (MCV, I, 16 n; PII, 201 n; ZE, XII, 381, n. 78). All three called attention to the agreement between the entry and our text with reference to the deferent and epicycles. But they failed to see that the "approach and withdrawal" of our text is identical with the "inequality of the diameter" in the entry; and that the value of 29 minutes given in both places establishes the correctness of S as against V.

This value varies but slightly from Ptolemy's. In his system, the inequality is produced by a small circle upon which the center of the eccentric revolves (HII, 252.26–253.6, 256.15–22; cf. Th 376.17–24). If we compare the radius of the small circle with the sum of the radii of the eccentric and epicycle (HII, 279.15-

25 by which I have measured the relative sizes of all the circles. The motion of the outer small circle performs two revolutions in a tropical year,[97] while the inner one completes four in the same time with twice the velocity in the opposite direction. By this composite motion the centers of the greater epicycle are carried along a straight line, just as I explained with regard to the librations in latitude.[98] Therefore, in the aforementioned positions of the earth with respect to the apse, the center of the greater epicycle is nearest to the center of the deferent; and it is most remote, when the earth is at a quadrant's distance[92] from these positions. When the earth is at the midpoints, that is, 45° from the points just mentioned, the center of the greater epicycle joins the center of the outer[99] small circle, and both centers coincide.[100] The amount of this with-

---

18), we get the ratio $1:27\frac{1}{2}$, while in the *Commentariolus* the corresponding ratio is $1 : 24$ ($29^m : 9^p24^m + 1^p41^m + 34^m$).

In *De rev.* Copernicus represents the inequality by adding an epicycle to the outer eccentric (Th 377.4-8, 18-23); so that, if we include this refinement, his arrangement for Mercury in *De rev.* is bieccentrepicyclic rather than eccentreccentric (Th 377.23-26). But he does not alter the amount of the inequality. For he puts the diameter of the epicycle at 380, where $r = 10,000$ (Th 382.23-27, 384.9-14). Then the amount of the inequality is 190 ($r = 10,000$), or $28\frac{1}{2}$ minutes, where $r = 25$.

[97] Müller failed to recognize *annus vertens* as the term for "tropical year" (see p. 46, above).

[98] See the closing paragraph of the section on "The Three Superior Planets."

[99] Müller omitted *exterioris* from his translation (ZE, XII, 382).

[100] Figure 25 may serve to clarify this motion in libration. In the initial position, the earth is at E_1 on the produced apse-line, the center of the greater epicycle is at A, the center of the inner small circle is at B_1, and the center of the outer small circle is at C. While the earth moves 45° from E_1 to E_2, the outer circle rotates through a quadrant, thereby moving the center of the inner circle from B_1 to B_2. But during this interval, the inner circle rotates through a semicircle, thereby bringing the center of the epicycle to C. As the earth moves 45° from E_2 to E_3, the center of the inner circle reaches B_3, and the center of the epicycle comes to D. As the earth moves from E_3 to E_4, the center of the inner circle goes to B_4, and the center of the epicycle to C. When the earth arrives at E_5, the center of the inner circle returns to B_1, and the center of the epicycle to A. While the earth completes the remaining semicircle E_5-E_6-E_7-E_8-E_1, the small circles repeat their previous motion. Therefore, whenever the earth is on the produced apse-line (E_1 or E_5), the center of the greater epicycle is nearest (A) to the center of the deferent. When the earth is at a quadrant's distance from the apse-

drawal and approach is 29 minutes[96] of one of the above-mentioned units. This, then, is the motion of Mercury in longitude.

Its motion in latitude is exactly like that of Venus, but always in the opposite hemisphere. For where Venus is in north latitude, Mercury is in south. Its deferent is inclined to the ecliptic at an angle of 7°.[101] The deviation, which is always south, never

line (E3 or E7), the center of the epicycle is most remote (D) from the center of the deferent. When the earth is at E2, E4, E6, or E8, the center of the epicycle coincides with C, the center of the outer small circle.

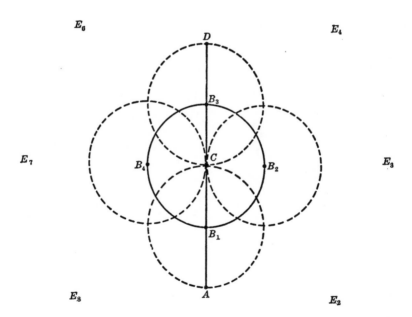

FIGURE 25

[101] In *De rev.* the angle is given as 6° 15′ for the declinations, and 7° for the reflexions (Th 424.23-27, 431.4-9). These were Ptolemy's values (HII, 536.20-22, 575.9-11).

exceeds ¾°.[102] For the rest, what was said about the latitudes of Venus may be understood here also, to avoid repetition.

Then Mercury runs on seven circles in all; Venus on five; the earth on three, and round it the moon on four; finally Mars, Jupiter, and Saturn on five each. Altogether, therefore, thirty-four circles suffice to explain the entire structure of the universe and the entire ballet of the planets.

[102] This was the traditional estimate (Th 433.20-25); but in *De rev.* Copernicus puts it at 51′ ± 18′ (Th 435.4-8; 440-41). Müller rendered this sentence by: "doch übersteigt die Ablenkung nach Süden nie den zwölften Teil eines Grades" (the southward deviation never exceeds ½₂°; ZE, XII, 382). This version omits *semper*, and puts *dodrantem* = ½₂. For ½₂ Copernicus wrote the usual word *uncia* (Th 159.28).

THE *LETTER AGAINST WERNER*

TO THE REVEREND BERNARD WAPOWSKI, Cantor and Canon of the Church of Cracow, and Secretary to His Majesty the King of Poland, from Nicholas Copernicus.

Some time ago, my dear Bernard, you sent me a little treatise on *The Motion of the Eighth Sphere* written by John Werner of Nuremberg. Your Reverence stated that the work was widely praised and asked me to give you my opinion of it. Had it been really possible for me to praise it with any degree of sincerity, I should have replied with a corresponding degree of pleasure. But I may commend the author's zeal and effort. It was Aristotle's advice that "we should be grateful not only to the philosophers who have spoken well, but also to those who have spoken incorrectly, because to men who desire to follow the right road, it is frequently no small advantage to know the blind alleys."[1] Faultfinding is of little use and scant profit, for it is the mark of a shameless mind to prefer the role of the censorious critic to that of the creative poet. Hence I fear that I may arouse anger if I reprove another while I myself produce nothing better. Accordingly I wished to leave these matters, just as they are, to the attention of others; and I intended to reply to your Reverence substantially along these lines, with a view to a favorable reception of my work. However, I know that it is one thing to snap at a man and attack him, but another thing to set him right and redirect him when he strays, just as it is one thing to praise, and another to flatter and play the fawner. Hence I see no reason why I should not comply with your request or why I should appear to hamper the pursuit and cultivation of these studies, in which you have a conspicuous place. Consequently, lest I seem to condemn the man gratuitously, I shall attempt to show as clearly as possible in what respects he errs regarding the motion of the sphere of the fixed stars and maintains an unsound position. Perhaps my criticism

[1] *Metaphysics* i minor.1 993b11-14; Copernicus departs considerably from the original and appears to be quoting from memory.

may even contribute not a little to the formation of a better understanding of this subject.

In the first place, then, he went wrong in his calculation of time. He thought that the second year of Antoninus Pius Augustus, in which Claudius Ptolemy drew up the catalogue of the fixed stars as observed by himself,[2] was A.D. 150,[3] when in fact it was A.D. 139. For in the *Great Syntaxis*, Book III, chapter i,[4] Ptolemy says that the autumnal equinox observed 463 years after the death of Alexander the Great fell in the third year of Antoninus.[5] But from the death of Alexander to the birth of Christ there are 323 uniform Egyptian years[6] and 130 days, because the interval between the beginning of the reign of Nabonassar and the birth of Christ is computed as 747 uniform years and 130 days.[7] This computation, I observe, is not questioned, certainly not by our author, as can be seen in his Proposition 22.[8] It is true that according to the Alfonsine

---

[2] The view that Ptolemy copied his star catalogue from that of Hipparchus corrected for precession was critically examined by J. L. E. Dreyer and rejected (*Monthly Notices of the Royal Astronomical Society*, LXXVII [1917], 528-39; LXXVIII [1918], 343-49); J. K. Fotheringham took the same position (LXXVIII [1918], 419-22). See also Armitage, *Copernicus*, p. 106.

[3] In Proposition 4 (not 3 as in PII, 173 n) of his *De motu octavae sphaerae tractatus primus* Werner dates an observation of the second year of Antoninus *anno dominicae incarnationis 150 incompleto*, "in the 150th year of the incarnation of our Lord." Copernicus does not call attention to a related error committed by Werner. The latter regarded February 22, 150, as the epoch of Ptolemy's catalogue of the fixed stars: "Therefore it is clear that Ptolemy established the true places of the fixed stars in the zodiac for the 22d day of February, according to the Roman calendar, A. D. 150." (Prop. 4); ". . . the era of Ptolemy, that is, 149 years, 53 days from the incarnation of the Lord . . ." (Prop. 10). But Ptolemy gives his epoch as the beginning of the reign of Antoninus (HII, 36.13-16).

[4] HI, 204.7-11.

[5] Siegmund Günther erred when he stated (MCV, II[1880], 5.15-19) that, according to Copernicus, Ptolemy equated the year 463 of Alexander with the second year of Antoninus.

[6] Of exactly 365 days, with 12 months of 30 days each and 5 additional days (cf. Th 172.29-173.12).

[7] Günther's "747 years and 140 days" (MCV, II, 6.3) is evidently a misprint, for ten lines below (*ibid.*, 6.13) he gives the correct number.

[8] Werner there states that "the interval between the years of Christ and Nebuchadnezzar is, according to the Alfonsine Tables, 747 uniform years, 131 days."

Tables there is one additional day. The reason for this discrepancy is that Ptolemy takes noon of the first day of the first Egyptian month Thoth as the starting point of the years reckoned from Nabonassar and Alexander the Great,[9] while Alfonso starts from noon of the last day of the preceding year,[10] just as we compute the years of Christ from noon of the last day of the month December.[11] Now the interval from Nabonassar to the death of Alexander the Great is given by Ptolemy, Book III, chap. viii,[12] as 424 uniform years; and Censorinus, relying on Marcus Varro, agrees with this estimate in his De die natali,[13] addressed to Quintus Caerellius.[14] This interval, subtracted from 747 years and 130 days, leaves a remainder of 323 years and 130 days as the period from the death of Alexander to the birth of Christ. Then from the birth of Christ to the aforementioned observation of Ptolemy there are 139 uniform years and 303 days.[15] Therefore it is clear that the

---

In De rev. Copernicus calls attention to the error, frequently made, of identifying Nebuchadnezzar with Nabonassar (Th 186.20-24).

[9] HI, 256.13-16.

[10] Libros del saber de astronomia, ed. Rico y Sinobas, IV, 120.

[11] Cf. Werner, Prop. 16, corollary 3: "Roman years . . . that is, years computed from the birth of the Savior and from noon of the last day of December."

[12] HI, 256.10-12. In H these lines are in chap. vii. Copernicus cites them from chap. viii, because he is using the translation of 1515, which divides Book III into ten chapters instead of nine. It makes a separate chapter of the penultimate paragraph of Heiberg's chap. v (HI, 251.10-252.8). L. A. Birkenmajer discovered Copernicus's copy of the 1515 translation (see his Mikołaj Kopernik, ch. x, and Bulletin international de l'académie des sciences de Cracovie, classe des sciences math. et naturelles [1909²], p. 24, n. 2).

[13] 21.9; Censorinus equates the year 986 of Nabonassar with the year 562 of Alexander.

[14] On the form of this name Curtze wrote (MCV, I, 25) a perplexing note, which Prowe repeats (PII, 174) with all its blunders. Hultsch's edition of Censorinus (Leipzig, 1867) reads Caerellius, not Caerellus (1.1, 15.1). The incunabular editions (GW 6,471-72) read Cerelius and Cerellius, not Cerillius. The praenomen is Quintus, not Gaius (PW s. v. Caerellius, No. 4; Prosopographia imperii Romani [2d ed.; ed. Groag and Stein, Berlin and Leipzig, 1933-], II, 30, No. 156).

[15] Günther's "139 years and 333 days" (MCV, II, 6.16) is a misprint. The observation was made in the early morning of the 9th day of Athyr (the third month in the Egyptian year), and is therefore assigned to the 68th astronomical day of the year. Then Copernicus reckons the interval from the death of Alex-

autumnal equinox observed by Ptolemy occurred 140[16] uniform years after the birth of our Lord, on the ninth day of the month Athyr; or 139 Roman years, September 25, the third year of Antoninus.[17]

Again, in the *Great Syntaxis,* Book V, chap. iii, Ptolemy counts 885 years and 203 days from Nabonassar to his observation of the sun and moon in the second year of Antoninus.[18] Therefore 138 uniform years and 73 days must have elapsed since the birth of Christ.[19] Hence the fourteenth day thereafter, that is, the ninth of Pharmuthi, on which Ptolemy observed

---

ander to the observation as 463^y 68^d;

subtract 323^y 130^d from the death of Alexander to the birth
$\overline{139^y\ 303^d}$ of Christ

[16] Since the first day of the Roman year fell on the 130th day of the Egyptian year, and $130 + 303 > 365$, the observation took place in the following Egyptian year.

[17] Since $\frac{1}{4} \times 139 = 34\frac{3}{4}$, subtract 35 days on account of leap-years: $303 - 35 = 268$. Hence the observation took place on the 268th day of the Roman year, or September 25. Copernicus assigns the observation to A. D. 140, although A. D. 139 is clearly correct (see emendation No. 12 on p. 254 in L. A. Birkenmajer, *Mikołaj Kopernik*). It was pointed out in n. 15, above, that he counted 463 years, 68 days, from Alexander to the observation. Yet the 1515 edition of the *Syntaxis* reads . . . *quod quidem fuit post mortem Alexandri in .463. anno . . .* (p. 28r), ". . . which was in the 463d year after the death of Alexander . . ." making the interval 462 years, 68 days. We are reduced to choosing between two alternatives: either Homer nodded, or the MSS of the *Letter against Werner* should read 138 years, 303 days, instead of 139 years, 303 days. Werner makes a commendable effort to avoid such ambiguity by attaching to a total number of years the adjective "complete" or "incomplete," so that $n + 1$ incomplete years $= n$ complete years. He writes: "Therefore it is clear that Ptolemy established the true places of the fixed stars in the zodiac for the 22d day of February, according to the Roman calendar, while the 150th year of our Lord was still incomplete" (Prop. 4); "It is clear, however, from Proposition 4 that Claudius Ptolemy established the true places of the fixed stars for 149 complete Roman years and 53 days from the beginning of the years of Christ" (Prop. 20).

Günther's statement that Copernicus assigns the observation under discussion to September 25, 138 (MCV, II, 6.18-19) is clearly erroneous.

[18] HI, 362.9-10, 19-21; the observation was made on the twenty-fifth of Phamenoth.

[19] 885^y 203^d from Nabonassar to observation
747 130 from Nabonassar to Christ
$\overline{138^y\ 73^d}$ from the birth of Christ to the observation.

Basiliscus in Leo,[20] was the 22d day of February in the 139th Roman year after the birth of Christ.[21] And this was the second year of Antoninus, which our author thinks was A.D. 150. Consequently his error consists of an excess of eleven years.

If anyone is still in doubt and, not satisfied by our previous criticism, desires a further test of this treatise, he should remember that time is the number or measure of the motion of heaven considered as "before" and "after." From this motion we derive the year, month, day, and hour. But the measure and the measured, being related, are mutually interchangeable.[22] Now since Ptolemy based his tables on fresh observations of his own, it is incredible that the tables should contain any sensible error or any departure from the observations that would make the tables inconsistent with the principles on which they rest. Consequently if anyone will take the positions of the sun and moon, which Ptolemy determined by the astrolabe in his examination of Basiliscus, in the second year of Antoninus, on the ninth day of the month Pharmuthi, 5½ hours after noon, and if he will consult Ptolemy's tables for these positions, he will find them, not 149 years after Christ, but 138 years, 88 days, 5½ hours, equal to 885 years after Nabonassar, 218 days, 5½ hours.[23] Thus is laid bare the error which frequently

[20] HII, 14.1-14; for Basiliscus see above, p. 76, n. 56.

[21] 138 years and 73 days + 14 days — 34 days (on account of leap-years) = 138 years and 53 days or February 22, 139. In *De rev.*, where Copernicus considers this observation without reference to any other, he gives the day as February 24 (Th 114.20-22). His computation may be reconstructed as follows. Pharmuthi 9 is the 219th day of the Egyptian year. Subtract 34 days on account of leap-years, and 130 days to get the equivalent day in the Roman year: 219 — (34 + 130) = 55th day of the Roman year, or February 24. Robert Schram's tables, *Kalendariographische und chronologische Tafeln* (Leipzig, 1908), give February 23.

[22] These reflections on the nature of time are an echo of Aristotle's views; see *Physics* iv.11 219b1-2, iv.12 220b14-16, iv.14 223b21-23, and vi.4 235a10-24.

[23] 747^y 130^d 0^h from Nabonassar to Christ
     $+138$ 88 $5½$ from Christ to observation
     $\overline{885^y \ 218^d \ 5½^h}$

We established in n. 12 (p. 95, above) that when Copernicus wrote the *Letter against Werner* he was using for his references to Ptolemy the translation of 1515. Consulting the tables for the sun in that work (pp. 29r, 33v), we get the following result:

vitiates our author's examination of the motion of the eighth sphere when he mentions time.

The hypothesis in which he expresses his belief that during the four hundred years before Ptolemy the fixed stars moved with equal motion only[24] involves a second error no less important than the first. To clarify this matter and make it more intelligible, attention should be directed, I think, to the propositions stated below. The science of the stars is one of those subjects which we learn in the order opposite to the natural order. For example, in the natural order it is first known that

$$
\begin{array}{rl}
810^{y} & 163° \ 4' \ 13'' \\
72^{y} & 342° \ 29' \ 42'' \\
3^{y} & 359° \ 16' \ 14'' \\
210^{d} & 206° \ 59' \ 0'' \\
8^{d} & 7° \ 53' \ 6'' \\
5^{h} & 0° \ 12' \ 19'' \\
\tfrac{1}{2}^{h} & 0° \ 1' \ 13'' \\
\hline
& 1{,}079° \ 55' \ 47'' \\
\end{array}
$$

initial position $\quad + \ 265° \ 15'$

$$
\begin{array}{rl}
& 1{,}345° \ 10' \\
& - \ 1{,}080° \\
\hline
& 265° \ 10' \\
\end{array}
$$

Gemini $\quad\quad 5° \ 30'$

mean place of sun $\quad 270° \ 40'$ or Pisces $\ 0° \ 40'$

anomaly $\qquad\qquad\qquad\qquad +2° \ 23'$

true place of sun $\qquad\qquad$ Pisces $\ 3° \ 3'$

And Ptolemy states that the true place of the sun was Pisces 3° 3' (HII, 14.14-16). There is no need to set down here the longer calculations required for the moon, inasmuch as Manitius gives them (*Ptolemäus Handbuch*, II, 397-98). For the method of using Ptolemy's solar tables, cf. A. Rome, *Commentaires de Pappus et de Théon d'Alexandrie sur l'Almageste*, I (Rome, 1931), xxxv-xxxvii.

[24] Werner, Prop. 6: "To prove that the motion of the fixed stars in the zodiac for approximately four hundred years before the era of Ptolemy was nearly uniform and equal. In many passages of his *Great Syntaxis* Ptolemy shows with reference to the motion of the stars that previous to him and to his observation of the fixed stars, they moved for about four hundred years only one degree in each century. Therefore if for four hundred years the motion of the fixed stars completed one degree in each century (*centenariis*, not *centenarios*, as Curtze [MCV, I, 26 n] and PII, 175-76 n), the consequence is that the motion of the fixed stars for four hundred years previous to Ptolemy was nearly uniform and equal." Prop. 8: "Therefore it is clear that the fixed stars moved only with equal motion, and lacked unequal motion; or if they had any unequal motion, it was very small and almost imperceptible."

the planets are nearer than the fixed stars to the earth, and then as a consequence that the planets do not twinkle. We, on the contrary, first see that they do not twinkle, and then we know that they are nearer to the earth.[25] In like manner, first we learn that the apparent motions of the planets are unequal, and subsequently we conclude that there are epicycles, eccentrics, or other circles by which the planets are carried unequally. I should therefore like to state that it was necessary for the ancient philosophers, first to mark with the aid of instruments the positions of the planets and the intervals of time, and then with this information as their guide, lest the inquiry into the motion of heaven remain interminable, to work out some definite planetary theory, which they seem to have found when the theory agreed in some harmonious manner with all[26] the observed and noted positions of the planets. The situation is the same with respect to the motion of the eighth sphere. However, by reason of the extreme slowness of this motion, the ancient mathematicians were unable to pass on to us a complete account of it. But if we desire to examine it,[27] we must follow in their footsteps and hold fast to their observations,[28] bequeathed to us like an inheritance. And if anyone on the contrary thinks that the ancients are untrustworthy in this regard, surely the gates of this art are closed to him. Lying before the entrance, he will dream the dreams of the disordered about the motion of the eighth sphere and will receive his deserts for supposing that he must support his own hallucination by defaming the ancients. It is well known that they observed all

[25] An echo of Aristotle's *De caelo* ii.8 290a17-24 and *Posterior Analytics* i.13 78a30-78b4.

[26] I take it that *omnibus* appears twice by dittography in the texts of Curtze (MCV, I, 27.7-8) and Prowe (PII, 176.15-16).

[27] Prowe reads incorrectly *cum* instead of *eum* (PII, 176.19).

[28] Curtze preferred *observationibus*, the reading of the Vienna MS, to *considerationibus*, the reading of the Berlin MS (MCV, I, 27 n). In his note, which Prowe follows (PII, 176 n), Curtze equates *consideratio* with "Betrachtung" (contemplation), and seems unaware of the use of *consideratio* in the technical sense of "observation" (e.g., Th 192.11, 28; 259.26-27; 261.8; 276.3; 337.25; 338.20-21, 29; 351.32-352.1; 357.19; 365.6-7; 366.2, 6; 367.17; 379.13; 385.6; p. 104, n. 45, below; for an example in the *Epitome* see p. 124, n. 62, below).

these phenomena with great care and expert skill, and bequeathed to us many famous and praiseworthy discoveries. Consequently I cannot be persuaded that in noting star-places they erred by ¼° or ⅓° or even ⅙°, as our author believes. But of this I shall say more below.

Another point must not be overlooked. In every celestial motion that involves an inequality, what we want above all is the entire period in which the apparent motion passes through all its variations. For an apparent inequality in a motion is what prevents the whole revolution and the mean motion from being measured by their parts. As Ptolemy and before him Hipparchus of Rhodes, in their investigation of the moon's path, divined with keen insight, in the revolution of an inequality there must be four diametrically opposite points, the points of extreme swiftness and slowness, and, at each end of the perpendicular, the two points of mean uniform motion. These points divide the circle into four parts, so that in the first quadrant the swiftest motion diminishes, in the second the mean diminishes, in the third the slowest increases, and in the fourth the mean increases.[29] By this device they could infer from the observed and examined motions of the moon in what portion of its circle it was at any specified time; and hence, when a similar motion recurred, they knew that a revolution of the inequality had been completed. Ptolemy explained this procedure more fully in the fourth book[30] of the *Great Syntaxis*.

This method should have been adopted also in studying the motion of the eighth sphere. But because it is extremely slow, as I have said, in thousands of years the unequal motion quite clearly has not yet returned upon itself; and we are not permitted to give a final statement forthwith in dealing with a motion that extends beyond many generations of men. Nevertheless it is possible to attain our goal by a reasonable conjecture; and we now have the assistance of some observations, added since Ptolemy, which agree with this explanation. For what has been determined cannot have innumerable explanations; just as, if a circumference is drawn through three given

---

[29] Cf. p. 112, n. 13, below.   [30] Chap. ii.

points not on a straight line, we cannot draw another circumference greater or smaller than the one first drawn.[31] But let me postpone this discussion to another occasion in order that I may return to the point where I digressed.

We must now see whether during the four hundred years before Ptolemy the fixed stars indeed moved, as our author says, with equal motion only. But let us not be mistaken in the meaning of terms. I understand by "equal motion," usually called also "mean motion," the motion that is half way between the slowest and the swiftest. We must not be deceived by the first corollary to the seventh proposition. There he says[32] that the motion of the fixed stars is slower when on his hypothesis the equal motion occurs, while the rest of the motion is more rapid and hence would at no time be slower than the equal motion. I do not know whether he is consistent in this regard when later on he uses the expression "much slower."[33] He derives his measure of the equal motion from the following uniformity: in the period from the earliest observers of the fixed stars, Aristarchus and Timocharis, to Ptolemy, and in equal periods of time, the fixed stars moved equal distances, namely, approximately 1° in a century. This rate is given quite clearly by Ptolemy,[34] and is repeated by our author in his seventh proposition.[35]

But being a great mathematician, he is not aware that at the points of equality, that is, the intersections of the ecliptic of the

---

[31] The foregoing passage, in a slightly altered form, was quoted by Tycho Brahe, as I pointed out in n. 15 on pp. 8-9, above. Then Brahe added: "From these remarks it is clear that Copernicus, who came to the science of astronomy with talents certainly equal to Ptolemy's, thought that it was not utterly useless to construct, from some carefully examined part of its motion, a probable conception of the entire motion of the eighth sphere" (*Tychonis Brahe opera omnia*, ed. Dreyer, IV, 292.20-23).

[32] . . . *motum fixorum siderum tardiorem existere* . . .

[33] Copernicus has placed his finger on a logical difficulty. For Werner the mean rate (Prop. 8: "Hence it can without difficulty be inferred that the fixed stars in their equal motion move only one degree in each century of uniform years. Corollary. Hence it is clear that the fixed stars in their equal motion complete one revolution in 36,000 uniform years.") is also the slowest (Prop. 13: "Therefore the motion of the fixed stars in Ptolemy's time was slower or slowest.").

[34] HII, 36.21-37.2.

[35] The reference should be to the sixth proposition.

tenth sphere with the circles of trepidation, as he calls them,[36] the motion of the stars cannot possibly appear more uniform than elsewhere.[37] The contrary is necessarily true: at those times the motion appears to change most, and least when the apparent motion is swiftest or slowest. He should have seen this from his own hypothesis and system and from the tables based on them, especially the last table which he drew up for the revolution of the entire equality or trepidation.[38]

In this table the apparent motion is found to be, according to the preceding calculation, only 49′ for the century following 200 B.C., and 57′ for the next century. During the first century A.D. the stars must have moved about 1°6′, and during the second about 1°15′. Thus in equal periods of time the motions were successively greater by a little less than ⅛°.[39] If you add the motion of the two centuries in either era, the total for the first interval will fall short of 2° by more than ⅛°, while the total for the second will exceed 2° by about ¼°.[40] Thus again in equal times the later motion will exceed the earlier by about 34′,[41] whereas our author had previously reported, trusting in Ptolemy, that the fixed stars moved 1° in a century. On the other hand, by the same law of the circles which he assumed, in the swiftest motion of the eighth sphere it happens that during 400 years a variation of scarcely 1′ is found in the apparent motion, as can be seen in the same table for the years

[36] Prop. 11: ". . . the apparent or unequal motion of the sphere of the fixed stars or of the eighth sphere is caused by the circumstance that the first points of Cancer and Capricornus of the ecliptic of the ninth sphere revolve on small circles. This revolution is called by Thābit and the Alfonsine Tables the forward and backward motion or trepidation of the eighth sphere. This trepidation proceeds sometimes in the order of the signs, sometimes in the contrary order. Hence the motion of the fixed stars is sometimes slow and sometimes rapid. It is clear, moreover, that the motion of the fixed stars is composed of the equal motion of the eighth sphere, and the trepidation or forward and backward motion of the ninth sphere on the small circles."

[37] In Prop. 13 Werner states: "Therefore it is clear that the first points of Cancer and Capricornus of the ninth sphere were, about Ptolemy's time, near the aforesaid intersections of the small circles with the ecliptic of the tenth sphere."

[38] This table is placed at the end of Prop. 30.

[39] The successive increases are 8′, 9′, 9′.

[40] 49′ + 57′ = 1° 46′; 1° 6′ + 1° 15′ = 2° 21′.

[41] 2° 21′ − 1° 46′ = 35′.

600–1000 A.D.; and similarly in the slowest motion, from 2060 B.C. for 400 years thereafter. Now the law governing an inequality is that, as was stated above,[42] in one semicircle of trepidation, the one that extends from extreme slowness to extreme swiftness, the apparent motion constantly increases; and in the other semicircle, the one that extends from extreme swiftness to extreme slowness, the motion, previously on the increase, constantly diminishes. The greatest increase and decrease occur at the points of equality, diametrically opposite to each other. Hence in the apparent motion for two continuous equal periods of time equal motions cannot be found, but one is greater or smaller than the other. An exception occurs only at the extremes of swiftness and slowness, where the motions to either side pass through equal arcs in equal times; beginning or ceasing to increase or decrease, they equal each other at those times by undergoing opposite changes.

Therefore it is clear that the motion during the four hundred years before Ptolemy was not at all mean, but rather the slowest. I see no reason why we should suppose any slower motion, for which we have not been able thus far to get any evidence. No observation of the fixed stars made before Timocharis has come down to us, and Ptolemy had none. Since the swiftest motion has already occurred, we are now as a consequence not in the same semicircle with Ptolemy. In our semicircle the motion diminishes, and no small part of it has already occurred.

Hence it should not be surprising that with these assumptions our author could not more nearly approach the recorded observations of the ancients; and that in his opinion they erred by ¼° or ⅕°, or even ½° and more. Yet nowhere does Ptolemy seem to have exercised greater care than in his effort to hand down to us a flawless treatment of the motion of the fixed stars. He could be successful only in that small portion of it from which he had to reconstruct the entire revolution. If an error, however imperceptible, entered that whole vast realm, it might have prodigious effects on the outcome. Therefore he seems to have joined Aristarchus to Timocharis of Alexandria, his contemporary, and Agrippa of Bithynia to Menelaus of

[42] Page 100.

Rome; in this way he would have most certain and unquestionable evidence when they agreed with each other, although separated by great distances. It is incredible that such great errors were made by these men or Ptolemy, who could deal with many other more difficult matters and, as the saying goes, put the finishing touches to them.

Finally, our author is nowhere more foolish than in his twenty-second proposition, especially in the corollary thereto. Wishing to praise his own work, he censures Timocharis with regard to two stars, namely, Arista Virginis,[43] and the star which is the most northerly of the three in the brow of Scorpio,[44] on the ground that for the former star Timocharis's calculation fell short, and for the latter was excessive.[45] But here our author commits a childish blunder. For both stars the difference in the distance, as determined by Timocharis and Ptolemy, is the same, namely, 4° 20' in approximately equal intervals of time; and hence the result of the calculation is practically the same. Yet our author disregards the fact that the addition of 4°7' to the place of the star which Timocharis found in 2° of Scorpio[46] cannot possibly produce 6°20' of Scorpio, the place where Ptolemy found the star.[47] Conversely when the same number is subtracted from 26°40', the place of Arista according to Ptolemy,[48] it cannot yield 22°20', as it

[43] Commonly called Spica; cf. p. 67, above.

[44] Scorpio 1 (HII, 108.18; Th 137.31), β Scorpii.

[45] Corollary to Prop. 22 (not 27, as in Curtze, MCV, I, 31 n. and Prowe, PII, 181 n): "This is clear from the observations [considerationibus] of Timocharis. In the case of the fixed star called Arista, they fall short of my calculation, but in the case of the star which is the most northerly of the three bright stars in the brow of Scorpio, they exceed my computation. However, if these observations [considerationes] made by Timocharis had both been true, they should equally fall short of my calculation, or equally exceed it. Therefore the trustworthiness of my tables is not less than that of the observations and discoveries of the ancients."   [46] HII, 32.20-33.1.

[47] HII, 109.18. Werner's remarks are: "The true motion of the fixed stars, in the interval between Timocharis and Ptolemy, will turn out to be 4° 7' 3" 28'''. If we add this difference to the true place of the fixed star which is the most northerly in the brow of Scorpio, to the place, that is, which Timocharis found in his observation, the result will be 6° 7' 3". But Ptolemy's tables place this star in 6° 20' of Scorpio" (Prop. 22).

[48] HII, 103.16. This statement enables us to correct the slip in Th 160.1; see n. 14 on p. 112, below.

should,[49] but it gives 22°32′.[50] Thus our author thought that in the one case the computation was deficient by the amount by which in the other case it was excessive, as though this irregularity were inherent in the observations, or as though the road from Athens to Thebes were not the same as the road from Thebes to Athens. Besides, if he had either added or subtracted the number in both cases, as parity of reasoning required, he would have found the two cases identical.

Moreover, between Timocharis and Ptolemy there were in reality not 443 years,[51] but only 432, as I indicated in the beginning.[52] Since the interval is shorter, the difference should be smaller; hence he departs from the observed motion of the stars not merely by 13′[53] but by ⅓°. Thus he imputed his own error to Timocharis, while Ptolemy barely escaped. And while he thinks that their reports are unreliable,[54] what else is left but to distrust his observations?

So much for the motion in longitude of the eighth sphere. From the foregoing remarks it can easily be inferred what we

[49] HII, 29.9-11.

[50] Not 22° 33′, because Werner is subtracting 4° 7′ 57″: "The difference will be 4° 7′ 57″, the true motion of the fixed stars for the 442 complete Roman years and 350 days between the observations of Ptolemy and Timocharis. If, finally, this value of 4° 7′ 57″ is subtracted from the true place of Arista as observed or calculated by Ptolemy, the remainder is 22° 32′ 3″ of Virgo, the true place of Arista in the zodiac, near the place found by Timocharis in his observation" (Prop. 22).

[51] Prop. 22: "Finally, between this observation of Timocharis and Ptolemy's investigation of the fixed stars there intervened 443 Roman years and 64 days."

[52] In his treatment of Werner's "first error," Copernicus established that Werner postdated Ptolemy by eleven years; cf. pp. 94-97, above.

[53] Prop. 22: "Therefore my tables would diminish the position of this star by 13′." Again, "However, my computation exceeds Timocharis's observation by 12′."

[54] Prop. 22, Corollary: "For this weakens not a little the reliability of the ancient observations of the fixed stars, since some of these observations exceed the computation based on the foregoing canons and tables, while certain of them fall short of this computation. Now if all the results of the ancient observations of the fixed stars coincided exactly with the truth, they should, with perfect propriety, all together fall short of the calculation based on the aforesaid tables, or they should all equally exceed it. But it has been shown above that the ancient observations partly fall short of, and partly exceed the calculation based on my tables." Yet with regard to the length of the year Werner is less confident: "For I do not venture to charge the ancient observers of the stars with any error" (Prop. 33).

must think about the motion in declination, which our author has complicated with two trepidations, as he calls them, piling a second one upon the first.[55] But since the foundation has now been destroyed, of necessity the superstructure collapses, being weak and incohesive. What finally is my own opinion concerning the motion of the sphere of the fixed stars? Since I intend to set forth my views elsewhere, I have thought it unnecessary and improper to extend this communication further. For it is enough if I satisfy your desire to have my judgment of this work, as you requested.[56] May your Reverence be of sound health and good fortune.   NICHOLAS COPERNICUS

To the Reverend Bernard Wapowski,   Frauenburg, June 3, 1524
Cantor and Canon of the Church of Cracow,
Secretary to His Majesty the King of Poland,
my highly esteemed lord and patron, etc.

[55] Prop. 18: "The first trepidation or forward and backward motion is a property of the ninth sphere and its small circles. This trepidation of the ninth sphere is called the first trepidation because, by reason of the variation in the maximum declination of the sun, a revolution or upward and downward movement on small circles must be assigned to the ecliptic of the tenth sphere also. This movement will, then, be named the second trepidation."

[56] We have already seen (cf. pp. 7-8, n. 14, above) how difficult it was for Tycho Brahe to obtain a copy of Werner's treatise on *The Motion of the Eighth Sphere*. His critical comment on it follows: "I have examined it, studied it thoroughly, and set it aside for a reason which I may briefly explain. Werner uses three stars as a basis for dealing with the rest, and from these three he attempts to construct complicated movements of the eighth sphere. He did not carefully observe the three stars in the heavens, as he should have, although he pretends to have done so (I wish, however, to say this with due respect to the memory of a man who was otherwise very learned, and who served the cause of mathematics admirably). Rather, he represented them as he pleased and adjusted them to fit his purpose. This is quite clear from the fact that he retains everywhere the ancient values for their latitudes and nevertheless, assuming his own motions in declination, works out changes in their longitude equal to the accepted account. These views cannot possibly be consistent. For the ancient determinations of the latitudes of these stars do not accord with what is in the heavens, except in the case of Spica alone, where only a single minute is lacking; and the accepted shifts in their longitude do not agree with the appearances. Hence it is clear how the rest of his argument, which he strives to erect not without keenness and subtlety of mind, turns out to be feeble and broken. I make no mention at present of the fact that neither Werner nor the great Copernicus noticed that the latitude of the stars changes in accordance with the shift in the obliquity of the ecliptic (as has been clearly established by me); nor did they explain the displacement in latitude by any hypothesis" (*Tychonis Brahe opera omnia*, ed. Dreyer, VII, 295.23-42; cf. II, 223.29-226.11).

THE *NARRATIO PRIMA*
OF RHETICUS

"Free in mind must be he who desires to have understanding."

ALCINOUS

TO THE ILLUSTRIOUS JOHN SCHÖNER, as to his own revered father, G. Joachim Rheticus sends his greetings.

On May 14th I wrote you a letter from Posen in which I informed you that I had undertaken a journey to Prussia,[1] and I promised to declare, as soon as I could, whether the actuality answered to report and to my own expectation. However, I have been able to devote scarcely[2] ten weeks to mastering the astronomical work of the learned man to whom I have repaired; for I had a slight illness and, on the honorable invitation of the Most Reverend Tiedemann Giese, bishop of Kulm, I went with my teacher to Löbau and there rested from my studies for several weeks.[3] Nevertheless, to fulfill my promises at last and gratify your desires, I shall set forth, as briefly and clearly as I can, the opinions of my teacher on the topics which I have studied.

First of all I wish you to be convinced, most learned Schöner, that this man whose work I am now treating is in every field of knowledge and in mastery of astronomy not inferior to Regiomontanus. I rather compare him with Ptolemy, not because I consider Regiomontanus inferior to Ptolemy, but because my teacher shares with Ptolemy the good fortune of completing, with the aid of divine kindness, the reconstruction of astronomy which he began, while Regiomontanus—alas, cruel fate—departed this life before he had time to erect his columns.

My teacher has written a work of six books in which, in imitation of Ptolemy, he has embraced the whole of astronomy,

---

[1] The basic study for the biography of Rheticus will be found in *Vierteljahrsschrift für Geschichte und Landeskunde Vorarlbergs*, neue Folge, II(1918), 5-46. For subsequent work consult *Forschungen zur Geschichte Vorarlbergs und Liechtensteins*, I(1920), 128-30; *Schriften des Vereines für Geschichte des Bodensees*, LV(1927), 122-37; and Martin Bilgeri, *Das Vorarlberger Schrifttum* (Vienna, 1936), pp. 64-70.

[2] Reading *vix* (Th 447.8) instead of *viri* (PII, 295.7).

[3] In the light of this remark, we must regard as incorrect Prowe's statement (PI², 395) that the *Narratio prima* was written at Löbau. Prowe himself declares that Rheticus's trip to Löbau kept him from his studies (PI², 428).

stating and proving individual propositions mathematically and by the geometrical method.

The first book contains the general description of the universe and the foundations by which he undertakes to save the appearances and the observations of all ages. He adds as much of the doctrine of sines and plane and spherical triangles as he deemed necessary to the work.

The second book contains the doctrine of the first motion[4] and the statements about the fixed stars which he thought he should make in that place.

The third book treats of the motion of the sun. And because experience has taught him that the length of the year measured by the equinoxes depends, in part, on the motion of the fixed stars, he undertakes in the first portion of this book to examine by right reason and with truly divine ingenuity the motions of the fixed stars and the mutations of the solstitial and equinoctial points.

The fourth book treats of the motion of the moon and eclipses; the fifth, the motions of the remaining planets; the sixth, latitudes.

I have mastered the first three books, grasped the general idea of the fourth, and begun to conceive the hypotheses of the rest. So far as the first two books are concerned, I have thought it unnecessary to write anything to you, partly because I have a special plan,[5] partly because my teacher's doctrine of the first motion does not differ from the common and received opinion,[6] save that he has so constructed anew the tables of declinations, right ascensions, ascensional differences, and the other tables belonging to this branch of the science that they can be brought by the method of proportional parts into agree-

---

[4] The apparent daily rotation of the heavens; see p. 41, above.

[5] Rheticus doubtless refers to his plan for writing a "Second Account." For an explanation why this "Second Account" was never written see p. 10, above.

[6] But in the common and received opinion the first motion was real; in Copernicus's system, apparent. Rheticus ignores the distinction, for it involves the motion of the earth. Throughout the first third of this *Account* he withholds all reference to Copernicus's principal alteration of astronomical theory, the shift from a stationary to a moving earth, and from geocentrism to heliocentrism (cf. below, pp. 135-36, n. 115).

ment with the observations of all ages. Therefore I shall set forth clearly to you, God willing, the subjects treated in the third book together with the hypotheses of all the remaining motions, so far as at present with my meager mental attainments I have been able to understand them.

The Motions of the Fixed Stars

My teacher made observations with the utmost care at Bologna, where he was not so much the pupil as the assistant and witness of observations of the learned Dominicus Maria;[7] at Rome, where, about the year 1500, being twenty-seven years of age more or less, he lectured on mathematics before a large audience of students and a throng of great men and experts in this branch of knowledge; then here in Frauenburg,[8] when he had leisure for his studies. From his observations of the fixed stars he selected the one which he made of Spica Virginis in 1525. He determined its distance from the autumnal point[9] as about $17°21'$, and its declination as not less than $8°40'$ south of the equator. Then comparing all the observations of previous writers with his own, he found that a revolution of the anomaly or of the circle of inequality had been completed and that the second revolution extends from Timocharis to our own time. Thereby he geometrically determined the mean motion of the fixed stars and the equations of their unequal motion.

Timocharis's observation of Spica in the 36th year[10] of the first Callippic cycle, when compared with his observation in the 48th year of the same cycle, shows us that the stars moved $1°$ in 72 years in that era.[11] From Hipparchus to

[7] Concerning whom Lino Sighinolfi has assembled some material, chiefly biographical, in his article "Domenico Maria Novara e Nicolò Copernico" (*Studi e memorie per la storia dell' università di Bologna*, V [1920], 211-35).

[8] Cf. Th 193, note to line 9.

[9] The first point of Libra (cf. Th 161.24-25).

[10] 295/4 B. C. A Callippic cycle contained 76 years (HII, 25.16-17; Th 159.11). See F. K. Ginzel, *Handbuch der mathematischen und technischen Chronologie* (Leipzig, 1906-14), II, 409-19; and J. K. Fotheringham in *Monthly Notices of the Royal Astronomical Society*, LXXXIV(1924), 387-92.

[11] HII, 28.11-30.17.

Menelaus they regularly completed 1° in 100 years.[12] My teacher therefore concluded that Timocharis's observations fell in the last quadrant of the circle of inequality,[13] in which the motion appears mean-diminishing, and that between Hipparchus and Menelaus the motion of inequality was slowest. A comparison of Menelaus's observations with Ptolemy's shows that the stars then moved 1° in 86[14] years. Therefore Ptolemy's

[12] Ptolemy accepts this estimate as the approximate value for the entire period from Hipparchus to himself (HII, 23.11-16); and he regards the rate of precession as constant (HII, 34.11-17).

[13] Copernicus held that the rate of precession varied. To represent the variation he constructs a "circle of inequality," (Fig. 26) in which *a* is the point of slowest

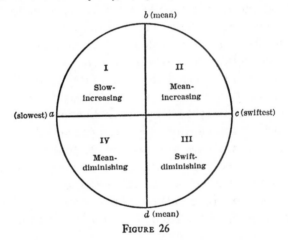

FIGURE 26

motion; *c*, the point of swiftest motion; *b* and *d*, the points of mean motion. The first quadrant *ab* is the quadrant of slow-increasing motion; the second quadrant *bc*, of mean-increasing motion; the third quadrant *cd*, swift-diminishing; the fourth quadrant *da*, mean-diminishing. See Th 169.25-170.4, and p. 100, above.

[14] This should be 96, as an examination of Menzzer's chart (p. 21 of his notes) shows. Menelaus determined the longitudinal distance of Spica from the summer solstice as 86° 15', and of β Scorpii from the autumnal equinox as 35° 55' (HII, 30.18-31.16, 33.3-24; Th 159.24-29). For Ptolemy, 40 years later, the corresponding values were 86° 40' and 36° 20' (HII, 103.16, 109.18; Th 159.29–160.4; *LXXXVI s.* in Th 160.1 is wrong, as can be seen in the *Letter against Werner* [cf. p. 104, above] and in the three editions of the *Syntaxis* available to Copernicus: 1515, p. 83r; 1528, p. 78r; 1538, p. 187; cf. Th 161.28-31). For both stars the motion is 25' in 40 years, or 1° in 96 years. It is more likely that an X has fallen out of LXXXXVI than that Rheticus made an error in his computations.

observations were made when the motion of anomaly was in the first quadrant, and the stars then moved with a slow-increasing[15] or -augmenting motion. Further, from Ptolemy to Albategnius, 66 years correspond to 1°;[16] a comparison of our observations with those of Albategnius shows that the stars in their unequal motion again completed 1° in 70 years;[17] and a comparison of the observation which I mentioned above with the others which my teacher made in Italy shows that the fixed stars in their unequal motion are once more passing through 1° in 100 years. Therefore it is clearer than sunlight that between Ptolemy and Albategnius the motion of inequality passed the first boundary of mean motion and the entire quadrant of mean-increasing motion, and about the time of Albategnius was in the region of swiftest motion. Between Albategnius and ourselves the third quadrant of unequal motion was completed (during this time the stars moved with a swift-diminishing motion) and the other boundary of mean motion was passed. In our era the anomaly has again entered the fourth quadrant of mean-diminishing motion, and hence the unequal motion is once more approaching the point of slowest motion.

To reduce these calculations to a definite system in which they would agree with all the observations, my teacher computed that the unequal motion is completed in 1,717 Egyptian years,[18] the maximum correction is about 70', the mean motion

---

[15] *addito* (Th 449.3) has dropped out of Prowe's text (PII, 298.14).

[16] Nallino, *Al-Battānī*, I, 124.32-33, 128.2-4. A translation of Albategnius's work into Latin was included in a book printed at Nuremberg in 1537 and was presumably available to Copernicus and Rheticus, as may be inferred from the latter's remark about Albategnius on p. 124, below. The volume opened with a treatise entitled in some copies *Rudimenta astronomica Alfragani*, and in others, *Brevis ac perutilis compilatio Alfragani*. I have been unable to consult this translation, but the relevant passage was excerpted by Menzzer (n. 81). Rheticus ignores the distinction made by Copernicus (Th 162.7-11) that the rate of precession was 1° in 66 years from Menelaus to Albategnius, and 1° in 65 years from Ptolemy to Albategnius. The distinction is based on the observations cited in Th 159.24-160.10.

[17] Copernicus states this rate as 1° in 71 years (Th 162.12-14); calculation from his data gives the fractional result 1° in 70¼ years.

[18] The changing rate of precession requires 1717 years to pass through the four quadrants of the circle of inequality.

of the stars in an Egyptian year is about 50″,[19] and the complete revolution of the mean motion will take 25,816 Egyptian years.[20]

General Consideration of the Tropical Year

This theory of the motions of the fixed stars is supported by the length of the year reckoned from the equinoctial points. It is quite clear why from Hipparchus[21] to Ptolemy there was a deficiency of ¹⁹⁄₂₀ of a day;[22] from Ptolemy to Albategnius, of about 7 days;[23] and from Albategnius to the observations which my teacher made in 1515, of about 5 days.[24] These discrepancies are not at all caused by a defect in the instruments, as was heretofore believed, but occur according to a definite and completely self-consistent law. Hence equality of motion must be measured, not by the equinoxes, but by the fixed stars, as observations of the motions not merely of the sun and moon but of the other planets as well testify with a remarkable unanimity of all ages.

[19] The mean rate of precession is about 50″ a year, or 1° in about 72 years; and the greatest difference between the mean equinox and the true equinox is about 70′ (Th 179.4-7).

(a) The slowest rate of precession is 1° in 100 years, or 36″ a year. The difference between the slowest rate and the mean rate is 14″ a year, and in 300 years (the three centuries before Menelaus) the maximum difference of 70′ between mean and true equinox is attained.

(b) The swiftest rate is 1° in 66 years, or slightly more than 54½″ a year. The difference between the swiftest rate and the mean rate is about 4½″ a year, and in 743 years (between Ptolemy and Albategnius) about ⅚ of the maximum difference of 70′ is attained; the remaining ⅙ accumulates because the rate of precession during the 620 years between Albategnius and Copernicus's observations in Italy is slightly more rapid than the mean rate.

Copernicus's estimate of the mean precession, about 50″ a year, agrees quite closely with the determination accepted at present. His belief in the cyclic variation of the rate of precession is of course erroneous.

[20] The complete passage of the stars around the celestial sphere requires 25,816 years.

[21] Michael Mästlin, editor of the fourth (1596) and fifth (1621) editions of the Narratio prima, correctly substituted "Hipparchus" for the older reading "Timocharis." Unfortunately the incorrect reading was revived in Th 449.24-25.

[22] HI, 203.22-204.18. The tropical year (t) is less than 365¼ days; the "deficiency" from year x to year y is $(y - x)$ (365¼ days — t).

[23] Nallino, Al-Battānī, I, 42.10-14.   [24] Th 193.20-21.

It is the accepted opinion that because from Timocharis to Ptolemy the stars moved very slowly the year was less than 365¼ days by only ⅟₃₀₀ of a day;[25] and from Ptolemy to Albategnius, because the stars moved rapidly, by ⅟₁₀₆ of a day.[26] If the observations of our age are compared with those of Albategnius, it is clear that the difference is ⅟₁₂₈ of a day.[27] Therefore a greater length of the tropical year apparently corresponds to a slow motion of the stars, a lesser length to a swift motion, and the lengthening of the year to a diminishing velocity; so that if the length of the tropical year in our era is accurately determined, it will again be almost the same as Ptolemy's value. Hence we must say that the equinoctial points, like the nodes of the moon,[28] move in precedence, and not that the stars move in consequence.[29]

We must accordingly imagine a mean equinox moving in precedence from the first star of Aries in the sphere of the fixed stars, and displacing them by its uniform motion. The true equinox deviates to either side of this mean equinox in an unequal and regular motion; but the radius of the distance between the true equinox and the mean equinox does not much exceed 70'. Thus a definite law governing the length of the tropical year has existed in all ages, and it can be ascertained today. It agrees very closely, moreover, and almost to the minute with the observations which all scholars have made of the fixed stars.

To offer you some taste of this matter, most learned Schöner, I have computed for you the true precession of the equinoxes at certain times of observation.

[25] HI, 205.9-14, 207.24–208.1.

[26] Albategnius's estimate of the length of the tropical year was $365^d14^m26^s$ (Nallino, *op. cit.*, I, 42.17). The difference between this value and 365¼ days ($= 365^d15^m$) is 34^s or ³⁴⁄₃₆₀₀ of a day. Albategnius expresses this difference as

$$\frac{3\frac{2}{6}}{360}$$

(*op. cit.*, I, 127.19-20). It is much closer to ⅟₁₀₆ than to ⅟₁₀₅ of a day, and Copernicus writes the more accurate fraction (Th 193.2-3). Hence I have followed Mästlin in changing our text from ⅟₁₀₅ to ⅟₁₀₆.

[27] Th 193.20-21, 194.4-5.

[28] See p. 73, above.

[29] Copernicus interprets precession as a motion of the equator.

Egyptian Year		*True Precession*		*Period*
		° ′		
B.C.	293	2	24	Timocharis
	127	4	3	Hipparchus
A.D.	138	6	40	Ptolemy
	880	18	10	Albategnius
	1076	19	37³⁰	Arzachel
	1525	27	21	present

Ptolemy's precession subtracted from the positions of the stars as given by Ptolemy leaves a remainder equal to the distance of the stars from the first star of Aries; then the addition of Albategnius's precession gives the true position of the observed star. A similar procedure is followed in all the other cases. The results thus obtained coincide to the utmost degree of exactness with the observations of all scholars, even where the minutes are noted, or are derived from recorded declinations or from the motion of the moon reduced to greater precision, as a comparison of our observations with those of the ancients shows us. For when the minutes are neglected, as you see, at least a part of a degree is cut off,[31] ½° or ⅓° or ¼°, etc. However, these results do not agree with the motions of the planetary apsides, and therefore an independent motion had to be assigned to them, as will be clear from solar theory.[32]

Realizing that equality of motion must be measured by the fixed stars, my teacher carefully investigated the sidereal year. He finds that it is 365 days, 15 minutes, and about 24 seconds[33] and that it has always been of this length from the time of

---

[30] Prowe states (PII, 300 n) that the first edition read incorrectly 12° 37′, and that the change to 19° 37′ was made by Mästlin. But the Basel edition of 1566 gave 19° 37′ (p. 198r); and that number is suspect, for it would make the rate of precession (*a*) between Albategnius and Arzachel too slow (1° 27′ in 196 years, or 1° in 135 years); and (*b*) between Arzachel and Copernicus too fast (7° 44′ in 449 years, or 1° in 58 years). To be consistent with the theory and the rest of the table, the true precession for Arzachel must be about 20° 57′.

[31] Reading *recidant* (Th 450.28) instead of *recitant* (PII, 301.9).

[32] See pp. 119-21, below.

[33] The minute of the text is a *minutum diei* = 1/60 of a day = 24ᵐ; in like manner, the second = 1/60 of a *minutum diei* = 24ˢ. The length of the sidereal year, then, is given here as about 365ᵈ6ʰ9ᵐ36ˢ. In *De rev.* it is given as 365ᵈ6ʰ9ᵐ40ˢ; cf. p. 67, above.

the earliest observations. For the fact that the Babylonians, according to Albategnius, assign 3 seconds more,[34] and Thābit 1 second less,[35] can be safely ascribed to either the instruments and observations, which, as you know, cannot have been entirely accurate, or to the inequality in the motion of the sun, or even to the circumstance that the ancients, having no sure theory of eclipses, neglected to take account of the solar parallax in their observations. In any case, this discrepancy over the entire period from the Babylonians to ourselves cannot be compared with the discrepancy of 22 seconds between Ptolemy and Albategnius.[36] That there necessarily was a deficiency of $1\frac{9}{20}$ of a day from Hipparchus to Ptolemy, and from Ptolemy to Albategnius of about 7 days, I have deduced, not without the greatest pleasure, most learned Schöner, from the foregoing theory of the motions of the stars and from my teacher's treatment of the motion of the sun, as you will see a little further on.[37]

The Change in the Obliquity of the Ecliptic

My teacher found that the cycle of maximum obliquity stands in the following relation: while the unequal motion of the fixed stars is once completed, half of the change in the obliquity occurs. He therefore concluded that the entire period of the change in the obliquity is 3,434 Egyptian years.[38]

[34] The Latin translation of Albategnius gave $365\frac{1}{4}^d + \frac{1}{131}^d = 365^d 15^m 27\frac{1}{2}^s$ (see above, p. 67, n. 25). However, the Arabic MS on which Nallino based his text reads (I, 40.28-29) $365\frac{1}{4}^d + \frac{1}{120}^d = 365^d 15^m 30^s$.

[35] Rheticus and Copernicus (Th 194.8-12) probably drew this information about Thābit from the *Epitome*, Bk. III, Prop. 2 (see above, p. 65, n. 19), which gave Thābit's value as $365^d 6^h 9^m 12^s$ ($= 365^d 15^m 23^s$). Proof that Rheticus used the *Epitome* is afforded by two references to it (p. 124, below) and by a quotation from it (pp. 133-34, below). For Thābit see George Sarton, *Introduction to the History of Science* (Baltimore, 1927–), I, 599-600.

[36] Rheticus is referring to the difference in their determinations of the length of the tropical year: Ptolemy $365^d 14^m 48^s$ (HI, 208.11-12)

Albategnius $365^d 14^m 26^s$ (See above, p. 115, n. 26)

$$22^s$$

[37] Pages 128-30, below.

[38] It was stated above (p. 113) that the period of the unequal motion of the fixed stars is 1,717 Egyptian years.

In the time of Timocharis, Aristarchus, and Ptolemy the change in the obliquity was very slow, so that they believed the maximum arc of declination to be invariable, always having the value of $^{11}\!/_{83}$ of a great circle.[39] After them, Albategnius announced the obliquity for his own era as about $23°35'$;[40] Arzachel, about 190 years after him, $23°34'$; and Prophatius Judaeus, 230 years later, $23°32'$. In our own era it appears not greater than $23°28\frac{1}{2}'$.[41] Accordingly it is clear that in the 400 years before Ptolemy the change in the obliquity was very slow. But from Ptolemy to Albategnius, a period of about 750 years, the obliquity decreased by $17'$, and from Albategnius to ourselves, a period of 650 years, by only $7'$. Hence it follows that the variation of the obliquity, like the deviation of the planets from the ecliptic,[42] is governed by a motion in libration or motion along a straight line. It is a property of such motion that in the middle the motion is quickest, and slowest at the ends. Then about the time of Albategnius the pole of the equator or of the ecliptic was approximately in the middle of this motion in libration, while at present it is near the second limit of slowest motion, where the poles approach each other most closely. But I stated above[43] that the motions of the fixed stars and the variation in the length of the tropical year are saved by the motion of the equator. Now the poles of the equator

[39] $^{11}\!/_{83} \times 360° = 47° \; 42' \; 40''$, which makes the obliquity $23° \; 51' \; 20''$ (HI, 68.4-6, 81.50).

[40] Nallino, *Al-Battānī*, I, 12.20-22; Menzzer, n. 87.

[41] Th 162.24-25, 171.31-172.4; cf. above, p. 64, n. 15. The foregoing statement about the history of the determinations of the obliquity is virtually identical with the scholion in Reinhold's 1542 edition of Peurbach's *Theoricae novae planetarum*, fol. e8r-v; cf. Boncompagni, *Bulletino di bibliografia e di storia delle scienze*, XX(1887), 594-95. Since the statement in our text is earlier than Reinhold's, but Reinhold's contains additional items, apparently they both drew from some common source. For Arzachel and Prophatius Judaeus see Sarton, *Introduction*, I, 758-59; II, 850-53. Copernicus believed that Prophatius obtained his value of $23° \; 32'$ by a direct determination; but it was rather a calculation from Arzachel's tables, according to Duhem (*Le Système du monde*, III, 311). J. Millàs i Vallicrosa has published Prophatius's translation of Arzachel's *Saphea* in *Don Profeit Tibbon, Tractat de l'assafea d'Azarquiel* (Barcelona, 1933).

[42] Cf. pp. 80-81, 84-85, above and pp. 180, 182-85, below.

[43] See above, p. 115, n. 29.

are the prolongations of the earth's axis, and it is from them that the altitude of the pole is measured. Let me in passing call your attention, most learned Schöner, to the sort of hypotheses or theories of motion that the observations require; but you will hear clearer evidence.

Furthermore, my teacher assumes that the minimum obliquity will be 23°28', and the difference between the minimum and the maximum, 24'. On this basis he geometrically constructed a table of[44] proportional minutes, from which the maximum obliquity of the ecliptic may be derived for all ages. Thus the proportional minutes were, in the time of Ptolemy 58, Albategnius 18, Arzachel 15, and in our own time 1.[45] If, using these figures, we take proportional parts of the 24' difference between minimum and maximum, we shall have a sure rule for the change in the obliquity.[46]

The Eccentricity of the Sun and the Motion of the Solar Apogee

Since every difficulty in the motion of the sun is connected with the variable and unstable length of the year, I must first speak of the change in the apogee and eccentricity, in order that all the causes of the inequality of the year may be enumerated. However, by the assumption of theories suitable to the purpose, my teacher shows that these causes are regular and certain.

When Ptolemy declared that the apogee of the sun was fixed,[47] he preferred accepting the common opinion to trusting his own observations, which differed perhaps but little from the common opinion. But it can be definitely established from

[44] In his 1621 edition (p. 99), Mästlin inserted "sixty."

[45] Thus in the case of Arzachel, $^{15}/_{60} \times 24' = 6' + 23° 28' = 23° 34'$. For Albategnius, the editions of our text put the number of proportional minutes at 24; I have emended this obviously incorrect number to 18.

[46] For modern astronomy the change in the obliquity is a progressive diminution. The evidence available to Copernicus warranted only the same conclusion (Th 76.27-28). But he believed that after the obliquity had decreased to 23° 28', it would increase to 23° 52', completing a cycle which would then be repeated.

[47] HI, 232.18-233.16; cf. n. 13, pp. 62-63, above.

his own account that about the time of Hipparchus, that is, 200 years before his own time, the eccentricity was 417[48] of the units of which 10,000 constitute the radius of the eccentric, and in his own time 414.[49] In the time of Arzachel (in whom Regiomontanus had great faith) the eccentricity was about 346, according to the maximum inequality.[50] But in our own time it is 323, since my teacher states that he finds the maximum inequality not greater than 1°50½′.[51]

Furthermore, carefully investigating the motions of the apsides of the sun and of the other planets, he learned, as you see from what has been said above,[52] that the apsides have independent motions in the sphere of the fixed stars. We are no more justified in attributing the apparent motions of the fixed stars and apsides, and the change in the obliquity to a single motion and a single cause than is one of your experts, who speak of the motions of the planets as self-moving, in attempting to produce the motions and appearances of each of the planets by one and the same device; or than anyone undertaking to defend the view that the foot, hand, and tongue exercise all their functions by means of the same muscle and by the same motive force. Therefore my teacher assigned two motions to the solar apogee, one mean and the other unequal, with which it moves in the eighth sphere. Moreover, since the true equinox moves with a regular unequal motion in the reverse

[48] HI, 233.5-8; Hipparchus determined the eccentricity as approximately ¹⁄₂₄ of the radius of the eccentric: ¹⁄₂₄ × 10,000 = 416⅔.

[49] HI, 236.15-18. Ptolemy's value for the eccentricity is slightly smaller than Hipparchus's; but since he believed the eccentricity to be constant (see above, p. 61, n. 9), he ignored the small difference between the two values, which he denotes as approximate (ἔγγιστα) in any case. Copernicus held that the eccentricity varies, and hence utilized the difference. Ptolemy's value would be more accurately expressed as 415 than as 414 (Th 209. n. to line 12).

[50] For the method of computing the eccentricity from the maximum inequality, see above, p. 61, n. 11. The information that Arzachel had put the maximum inequality at 1° 59′ 10″ (cf. Th 212.15-16) was obtained by Copernicus and Rheticus from the *Epitome*, Book III, Prop. 13. By the Table of Chords (Th 44.18-19), this inequality would correspond to an eccentricity of 346 (cf. Th 210.1-6).

[51] Th 211.16-19; 212.16; 224.8, 37.

[52] Page 116.

order of the signs, the apogees of the sun and of the other planets, like the fixed stars,[53] are displaced eastward. Consequently, to harmonize the observations of all ages in a consistent law, my teacher was compelled to distinguish these three motions.

To understand this analysis, assume a maximum eccentricity of 417 units, and a minimum of 321. Let the difference, 96, be the diameter of a small circle, on whose circumference the center of the eccentric moves from east to west. The distance from the center of the universe, then, to the center of the small circle will be 369 units. You will recall that 10,000 of these units constitute the radius of the eccentric. This is the device which my teacher derived from the three above-mentioned eccentricities, in a manner closely resembling the surely divine discovery by which the uniform motions of the moon are determined from three lunar eclipses.[54]

My teacher further established that the velocity with which the center of the eccentric revolves is the same as that with which each value of the changing obliquity recurs. This discovery is indeed worthy of the highest admiration, since it is achieved with such great and remarkable agreement.

The eccentricity was greatest about 60 B.C., when the declination of the sun was also at its maximum. The eccentricity has decreased, moreover, in accordance with this single law, similar to no other. This and other[55] like sports of nature often bring me great solace in the fluctuating vicissitudes of my fortunes, and gently soothe my troubled mind.

The Kingdoms of the World Change with the Motion of the Eccentric

I shall add a prediction. We see that all kingdoms have had their beginnings when the center of the eccentric was at some special point on the small circle. Thus, when the eccentricity of the sun was at its maximum, the Roman government be-

---

[53] See p. 115, above.

[54] HI, 265.16-19, 268.3-12; Th 236.15-17, 28-32; 246.3-8.

[55] Reading *alii* (Th 453.7) instead of *alibi* (PII, 305.1).

came a monarchy; as the eccentricity decreased, Rome too declined, as though aging, and then fell. When the eccentricity reached the boundary and quadrant of mean value, the Mohammedan faith was established; another great empire came into being and increased very rapidly, like the change in the eccentricity. A hundred years hence, when the eccentricity will be at its minimum, this empire too will complete its period. In our time it is at its pinnacle from which equally swiftly, God willing, it will fall with a mighty crash. We look forward to the coming of our Lord Jesus Christ when the center of the eccentric reaches the other boundary of mean value, for it was in that position at the creation of the world. This calculation does not differ much from the saying of Elijah, who prophesied under divine inspiration that the world would endure only 6,000 years,[56] during which time nearly two revolutions are completed. Thus it appears that this small circle is in very truth the Wheel of Fortune, by whose turning the kingdoms of the world have their beginnings and vicissitudes. For in this manner are the most significant changes in the entire history of the world revealed, as though inscribed upon this circle. Moreover, I shall soon, God willing, hear from your own lips how it may be inferred from important conjunctions and other learned prognostications, of what nature these empires were destined to be, whether governed by just or oppressive laws.[57]

[56] From Rheticus's language it appears that he attributed the dire prophecy to the prophet Elijah. But the Old Testament contains no such prediction by Elijah; however, the late Prof. Ralph Marcus kindly called my attention to the following passage in the Babylonian Talmud: "The Tanna debe Eliyyahu teaches: The world is to exist six thousand years" (Babylonian Talmud, English translation, ed. Isidore Epstein [London, 1935–], Sanhedrin, Vol. II [= Nezikin, Vol. VI], p. 657).

[57] Rheticus again displays his devotion to astrological superstition in the Preface to Werner's De triangulis sphaericis. He there declares: "The changes in empires depend upon celestial phenomena. Lands formerly distinguished for their culture, fertile soil, and possessions now lie barren and desolate, inhabited by barbarians, oppressed by tyranny . . . The fiercest nations become civilized, unproductive land is brought under cultivation, from heaven are sent down new forms of earth, culture, and physical type of man. And we see that at intervals of about three hundred and fifty years there always occurs some significant change in the sub-

Now while the center of the eccentric descends toward the center of the universe, the center of the small circle, it is clear, moves in the order of the signs about 25″ each Egyptian year. And starting from the point of its greatest distance from the center of the universe, the center of the eccentric moves in precedence. Hence the inequality arising from the motion of the anomaly for any specified time is subtracted from the mean motion, until a semicircle is completed; but in the other semicircle it is added, in order to obtain the true[58] motion of the apogee. Now the greatest difference between the true and mean apogee is deduced, in the proper geometrical manner, from the above-mentioned data as 7°24′;[59] the other differences are determined, in the customary way, from the position of the center of the eccentric on the small circle. The unequal motion is known, since three positions are given. With regard to the mean motion there is some doubt, since we do not have for these three positions the true place of the solar apogee on the ecliptic. The doubt arises from the disagreement between

lunar world, corresponding to some important alteration in the motion of the sphere of the fixed stars" (*Abh. zur Gesch. d. math. Wiss.*, XXIV, 1, fol. a2v). Later in the same Preface he asserts: "As far as the stars are concerned, I have no doubt that for the Turkish empire there is impending disaster, momentous, sudden, and unforeseen, since the influence of the fiery Triangle is approaching, and the strength of the watery Triangle is declining. Moreover, the anomaly of the sphere of the fixed stars is nearing its third boundary. Whenever it reaches any such boundary, there always occur the most significant changes in the world and in the empires, as history makes clear" (*ibid.*, fol. a5r).

In a letter to Tycho Brahe, Christopher Rothmann censures Rheticus and asks: "How can the variation in the eccentricity of the sun produce a change of empires?" (*Tychonis Brahe opera omnia*, ed. Dreyer, VI, 160.28-29). I know of no evidence indicating that Copernicus shared the astrological views of Rheticus. Dreyer would perhaps not have advanced this suggestion (*Planetary Systems*, p. 333), had he been familiar with the aforementioned Preface by Rheticus.

Schöner's *Opera mathematica* appeared at Nuremberg in 1551, and again in 1561. The first paper is an introduction to judicial astrology (*Isagoge astrologiae iudiciariae*), and the second is a fearfully thorough essay in genethlialogy (*De iudiciis nativitatum*).

[58] Reading *verus* (Th 453.35) instead of *versus* (PII, 306.3).

[59] In *De rev.* Copernicus puts the greatest difference at about 7½° (Th 223. 5-8); while an earlier passage gives 7° 28′ (Th 221.3-5), Rheticus has chosen to follow Copernicus's tables, which give 7° 24′ (Th 224.8-10).

Albategnius and Arzachel, pointed out by Regiomontanus in the *Epitome*, Book III, Proposition 13.[60]

Albategnius is too free in his treatment of the inner secrets of astronomy, as can be seen in many passages. Did he commit this fault in his determination of the solar apogee? Let us grant that he had the correct time of the equinox. Nevertheless, it is impossible, as Ptolemy states,[61] by means of instruments to determine with precision the times of the solstices. For a single minute of declination, which of course easily escapes the eye, may deceive us in this matter by about 4°, to which four days correspond. How was Albategnius able to determine the position of the solar apogee? If he used the method of intermediate positions on the ecliptic, explained by Regiomontanus in the *Epitome*, Book III, Proposition 14, he failed to employ a more trustworthy procedure. He is therefore himself to blame for going astray, since he selected eclipses occurring not near the apogee, but near the middle longitudes of the eccentric of the sun, where the solar apogee, even if mistakenly located 6° from its true position, could produce no noticeable error in eclipses.

According to Regiomontanus,[62] Arzachel boasts that he made 402 observations, and determined from them the position of the apogee. We grant that by this diligence he found the true eccentricity. But since it is not clear that he took into account lunar eclipses occurring near the apsides of the sun, it is apparent that we must no more accept his[63] determination of the higher apse than that of Albategnius.

[60] "Albategnius determined the eccentricity as 2° 4′ 45″, and the arc BH as 7° 43′. Arzachel, however . . . found the same eccentricity as Albategnius, but his value for the arc BH was 12° 10′. This is certainly surprising, since Arzachel lived after Albategnius." The arc BH is the distance from the apogee to the summer solstice.

[61] HI, 196.21–197.11.

[62] *Epitome*, Book III, Prop. 13: "Arzachel, 193 years after Albategnius, made 402 observations [*considerationes*] of the four points midway between the equinoctial and solstitial points, and found BH to be 12° 10′." It should be noted, with reference to the equivalence of *consideratio* and *observatio* (see above, p. 99, n. 28), that in citing this passage Rheticus altered *considerationes* to *observationes*.

[63] Reading *ei* (Th 454.20) instead of *eis* (PII, 306.32).

Now you see what great effort my teacher had to put forth
to determine the mean motion of the apogee. For nearly 40
years in Italy and here in Frauenburg, he observed eclipses
and the motion of the sun. He selected the observation by
which he established that in A.D. 1515 the solar apogee was at
6⅔° of Cancer.[64] Then examining all the eclipses in Ptolemy
and comparing them with his own very careful observations, he
concluded that the mean annual[65] motion of the apogee with
reference to the fixed stars was about 25″,[66] and with reference
to the mean equinox about 1′15″.[67] Through this result it is
established, by applying the true precession to both the mean
and the unequal motions, that the true[68] position of the apogee
was in the time of Hipparchus 63° from the true equinox,
Ptolemy 64½°, Albategnius 76½°, and Arzachel 82°, while
in our time all the calculations agree with experience. These
figures are surely more satisfactory than the Alfonsine, which
put the solar apogee at 12° of Gemini in the time of Ptolemy,
and at the beginning of Cancer in our time.[69] We are 2° closer
than the Alfonsine Tables to the estimate of Arzachel.[70] Albategnius's computation of the position of the apogee exceeds the
Alfonsine by 1°, while we, for a good reason, fall short of his
figure by 6°.[71] For my teacher cannot depart from Ptolemy
and from his own observations, not only because he made and

[64] Th 210.10–211.26.

[65] Reading *annuum* (Th 454.26) instead of *annum* (PII, 307.8).

[66] Th 221.32–222.3.

[67] The mean annual motion of the equinox (mean precession) is about 50″
(see pp. 113–14, above; cf. Th 172.14–17), and it is retrograde (see p. 115,
above). The motion of the apogee is direct (see p. 123, above). Hence, to obtain
the motion of the apogee relative to the equinox, the two mean annual rates must
be added: 25″ + 50″ = 1′ 15″.

[68] Reading *verus* (Th 454.29) instead of *versus* (PII, 307.12).

[69] That is, at 72° for Ptolemy's time, and at 90° for Copernicus's time.

[70] As we saw above (notes 60 and 62 on p. 124), the *Epitome* stated that
Arzachel found the apogee 12° 10′ from the summer solstice = 77° 50′ from the
equinox; cf. Th 210.5–8.

[71] Albategnius located the apogee 7° 43′ from the summer solstice = 82° 17′
from the equinox; cf. above, p. 124, n. 60 and Nallino, *Al-Battānī*, I, 44.29–33.
The version of the Alfonsine Tables to which Rheticus refers evidently contained
the following values: for Ptolemy's time, 72°; Albategnius's, 81°; Arzachel's,
72° (or 84°?); Copernicus's, 90°.

noted his own observations with his own eyes, but also because he knows that Ptolemy, working with the utmost care and making use of eclipses, accurately investigated the motions of the sun and the moon and established them correctly, so far as he could. We are compelled, nevertheless, to differ from him by about 1°,[72] as the motion of the apogee has made clear to us. For Ptolemy regarded the apogee as fixed and therefore showed little care in his treatment of this topic.

You have the opinion of my teacher regarding the motion of the sun. He has accordingly drawn up tables in which he collects for any specified time the true position of the solar apogee, the true eccentricity, the true inequalities, the uniform motions of the sun with reference to the fixed stars and to the mean equinoxes, and hence the true position of the sun corresponding to the observations of all ages. Clearly the tables of Hipparchus, Ptolemy, Theon, Albategnius, and Arzachel, and the Alfonsine Tables, which are to some extent a composite of the others, are temporary only and can endure at most 200 years, until, that is, the discrepancy in the length of the year, eccentricity, inequality, etc., becomes evident, a thing which occurs in the motions and appearances of the other planets for a similar definite reason. Not undeservedly, therefore, could the astronomy of my teacher be called perpetual, as the observations of all ages testify, and the observations of posterity will doubtless confirm. But he calculates his motions and the positions of the apsides from the first star of Aries,[73] since equality of motion is measured by the fixed stars. Then by adding the true precession, he computes and determines the distance in each age of the true positions of the planets from the true equinox.

If such an account of the celestial phenomena had existed a

[72] Hipparchus found the apogee 24½° from the summer solstice (HI, 233.8-10) = 65½° from the equinox, and Ptolemy accepted his determination (HI, 237.6-11).

[73] γ Arietis (Th 130.6-7), not α Arietis (Th 130.22) as Berry thought (A Short History of Astronomy, p. 110 n); cf. above, p. 63, n. 13; Rudolf Wolf, Geschichte der Astronomie (Munich, 1877), p. 240; Dreyer, Planetary Systems, p. 330; Armitage, Copernicus, pp. 105-6.

little before our time, Pico would have had no opportunity, in his eighth and ninth books,[74] of impugning not merely astrology but also astronomy. For we see daily how markedly common calculation departs from the truth.

Special Consideration of the Length of the Tropical Year

In improving the calendar most scholars enumerate various lengths of the year as computed by writers. But they do this in a confused way and come to no conclusion—surely a remarkable procedure for such great mathematicians.

From what has been said above, however, most learned Schöner, you see the four causes of the unequal motion of the sun as measured by the equinoxes: the inequality of the precession of the equinoxes, the inequality of the motion of the sun in the ecliptic, the decrease of the eccentricity, and last, the motion of the apogee for a twofold reason. By virtue of the same causes, the tropical year cannot be equal.

We may readily pardon Ptolemy for measuring equality of motion by the equinoxes,[75] since he held that the fixed stars move in consequence,[76] the position of the apogee is fixed,[77] and the eccentricity of the sun does not decrease.[78] How others would excuse themselves, I do not know. Let us even grant them that the stars and the solar apogee have the same motion in consequence; that therefore time measured by the true equinox in reality does not change; and that the entire inequality (though to assert this in our time would be most absurd) is caused by the defect in the instruments, since the motion of the solar apogee produces only a slight change in the length of the year. Nevertheless, it will not therefore follow that the sun regularly returns to the true equinox always in equal times, as we say that the moon regularly increases its distance from the mean apogee of the epicycle, and returns to the same position in equal times—a statement quoted by the

[74] Pico della Mirandola, *Disputationes adversus astrologos*, Books VIII–IX (pp. 457-82 in the Venice, 1498, edition of Pico's *Opera omnia*).

[75] See above, p. 65, n. 18.

[76] HI, 193.14-16; cf. above, p. 63, n. 13.

[77] See p. 119, above. [78] See above, p. 120, n. 49.

learned Marcus Beneventanus[79] from the Alfonsine Tables. For since we surely cannot deny that the eccentricity of the sun changes, how can they assert that the variation of the angle of anomaly from the mean motion does not alter the length of the tropical year? I heartily congratulate the state and all scholars, whom the work of my teacher will advantage, that we have a sure understanding of the inequality of the year.

But that you may the more readily grasp all these ideas, most learned Schöner, I set them forth numerically before your eyes, in order that I may at length fulfill the pledge I made above.[80]

Let the sun be at the mean vernal equinoctial point, which was 3°29′ west of the first star of Aries at the time of the observation of the autumnal equinox made by Hipparchus in 147 B.C.[81] Let the sun move from this point in the eighth sphere and return to it in a sidereal year (365 days, 15 minutes, and about 24 seconds).[82] However, because the mean equinox in a sidereal year moves about 50″ in the direction opposite to that of the sun, the result[83] is that the sun reaches the new position of the mean vernal equinoctial point before it reaches the starting point, where the sun and the mean equinox had occupied the same position on the ecliptic. Therefore the year as measured by the mean equinox is shorter than the sidereal year,[84] and is computed to be, on the basis of our hypotheses, 365 days, 14 minutes, and about 34 seconds.[85] Now if, for the year measured by the mean equinox, we inquire what the excess[86] in days and fractions of days amounted to in the

---

[79] For a brief account of the life and work of Marco da Benevento see L. Birkenmajer in *Bulletin international de l'académie des sciences de Cracovie, classe des sciences math. et naturelles* (1901), pp. 63-71; and A. Birkenmajer in *Philosophisches Jahrbuch*, XXXVIII(1925), 336-44.

[80] Page 117.   [81] HI, 195.17-20, 204.1-6.

[82] See above, p. 116, n. 33.

[83] Reading *fit* (Th 456.17) instead of *sit* (PII, 310.11).

[84] See p. 46, above.

[85] That is, $365^d 5^h 49^m 36^s$. Newcomb's determination (1900) is $365^d.24219879$ = $365^d 5^h 48^m 46^s$ (*American Ephemeris for 1940*, p. xx).

[86] That is, the length of time by which the tropical year exceeds the Egyptian year of exactly 365 days.

285[87] years between Hipparchus and Ptolemy, we shall find that it was about 69^d 9^m.[88] Then there would be a deficiency of 2^d 6^m,[89] if we assumed that each year exceeded 365 days by ¼ of a day. Let us therefore consider the remaining causes, until we find a deficiency of only ¹⁹⁄₂₀ of a day.

At the time of Hipparchus's observation, the true equinox was about 21′ of the starry ecliptic west of the mean equinox, and the sun was then in the same position as the mean equinox. But in the time of Ptolemy the true equinox was about 47′ east of the mean equinox. Therefore when the sun in the time of Ptolemy arrived at the point 21′ west of the mean equinox, where the true equinoctial point had been in the time of Hipparchus, the equinox did not occur. Nor did it occur when the sun reached the mean equinox. But after it had moved 47′ beyond the mean equinox, it came to the center of the earth, as Pliny says,[90] that is, to the true equinoctial point. The sun, then, had to pass through 1°8′,[91] an arc which it completed in its true motion in 1^d 8^m. Retaining this as a side, I ask how much the angle of anomaly decreased in this instance, and I find that about 1 minute of a day corresponds to it. Thus it is clear that to the excess computed for the year as measured by the mean equinox, there is an addition of 1^d 9^m.[92] Ptolemy correctly stated,[93] then, that between his own observation and that of Hipparchus, from true equinox to true equinox, there were 285^y 70^d 18^m. Therefore there was a deficiency of 57 minutes of a day, the result of subtracting 1^d 9^m from 2^d 6^m, the deficiency which appeared above for the year as measured by the mean equinox.

Let us now consider the deficiency of 7 days between

[87] In the table on p. 116, above, the interval between Hipparchus and Ptolemy is 265 years, because Hipparchus is assigned to 127 B.C. The present passage uses an observation made by Hipparchus in 147 B.C.

[88] 285 × 14^{m}34^s = 69^{d}11½^m.

[89] 285 × ¼^d = 71^{d}15^m
— 69 9
$\overline{2^d\ 6^m}$

[90] *Natural History* ii.19(17).81.

[91] 21′ + 47′ = 1° 8′.   [92] 69^{d}9^m + 1^{d}9^m = 70^{d}18^m.   [93] HI, 204.11-16.

Ptolemy and Albategnius. The situation is clear because the interval of time, 743 years, is greater, and hence all the causes will be more obvious. In the time of Ptolemy the mean equinox was about 7°28′ west of the first star of Aries.[94] But since the mean equinox moved from that position, as has been explained, in the direction opposite to that of the sun, the result is that between Ptolemy and Albategnius there was an excess of about 180^d 14^m for the year as measured by the mean equinox.[95] Then there will be a deficiency of 5^d 31^m, if we compare the year as measured by the mean equinox with the result obtained by adding a day every four years.[96] Whereas in the time of Ptolemy the true equinox was 47′ east of the mean equinox, in the time of Albategnius it was 22′ west of the mean equinox. Therefore the sun reached the true equinox before it reached the mean equinox or the former position of the true equinox,[97] in contrast with our previous example. Hence the time corresponding to 1°9′[98] will be subtracted from the excess for the year as measured by the mean equinox, and added to the deficiency of 5^d 31^m. We must deal in the same way with the variation in the angle of anomaly caused by the decrease in the eccentricity, to which 30 minutes of a day correspond. Then the change in the angle of anomaly, and the unequal motion of precession, combined with the other two causes of the unequal motion of the sun, produce a further deficiency of 1^d 30^m to be subtracted from the excess for the year as measured by the mean equinox. Hence the true excess from the time of Ptolemy to the time of Albategnius's observation becomes 178^d 44^m.[99] But the addition of this further deficiency to 5^d 31^m shows that the total deficiency was 7^d 1^m.[100] Q.E.D.

[94] For it had moved 3° 59′ from its position at the time of Hipparchus: 285 × 50″ = 3° 57½′.

[95] 743 × 14^{m}34^s = 180^{d}23^m.

[96] 743 × ¼^d = 185^{d}45^m
   − 180 14
   ————
   5^{d}31^m

[97] Prowe states (PII, 312 n) that Mästlin substituted *aequinoctium* for the older reading *aequinoctialem*. But both of Mästlin's editions show *aequinoctialem*.

[98] 47′ + 22′ = 1° 9′.

[99] 180^{d}14^m − 1^{d}30^m = 178^{d}44^m.   [100] 5^{d}31^m + 1^{d}30^m = 7^{d}1^m.

It was a difficult task to recover by this analysis the motions of the fixed stars and of the sun, and through the computation of these motions to attain a correct understanding of the length of the tropical year. A boundless kingdom in astronomy has God granted to my learned teacher. May he, as its ruler, deign to govern, guard, and increase it, to the restoration of astronomic truth. Amen.

I intended to report briefly to you, most learned Schöner, the entire treatment of the motions of the moon and of the remaining planets, as well as of the fixed stars and sun, in order that you might understand what benefits to students of mathematics and to all posterity will flow from the writings of my teacher, as from a most plentiful spring. But when I saw that my book was already growing excessively long, I decided to compose a special "Account"[101] of these topics. However, the material which I thought must precede and prepare the way, as it were, I shall set forth at this point. And I shall interweave with my teacher's hypotheses for the motions of the moon and of the remaining planets certain general considerations, in order that you may conceive greater hope for the entire work, and understand why he was compelled to assume other hypotheses or theories.

Having stated at the beginning of this *Account*[102] that my teacher in writing his book imitated Ptolemy, I see that there is practically nothing left for me to take up with you in reference to his method of improving the motions. For Ptolemy's tireless diligence in calculating, his almost superhuman accuracy in observing, his truly divine procedure in examining and investigating all the motions and appearances, and finally his completely consistent method of statement and proof cannot be sufficiently admired and praised by anyone to whom Urania is gracious.

In one respect, however, a burden greater than Ptolemy's confronts my teacher. For he must arrange in a certain and consistent scheme or harmony the series and order of all the

[101] Cf. above, p. 110, n. 5.
[102] Pages 109-10.

motions and appearances, marshalled on the broad battlefield of astronomy by the observations of 2,000 years, as by famous generals. Ptolemy, on the other hand, had the observations of the ancients, to which he could safely entrust himself, for scarcely a quarter of this period. Time, the true god and teacher of the laws of the celestial state, discloses the errors of astronomy to us. For an imperceptible or unnoticed error at the foundation of astronomical hypotheses, principles, and tables is revealed or greatly increased by the passage of time. Therefore my teacher must not so much restore astronomy as build it anew.

Ptolemy was able to harmonize satisfactorily most of the hypotheses of the ancients—Timocharis, Hipparchus, and others—with every inequality in the motions known to him from so small an elapsed period of observation. Therefore he quite rightly and wisely—a praiseworthy action—selected those hypotheses which seemed to be in better agreement with reason and our senses, and which his greatest predecessors had employed.[103] Nevertheless, the observations of all scholars and heaven itself and mathematical reasoning convince us that Ptolemy's[104] hypotheses and those commonly accepted do not suffice to establish the perpetual and consistent connection and harmony of celestial phenomena and to formulate that harmony in tables and rules. It was therefore necessary for my teacher to devise new hypotheses, by the assumption of which he might geometrically and arithmetically deduce with sound logic systems of motion like those which the ancients and Ptolemy, raised on high, once perceived "with the divine eye of the soul,"[105] and which careful observations reveal as existing in the heavens to those today who study the remains of the ancients. Surely students hereafter will see the value of Ptolemy and the other ancient writers, so that they will recall these men who have been until now excluded from the schools, and restore them, like returned exiles, to their ancient place of honor. The

[103] Reading *fuerant* (Th 458.27) instead of *fuerunt* (PII, 314.6).
[104] Reading *Ptolemaei* (Th 458.28) instead of *Ptolemaeo* (PII, 314.8).
[105] A Greek phrase quoted from the pseudo-Aristotelian *De mundo* 391a15.

poet says: "No one desires the unknown."[106] Hence it is not strange that heretofore Ptolemy together with all antiquity has lain ignored in obscurity, as doubtless you, most excellent Schöner, together with other good and learned men have often grieved.

General Considerations Regarding the Motions of the Moon, Together with the New Lunar Hypotheses

The theory of eclipses all by itself seems to maintain respect for astronomy among uneducated people; yet we see daily how much it differs nowadays from common calculation in the prediction of both the duration and extent of eclipses. In constructing astronomical tables we should not, as we see certain writers doing,[107] reject the precise observations of Ptolemy and other excellent authorities as false and untrustworthy, unless the passage of time discloses to us that some manifest error has crept in. For what is more human than sometimes to be mistaken and deceived even by the appearance of truth, especially in these difficult, abstruse, and by no means obvious matters?

In his exposition of the motion of the moon, my teacher assumes such theories and schemes of motion as make it clear that the eminent ancient philosophers were not at all blind in their observations. Just as we showed above that the increase and decrease of the tropical year are regular, so, from a careful investigation of the motions of the sun and moon, it is possible to deduce for each age the true distances of the sun, moon, and earth from one another, or the reason why the diameters of the sun, moon, and earth's shadow have been found different at different times, and thus, in addition, to attain a correct understanding of the solar and lunar parallax.

In the *Epitome*, Book V, Proposition 22, Regiomontanus says: "But it is noteworthy that the moon does not appear so great at quadrature, when it is in the perigee of the epicycle, whereas, if the entire disk were visible, it should appear four

---

[106] Ovid *Ars amatoria* iii.397.

[107] Doubtless Rheticus intended to include Werner in the group; cf. Copernicus's sharp protests in the *Letter against Werner*, pp. 99-100, 103-5, above.

times its apparent size at opposition, when it is in the apogee of the epicycle."[108] This difficulty was noticed by Timocharis and Menelaus, who always use the same lunar diameter[109] in their observations of the stars. But experience has shown my teacher that the parallax and size of the moon, at any distance from the sun, differ little or not at all from those which occur at conjunction and opposition, so that clearly the traditional eccentric cannot be assigned to the moon. He supposes therefore that the lunar sphere encloses the earth together with[110] its adjacent elements, and that the center of the deferent is the center of the earth, about which the deferent revolves uniformly, carrying the center of the lunar epicycle.

The second inequality, which appears in the distance of the moon from the sun, he saves as follows. He assumes that the moon moves on an epicycle of an epicycle of a concentric; that is, to the first epicycle, which in general is in evidence at conjunction and opposition, he joins a second small[111] epicycle which carries the moon; and he shows that the ratio of the diameter of the first epicycle to the diameter of the second is as 1,097:237. The scheme of motions is as follows. The inclined circle has the same motion as heretofore, save that its equal periods are measured by the fixed stars. The deferent, which is concentric, moves regularly and uniformly about its own center (which is also the center of the earth), at the same time rotating uniformly and regularly from the line of mean motion of the sun. The first epicycle also revolves uniformly about its own center; in its upper circumference it carries the center of the small second epicycle in precedence, in its lower circumference, in consequence.[112] My teacher computes this uniform and regular motion from the true apogee. This point

[108] The quotation is substantially correct. But the original has *aux* and *oppositum augis*, for which Rheticus substitutes *apogium* and *perigium* (cf. above, p. 34, n. 117).

[109] ½° (Th 235.8-11); the observations referred to are cited in HII, 25.15-34.8.

[110] Prowe's text (PII, 315.29) omits *cum* between *terram* and *adiacentibus* (Th 459.30).

[111] Reading *parvum* (Th 459.35) for *parum* (PII, 316.3).

[112] Cf. above, pp. 68-69, n. 28.

is marked on the upper circumference of the first epicycle by a line drawn from the center of the earth through the center of the first epicycle to its circumference. Starting from the small epicycle's true apogee, which is indicated on its circumference by a line drawn from the center of the first epicycle through the center of the second epicycle, the moon also moves regularly and uniformly on the circumference of the small second epicycle. The rule governing this motion is the following: the moon revolves twice on its epicycle[113] in one period of the deferent, so that at every conjunction and opposition the moon is found in the perigee of the small epicycle, but at the quadratures in its apogee. This is the device or hypothesis by which my teacher removes all the aforementioned incongruities, and which satisfies all the appearances, as he clearly shows, and as can be inferred also from his tables.

Furthermore, most learned Schöner, you see that here in the case of the moon we are liberated from an equant by the assumption of this theory, which, moreover, corresponds to experience and all the observations. My teacher dispenses with equants for the other planets as well, by assigning to each of the three superior planets only one epicycle and eccentric; each of these moves uniformly about its own center, while the planet revolves on the epicycle in equal periods with the eccentric. To Venus and Mercury, however, he assigns an eccentric on an eccentric. The planets are each year observed as direct, stationary, retrograde, near to and remote from the earth, etc.[114] These phenomena, besides being ascribed to the planets, can be explained, as my teacher shows, by a regular motion of the spherical earth; that is, by having the sun occupy the center of the universe, while the earth revolves instead of the sun on the eccentric,[115] which it has pleased him to name the great

---

[113] While discarding an unnecessary emendation of Mästlin's, Prowe's text (PII, 316.20) inserts *parvo*, for which there is no warrant in the Basel edition of 1566 (p. 201v).

[114] Reading *etc.* with the editions of 1566 (p. 202r), 1596 (p. 110), and 1621 (p. 107), instead of *et cum* (PII, 317.10; Th 460.24).

[115] This is the first indication in the *Narratio prima* that the astronomy of Copernicus involves heliocentrism and a moving earth. Rheticus evidently deemed

circle. Indeed, there is something divine in the circumstance that a sure understanding of celestial phenomena must depend on the regular and uniform motions of the terrestrial globe alone.

The Principal Reasons Why We Must Abandon the Hypotheses of the Ancient Astronomers

In the first place, the indisputable precession of the equinoxes, as you have heard, and the change in the obliquity of the ecliptic persuaded my teacher to assume that the motion of the earth could produce most of the appearances in the heavens, or at any rate save them satisfactorily.

Secondly, the diminution of the eccentricity of the sun is observed, for a similar reason and proportionally, in the eccentricities of the other planets.

Thirdly, the planets evidently have the centers of their deferents in the sun, as the center of the universe. That the ancients, not to mention the Pythagoreans for the moment, were aware of this fact is sufficiently clear for example from Pliny's statement, following undoubtedly the best authorities, that Venus and Mercury do not recede further from the sun than fixed, ordained limits because their paths encircle the sun;[116] hence these planets necessarily share the mean motion of the sun. As Pliny says,[117] the course of Mars is hard to trace. In addition to the other difficulties in the correction of its motion, Mars unquestionably shows a parallax sometimes greater

---

it advisable, before introducing these ideas, to paint the portrait of Copernicus as a great astronomer, who made careful observations and painstaking calculations, who studied thoroughly the work of his predecessors and respected, in particular, the authority of Ptolemy. The cautiousness of Rheticus stands in striking contrast to the forthright procedure of Copernicus in the *Commentariolus* (cf. pp. 57-59, above).

[116] *Natural History* ii.17(14).72. It is more likely that Pliny's *conversas absidas* meant simply "different courses," i.e., orbits unlike those of the superior planets; cf. Rackham's translation in the Loeb Classical Library (London, 1938). Rheticus's understanding of the passage was governed by Th 27.18-25. For Kepler's comments on this obscure section in Pliny and on Copernicus's interpretation of it see his *Opera*, ed. Frisch, I, 271-72.

[117] *Natural History* ii.17(15).77.

than the sun's, and therefore it seems impossible that the earth should occupy the center of the universe. Although Saturn and Jupiter, as they appear to us at their morning and evening rising, readily yield the same conclusion, it is particularly and especially supported by the variability of Mars when it rises. For Mars, having a very dim light, does not deceive the eye as much as Venus or Jupiter, and the variation of its size is related to its distance from the earth. Whereas at its evening rising Mars seems to equal Jupiter in size, so that it is differentiated only by its fiery splendor, when it rises in the morning just before the sun and is then extinguished in the light of the sun, it can scarcely be distinguished from stars of the second magnitude. Consequently at its evening rising it approaches closest to the earth, while at its morning rising it is furthest away; surely this cannot in any way occur on the theory of an epicycle. Clearly then, in order to restore the motions of Mars and the other planets, a different place must be assigned to the earth.

Fourthly, my teacher saw that only on this theory could all the circles in the universe be satisfactorily made to revolve uniformly and regularly about their own centers, and not about other centers—an essential property of circular motion.

Fifthly, mathematicians as well as physicians must agree with the statements emphasized by Galen here and there: "Nature does nothing without purpose"[118] and "So wise is our Maker that each of his works has not one use, but two or three or often more."[119] Since we see that this one motion of the earth satisfies an almost infinite number of appearances, should we not attribute to God, the creator of nature, that skill which we observe in the common makers of clocks? For they carefully avoid inserting in the mechanism any superfluous wheel or any whose function could be served better by another with a slight change

---

[118] *De usu partium* x.14 (ed. Helmreich, Leipzig, 1907–9, II, 109.2). Rheticus quotes the Greek text, the first edition (Venice, 1525) of which was available to him, as was also the Basel edition of 1538. The words quoted appear in the 1525 edition, Vol. I, fol. H7v.47 (= fol. 63v of the separate pagination for the *De usu partium*).

[119] *Ibid.* x.15 (ed. Helmreich, II, 111.5-8; 1525 ed., Vol. I, fol. H8r.14-16). Rheticus accommodates the quotation to the structure of his own sentence.

of position. What could dissuade my teacher, as a mathematician, from adopting a serviceable theory of the motion of the terrestrial globe, when he saw that on the assumption of this[120] hypothesis there sufficed, for the construction of a sound science of celestial phenomena, a single eighth sphere, and that motionless, the sun at rest in the center of the universe, and for the motions of the other planets, epicycles on an eccentric or eccentrics on an eccentric or epicycles on an epicycle? Moreover, the motion of the earth in its circle produces the inequalities of all the planets except the moon; this one motion alone seems to be the cause of every apparent inequality at a distance from the sun, in the case of the three superior planets, and in the neighborhood of the sun, in the case of Venus and Mercury. Finally, this motion makes it possible to satisfy each of the planets by only one deviation in latitude of the deferent of the planet. Hence it is particularly the planetary motions that require such hypotheses.

Sixthly and lastly, my teacher was especially influenced by the realization that the chief cause of all the uncertainty in astronomy was that the masters of this science (no offense is intended to divine Ptolemy, the father of astronomy) fashioned their theories and devices for correcting the motion of the heavenly bodies with too little regard for the rule which reminds us that the order and motions of the heavenly spheres agree in an absolute system. We fully grant these distinguished men their due honor, as we should. Nevertheless, we should have wished them, in establishing the harmony of the motions, to imitate the musicians who, when one string has either tightened or loosened, with great care and skill regulate and adjust the tones of all the other strings, until all together produce the desired harmony, and no dissonance is heard in any. If Albategnius, to speak of him for the moment, had followed this precept in his work, we should doubtless have today a surer understanding of all the motions. For it is likely that the Alfonsine Tables drew heavily from him; and since this one rule was neglected, we should have had to face at some time, if we intend to speak the truth, the collapse of all astronomy.

[120] Reading *tali* (Th 461.31) instead of *talia* (PII, 319.13).

Under the commonly accepted principles of astronomy, it could be seen that all the celestial phenomena conform to the mean motion of the sun and that the entire harmony of the celestial motions is established and preserved under its control. Hence the sun was called by the ancients leader, governor of nature, and king. But whether it carries on this administration as God rules the entire universe, a rule excellently described by Aristotle in the *De mundo*,[121] or whether, traversing the entire heaven so often and resting nowhere, it acts as God's administrator in nature, seems not yet altogether explained and settled. Which of these assumptions is preferable, I leave to be determined by geometers and philosophers (who are mathematically equipped). For in the trial and decision of such controversies, a verdict must be reached in accordance with not plausible opinions but mathematical laws (the court in which this case is heard). The former manner of rule has been set aside, the latter adopted. My teacher is convinced, however, that the rejected method of the sun's rule in the realm of nature must be revived, but in such a way that the received and accepted method retains its place. For he is aware that in human affairs the emperor need not himself hurry from city to city in order to perform the duty imposed on him by God; and that the heart does not move to the head or feet or other parts of the body to sustain[122] a living creature, but fulfills its function through other organs designed by God for that purpose.

Now my teacher concluded that the mean motion of the sun must be the sort of motion that is not only established by the imagination, as in the case of the other planets, but is self-caused, since it appears to be truly "both choral dancer and choral leader." He then showed that his opinion was sound and not inconsistent with the truth, for he saw that by his hypotheses the efficient cause of the uniform motion of the sun could be geometrically deduced and proved. Hence the mean motion of the sun would necessarily be perceived in all the

---

[121] Chap. vi. This work is now athetized; see Wilhelm Capelle, *Neue Jahrbücher für das klassische Altertum*, XV(1905), 532.

[122] Reading *conservationem* (Th 462.37) instead of *conversationem* (PII, 321.4).

motions and appearances of the other planets in a definite manner, as appears in each of them. Thus the assumption of the motion of the earth on an eccentric provides a sure theory of celestial phenomena, in which no change should be made without at the same time re-establishing the entire system, as would be fitting, once more on proper ground. While we were unable from our common theories even to surmise this rule by the sun in the realm of nature, we ignored most of the ancient encomia of the sun as poetry. You see, then, what sort of hypotheses for saving the motions my teacher had to assume under these circumstances.

Transition to the Explanation of the New Hypotheses for the Whole of Astronomy

I interrupt your thoughts, distinguished sir, for I am aware that while you listen to the reasons, investigated by my teacher with remarkable learning and great devotion, for revising the astronomical hypotheses, you thoughtfully consider what foundation may finally prove to be suitable for the hypotheses of astronomy reborn. But the men who endeavor to pull all the stars together around in the ether in accordance with their own opinion, as though they had put chains upon them, merit pity rather than resentment, in your judgment as in that of other true mathematicians and all good men. You are not unacquainted with the importance to astronomers of hypotheses or theories, and with the difference between a mathematician and a physicist.[123] Hence you agree, I feel, that the results to which the observations and the evidence of heaven itself lead us again and again must be accepted, and that every difficulty must be faced and overcome with God as our guide and mathematics and tireless study as our companions.

Accordingly, anyone who declares that he must be mindful of the highest and principal end of astronomy will be grateful

[123] Presumably a reference to Aristotle's *Physics* 193b22-23, and to Simplicius's *Commentary on Aristotle's Physics*, second comment on Book ii.2 (*Commentaria in Aristotelem Graeca*, Vol. IX, ed. H. Diels, Berlin, 1882, pp. 290-93); the first edition of Simplicius's *Commentary* (Venice, 1526) was available to Rheticus and Schöner.

with us to my teacher and will consider as applicable to himself
Aristotle's remark: "When anyone shall succeed in finding
proofs of greater precision, gratitude will be due to him for
the discovery."[124] The examples of Callippus and Aristotle[125]
assure us that, in the effort to ascertain the causes of the phenomena, astronomy must be revised as unequal motions of the
heavenly bodies are encountered. Hence I may hope that Averroes, who played the role of the severe Aristarchus[126] to Ptolemy, would not receive the hypotheses of my teacher harshly,
if only he would examine natural philosophy patiently. In my
opinion, Ptolemy was not so bound and sworn to his own hypotheses that, were he permitted to return to life, upon seeing
the royal road blocked and made impassable by the ruins of so
many centuries, he would not seek another road over land and
sea to the construction of a sound science of celestial phenomena,
since he could not rise through the air and open sky to the
desired goal. For what else shall I say of the man who wrote
the following words:

Propositions assumed without proof, if once they are perceived to be in
agreement with the phenomena, cannot be established without some
method and reflection; and the procedure for apprehending them is
hard to explain, since in general, of first principles, there naturally is
either no cause or one difficult to set forth.[127]

How modestly and wisely Aristotle speaks on the subject of
the celestial motions can be seen everywhere in his works. He
says in another connection: "It is the mark of an educated man
to look for precision in each class of things just so far as the

[124] *De caelo* ii.5 287b34–288a1 (J. L. Stocks's rendering in the Oxford translations of the works of Aristotle, 1930, Vol. II).

[125] *Metaphysics* xii.8 1073b32–1074a5.

[126] For Averroes as an adversary of the Ptolemaic astronomy see pp. 194-95,
below, and Duhem, *Le Système du monde*, II, 133-39. In a long article devoted to
the relation between Copernicus and the astronomer Aristarchus of Samos, Brachvogel failed to recognize that the Aristarchus of our passage is unquestionably Aristarchus of Alexandria, the severe critic of Homer, not Aristarchus of Samos (ZE,
XXV[1933–35], 703, n. 15).

[127] HII, 212.11-16. Rheticus presented to Copernicus (PI², 411) a copy of the
first edition of the Greek text of the *Syntaxis* (Basel, 1538); the passage quoted
begins at the foot of the page numbered (incorrectly) 219.

nature of the subject admits."[128] Now in physics as in astronomy, one proceeds as much as possible from effects and observations to principles. Hence I am convinced that Aristotle, who wrote careful discussions of the heavy and the light, circular motion, and the motion and rest of the earth,[129] if he could hear the reasons for the new hypotheses, would doubtless honestly acknowledge what he proved in these discussions, and what he assumed as unproved principle. I can therefore well believe that he would support my teacher, inasmuch as the well-known saying attributed to Plato[130] is certainly correct: "Aristotle is the philosopher of the truth." On the other hand, were he to burst forth in harsh language, it would be only to lament bitterly, I am persuaded, the condition of this most beautiful part of philosophy in the following terms: "It has been said very well by Plato[131] that 'geometry and the studies that accompany it dream about being, but the clear waking vision of it is impossible for them as long as they leave the assumptions which they employ undisturbed and cannot give any account of them'"; and he would add: "We must be deeply grateful to the immortal gods for the knowledge of such a theory of the phenomena."

But since these remarks are less appropriate here than in a certain other treatise,[132] I shall proceed to set forth the remaining hypotheses of my teacher in an open and orderly manner, in an endeavor to throw some light on my previous statements.

The Arrangement of the Universe

Aristotle says: "That which causes derivative truths to be true is most true."[133] Accordingly, my teacher decided that he

---

[128] *Nicomachean Ethics* 1094b23-25 (W. D. Ross's translation, Oxford, 1925).

[129] *De caelo* i.2-4; ii.3, 13-14; *Physics* viii.8-9.

[130] The authentic works of Plato contain no reference to the philosopher Aristotle.

[131] *Republic* vii.13 533B-C, slightly altered (P. Shorey's translation, Loeb Classical Library, London, 1930–35).

[132] Probably Rheticus has in mind his projected "Second Account."

[133] *Metaphysics* i minor.1 993b26-27 (W. D. Ross's translation, Oxford, 1928). Although Rheticus usually quotes from Greek authors in the original language, in the present instance he is using Bessarion's Latin translation of the *Metaphysics*.

must assume such hypotheses as would contain causes capable of confirming the truth of the observations of previous centuries, and such as would themselves cause, we may hope, all future astronomical predictions of the phenomena to be found true.

First, surmounting no mean difficulties, he established by hypothesis that the sphere of the stars, which we commonly call the eighth sphere, was created by God to be the region which would enclose within its confines the entire realm of nature, and hence that it was created fixed and immovable as the place of the universe. Now motion is perceived only by comparison with something fixed; thus sailors on the sea, to whom

> land is no longer
> Visible, only the sky on all sides and on all sides the water[134]

are not aware of any motion of their ship when the sea is undisturbed by winds, even though they are borne along at such high speed that they pass over several long miles in an hour. Hence this sphere was studded by God for our sake with a large number of twinkling stars, in order that by comparison with them, surely fixed in place, we might observe the positions and motions of the other enclosed spheres and planets.

Then, in harmony with these arrangements, God stationed in the center of the stage His governor of nature, king of the entire universe, conspicuous by its divine splendor, the sun

> To whose rhythm the gods move, and the world
> Receives its laws and keeps the pacts ordained.[135]

The other spheres are arranged in the following manner. The first place below the firmament or sphere of the stars falls to the sphere of Saturn, which encloses the spheres of first Jupiter, then Mars; the spheres of first Mercury, then Venus

---

[134] Vergil *Aeneid* iii.192-93.

[135] Giovanni Gioviano Pontano, *Urania* or *De stellis* i.240-41 (Florence, 1514, p. 71); Pontano's poems were reprinted in *Ioannis Ioviani Pontani carmina* (ed. Benedetto Soldati, Florence, 1902), where the quoted lines appear on p. 10. Copernicus owned a copy of the selection of Pontano's prose works which was printed at Venice in 1501 (PI², 417).

surround the sun; and the centers of the spheres of the five planets are located in the neighborhood of the sun. Between the concave surface of Mars' sphere and the convex of Venus', where there is ample space, the globe of the earth together with its adjacent elements, surrounded by the moon's sphere, revolves in a great circle which encloses within itself, in addition to the sun, the spheres of Mercury and Venus, so that the earth moves among the planets as one of them.

As I carefully consider this arrangement of the entire universe according to the opinion of my teacher, I realize that Pliny set down an excellent and accurate statement when he wrote: "To inquire what is beyond the universe or heaven, by which all things are overarched, is no concern of man, nor can the human mind form any conjecture concerning this question." And he continues: "The universe is sacred, without bounds, all in all; indeed, it is the totality, finite yet similar to the infinite, etc."[136] For if we follow my teacher, there will be nothing beyond the concave surface of the starry sphere for us to investigate, except insofar as Holy Writ has vouchsafed us knowledge, in which case again the road will be closed to placing anything beyond this concave surface. We will therefore gratefully admire and regard as sacrosanct all the rest of nature, enclosed by God within the starry heaven. In many ways and with innumerable instruments and gifts He has endowed us, and enabled us[137] to study and know nature; we will advance to the point to which He desired us to advance, and we will not attempt to transgress the limits imposed by Him.

That the universe is boundless up to its concave surface, and truly similar to the infinite is known, moreover, from the fact that we see all the heavenly bodies twinkle, with the exception of the planets including Saturn, which, being the nearest of them to the firmament, revolves on the greatest circle. But this conclusion follows far more clearly by deduction from the hypotheses of my teacher. For the great circle which carries the

---

[136] Abridged from *Natural History* ii.1.1-2.

[137] Prowe's text (PII, 326.18) omits *nos* between *idoneos* and *effecit* (Th 466.20).

earth has a perceptible ratio to the spheres of the five planets, and hence every inequality in the appearances of these planets is demonstrably derived from their relations to the sun. Every horizon on the earth, being a great circle of the universe, divides the sphere of the stars into equal parts. Equal periods in the revolutions of the spheres are shown to be measured by the fixed stars. Consequently it is quite clear that the sphere of the stars is, to the highest degree, similar to the infinite, since by comparison with it the great circle vanishes, and all the phenomena are observed exactly as if the earth were at rest in the center of the universe.

Moreover, the remarkable symmetry and interconnection of the motions and spheres, as maintained by the assumption of the foregoing hypotheses, are not unworthy of God's workmanship and not unsuited to these divine bodies. These relations, I should say, can be conceived by the mind (on account of its affinity with the heavens) more quickly than they can be explained by any human utterance, just as in demonstrations they are usually impressed upon our minds, not so much by words as by the perfect and absolute ideas, if I may use the term, of these most delightful objects. Nevertheless it is possible, in a general survey of the hypotheses, to see how the inexpressible harmony and agreement of all things manifest themselves.

For in the common hypotheses there appeared no end to the invention of spheres; moreover, spheres of an immensity that could be grasped by neither sense nor reason were revolved with extremely slow and extremely rapid motions. Some writers stated that the daily motion of all the lower spheres is caused by the highest movable sphere;[138] but when a great storm of controversy raged over this question, they could not explain why a higher sphere should have power over a lower. Others, like Eudoxus[139] and those who followed him, assigned to each planet a special sphere, the motion of which caused the

[138] See Dreyer, *Planetary Systems*, p. 91.

[139] See Aristotle *Metaphysics* xii.8 1073b17-26; and Simplicius's *Commentary on Aristotle's De caelo* (*Commentaria in Aristotelem Graeca*, Vol. VII, ed. Hei-

planet to revolve about the earth once in a natural day. Moreover, ye immortal gods, what dispute, what strife there has been until now over the position of the spheres of Venus and Mercury, and their relation to the sun. But the case is still before the judge. Is there anyone who does not see that it is very difficult and even impossible ever to settle this question while the common hypotheses are accepted? For what would prevent anyone from locating even Saturn below the sun, provided that at the same time he preserved the mutual proportions of the spheres and epicycle, since in these same hypotheses there has not yet been established the common measure of the spheres of the planets, whereby each sphere may be geometrically confined to its place? I refrain from mentioning here the vast commotion which those who defame this most beautiful and most delightful part of philosophy have stirred up on account of the great size of the epicycle of Venus, and on account of the unequal motion, on the assumption of equants, of the celestial spheres about their own centers.

However, in the hypotheses of my teacher, which accept, as has been explained, the starry sphere as boundary, the sphere of each planet advances uniformly with the motion assigned to it by nature and completes its period without being forced into any inequality by the power of a higher sphere. In addition, the larger spheres revolve more slowly, and, as is proper, those that are nearer to the sun, which may be said to be the source of motion and light, revolve more swiftly. Hence Saturn, moving freely in the ecliptic, revolves in thirty years, Jupiter in twelve, and Mars in two. The center of the earth measures the length of the year by the fixed stars. Venus passes through the zodiac in nine months, and Mercury, revolving about the sun on the smallest sphere, traverses the universe in eighty days. Thus there are only six moving spheres which revolve about

berg, Berlin, 1894, 493.11-15; 494.12-18, 23-26; 495.8-9, 17-22; 496.15-19). The first edition of Simplicius's *Commentary* (Venice, 1526) was available to Rheticus; it was a Greek version, done by Bessarion or one of his circle, of William of Moerbeke's Latin translation (*Sitzungsberichte der Akademie der Wissenschaften zu Berlin*, 1892, pp. 74-75).

the sun, the center of the universe. Their common measure is the great circle which carries the earth, just as the radius of the spherical earth is the common measure of the circles of the moon, the distance of the sun from the moon, etc.

Who could have chosen a more suitable and more appropriate number than six? By what number could anyone more easily have persuaded mankind that the whole universe was divided into spheres by God the Author and Creator of the world? For the number six is honored beyond all others in the sacred prophecies of God and by the Pythagoreans and the other philosophers. What is more agreeable to God's handiwork than that this first and most perfect work should be summed up in this first and most perfect number?[140] Moreover, the celestial harmony is achieved by the six aforementioned movable spheres. For they are all so arranged that no immense interval is left between one and another; and each, geometrically defined, so maintains its position that if you should try to move any one at all from its place, you would thereby disrupt the entire system.

But now that we have touched on these general considerations, let us proceed to an exposition of the circular motions which are appropriate to the several spheres and to the bodies that cleave to and rest upon them. First we shall speak of the hypotheses for the motions of the terrestrial globe, on which we have our being.

The Motions Appropriate[141] *to the Great Circle and Its Related Bodies. The Three Motions of the Earth: Daily, Annual, and the Motion in Declination*

Following Plato and the Pythagoreans, the greatest mathematicians of that divine age, my teacher thought that in order

---

[140] This passage, in which Rheticus reveals his acceptance of number mysticism, finds no parallel in the works of Copernicus; cf. above, p. 122, n. 57. For an excellent discussion of the metamathematical superstructure, erected in the early modern period on the basis of Pythagorean and Platonic philosophy, see Edward W. Strong, *Procedures and Metaphysics* (Berkeley, Calif., 1936), chap. viii.

[141] Reading *competant* (Th 468.9) instead of *computant* (PII, 329.23).

to determine the causes of the phenomena circular motions must be ascribed to the spherical earth.[142] He saw (as Aristotle also points out[143]) that when one motion is assigned to the earth, it may properly have other motions, by analogy with the planets. He therefore decided to begin with the assumption that the earth has three motions, by far the most important of all.

For in the first place, having assumed the general arrangement of the universe described above, he showed that, enclosed by its poles within the lunar sphere, the earth, like a ball on a lathe, rotates from west to east, as God's will ordains; and that by this motion, the terrestrial globe produces day and night and the changing appearances of the heavens, according as it is turned toward the sun. In the second place, the center of the earth, together with its adjacent elements and the lunar sphere, is carried uniformly in the plane of the ecliptic by the great circle, which I have already mentioned more than once,[144] in the order of the signs. In the third place, the equator and the axis of the earth have a variable inclination to the plane of the ecliptic and move in the direction opposite to that of the motion of the center, so that on account of this inclination of the earth's axis and the immensity of the starry sphere, no matter where the center of the earth may be, the equator and the poles of the earth are almost invariably directed to the same points in the heavens. This result will ensue if the ends of the earth's axis, that is, the poles of the earth, are understood to move daily in precedence a distance almost exactly equal to the motion of the center of the earth in consequence on the great circle, and to describe about the axis and poles of the great circle or ecliptic small circles equidistant from them.

But to these motions we should add, in the opinion of my teacher, two librations of the poles of the earth and the two motions, the one uniform and the other unequal, with which

---

[142] Whether Plato held that the earth is at rest or in motion is much disputed; see Thomas L. Heath, *Aristarchus of Samos* (Oxford, 1913), pp. 174-85.

[143] *De caelo* ii.14 296a24-296b3.

[144] Pages 135-36, 144, 145, 147.

the center of the great circle advances in the ecliptic;[145] let us also recall what was said above[146] concerning the motions of the moon about the center of the earth. We shall then have, most learned Schöner, a true system of hypotheses for deducing in its entirety what the moderns call the doctrine of the first motion, which at present is derived from all sorts of motions of the starry sphere; and for determining the causes of the motions and phenomena of the sun and moon, as they have been carefully observed by scholars for the past two thousand years. I may merely mention, since I shall have occasion to deal with the topic more fully below,[147] that the motion of the great circle unquestionably affects the appearances of the other five planets. With so few motions and, as it were, with a single circle is so vast a subject comprehended.

In the doctrine of the first motion nothing need be changed. For, utilizing the properties of things which are interrelated, we shall determine the maximum obliquity and in the same way investigate the declinations of the remaining parts of the ecliptic, right ascensions, the theory of shadows and gnomons in all regions of the earth, the lengths of days, oblique ascensions, the rising and setting of the stars, etc. However, our hypotheses differ from those of antiquity in that in ours, as opposed to the views[148] of the ancients, no circle except the ecliptic is properly described by the imagination on the starry sphere. The other circles, to wit, the equator, the two tropics, arctics and antarctics, horizons, meridians, and all the others connected with the doctrine of the first motion, e.g., vertical circles, parallels of altitude, colures etc. are properly traced upon the globe of the earth, and transferred by a certain relation to the heavens.

In addition to the apparent daily revolution about the earth, which the sun shares with all the stars and the other planets, there are those phenomena related to the sun which Ptolemy and the moderns have attributed to the sun's own motions and also those which are observed to occur in connection with the

---

[145] See pp. 121, 123, above.   [146] Pages 134-35.   [147] Pages 168-85.
[148] Reading *praescriptum* (Th 469.9) instead of *praeceptum* (PII, 331.12).

shift of the solstitial and equinoctial points, the distance of the stars from them, and the motion of the apogee among the fixed stars. All these phenomena present themselves to our eyes as if the sun and the sphere of the stars move. For the way in which, according to common belief, these bodies emerge in the east or rise, gradually climb above the horizon until they reach the meridian, from which they descend in like manner, and then traverse the lower hemisphere, daily completing their diurnal revolutions, is caused, clearly enough, by the first motion which my teacher, in company with Plato,[149] assigns to the earth.

The sun seems to us to move in the order of the signs, and we persuade ourselves that by this motion it describes the ecliptic and determines the length of the year. But these phenomena can be produced by the second motion which my teacher attributes to the earth. For as the earth moves on the great circle and comes to a position between the constellation Libra and the sun, those of us who suppose the earth to be at rest think that the sun is in the constellation Aries, because a line drawn from the center of the earth through the sun to the sphere of the stars strikes that constellation. Then, as the earth advances to Scorpio, the sun seems to be in Taurus, and so to traverse the zodiac.[150] I assert, however, that with the sun at rest this motion is properly the earth's. And the sidereal year is the time in which the center of the earth or, in appearance, of the sun completes a single revolution from a star to the same star.

The third motion of the earth produces the regular, cyclic changes of season on the whole earth; for it causes the sun and the other planets to appear to move on a circle oblique to the equator, and the sun to appear to the several regions of the earth exactly as it would if the earth were by hypothesis at the center of the universe and the planets moved on an oblique circle. For on account of the above-mentioned motion of its

[149] Cf. above, p. 148, n. 142.
[150] Prowe's text (PII, 332.9) inserts *totum*, for which there is no warrant in the Basel edition of 1566 and Mästlin's editions of 1596 and 1621.

poles, the plane of the equator, in comparison with the sun, turns away from the plane of the ecliptic and returns toward it, or as the Greeks say λοξεύεται καὶ ἐγκλίνει.[151] Hence the same inclination of the equator to the ecliptic recurs at almost the same points on the ecliptic, and the poles of the daily rotation are always in very nearly the same spot on the starry sphere.

Now when the equator attains its greatest inclination to the plane of the ecliptic, that is, to the sun, the line drawn from the center of the sun to the center of the earth cuts a cone in the globe of the earth as it performs its daily rotation, thereby describing the tropics. Furthermore, when the plane of the equator returns to the plane of the ecliptic, that is, to the sun, all over the earth the equinox occurs, since the line of which I just spoke divides the globe of the earth along the equator into two hemispheres. But the other parallels of latitude are marked on the earth according as the motions of the equator away from and toward the sun (or to use Ptolemy's terms λόξωσις καὶ ἔγκλισις) are combined. The arctics and antarctics are described by their points of contact with the horizons.[152] The poles of the ecliptic, in the opinion of my teacher, describe the polar circles equidistant from the poles of the equator. The great circle of the earth's globe which passes through the poles of the equator and the aforesaid equidistant poles of the ecliptic is the solstitial colure; and another great circle, intersecting the first in the poles of the equator at spherical right angles, is the equinoctial colure. And it is to be understood that in this manner the circles of any point at all and any other circles whatsoever are readily traced upon the earth and thence transferred to the overarching heavens.

Moreover, in obedience to the command given by the observations, the globe of the earth has risen to the circumference of the eccentric, while the sun has descended to the center of the universe. Now, in the common hypotheses the center of the eccentric was situated in our age between the center of the entire universe (which in these hypotheses was also the center

---

[151] For example, *Syntaxis* xiii.1 (HII, 528.11-16) and xiii *passim*.
[152] Cf. Th 74.23-28.

of the earth) and the constellation Gemini. Conversely, in my
teacher's hypotheses the center of the great circle, which I referred to in the beginning of this *Account*[153] as the center of
the eccentric, is found between the sun, which is the center of
the universe according to my teacher, and the constellation
Sagittarius; and the diameter of the great circle that passes
through the center of the earth represents the line of mean
motion of the sun. Since the line drawn from the center of the
earth through the center of the sun to the ecliptic determines
the true place of the sun, it is not difficult to see how in the
system of Ptolemy and the moderns the sun is conceived to
move unequally in the ecliptic and how the angle of inequality
from the mean motion is investigated geometrically. When the
earth is in the higher apse of the great circle, the sun is thought
to be at the apogee on the eccentric, and, conversely, when the
earth is in the lower apse, the sun seems to be in perigee.

But the manner in which the fixed stars appear to alter their
distance from the equinoctial and solstitial points, and the
greatest obliquity of the sun to vary, etc. (my treatment of
these topics at the beginning of the *Account* is drawn from
Book III of my teacher's work) has been shown by him to
depend on the motion in declination, which I have set forth in
a general way, and on two mutually interacting librations. From
the poles which were referred to just above as the equidistant
poles of the ecliptic, in both hemispheres let $23°40'$[154] of a
great circle be measured off, and let two points be marked there
in order to designate the poles of the mean equator. Let the
two colures be drawn in the proper manner to indicate the mean
solstices and equinoxes. For purposes of study, let these points
be imagined and indicated on a small sphere which encloses
the globe of the earth and which, by its uniform motion, produces the third motion assigned to the earth.

Now, with the center of the earth between the sun and the
constellation Virgo, let the mean equator be inclined or oblique
to the sun, and let the line of the true place of the sun pass

---

[153] Page 121.

[154] The mean value between the maximum of $23° 52'$ and minimum of $23° 28'$.

through the common intersection of the plane of the ecliptic, the mean equator, and the mean equinoctial colure. Let the mean vernal equinox and true vernal equinox occur simultaneously where required by the scheme of motions, as will be crystal clear from what follows. The center of the earth advances from its position 59′8″11‴[155] each day with uniform motion as reckoned by the fixed stars. In addition to this motion of the center of the earth, let the mean vernal point move an equal distance in precedence; and since it moves at a slightly faster rate, let it describe an angle greater by about 8‴. This is the reason why I said just above that the motion in declination is almost exactly equal to the uniform motion of the center of the earth as reckoned by the fixed stars. But there is a continual increase in the angle made by the vernal point of the mean equator as compared with the center of the earth (in accordance with the rule given above). Hence, before the center of the earth finally returns to the point on the ecliptic whence it set out, the line of the true place of the sun reaches the mean equinox, and the stars seem to us to move with a mean or uniform motion in consequence, to the amount of the precession. This precession, as I stated in the beginning,[156] is about 50″[157] in an Egyptian year, and in 25,816 Egyptian years it performs a complete revolution. Thus it is clear what the mean equinox is, what the mean precession is, and how these phenomena can be presented to the eyes as though by a mechanical device.

Librations

Let there be[158] a straight line $a\,b$ of finite length, for example 24′, divided at d into two equal parts. Then with the point of the compass placed at d, describe a circle $c\,e$ with the radius $d\,c$ directed to a and 6′ long (that is, a quarter of the entire length). Construct a second circle of the same size in this

---

[155] Although PII, 334.14 and Th 471.12 give 59′ 8″ 2‴, the correct reading is unquestionably 59′ 8″ 11‴ (cf. Th 196.9-11, 198.5). The error undoubtedly arose because 11 was interpreted as a Roman numeral, whereas it was Arabic (cf. 1566 ed., p. 205v; 1596 ed., p. 124; and 1621 ed., p. 123).

[156] Pages 113-14.   [157] 8‴ × 365 = 48⅔″.   [158] Reading *Sit* (Th 471.27).

figure;[159] and let the two small circles (to use this term for the moment) be so placed that each is attached to the circumference of the other and can move freely about its own center. Call that circle the first which carries the other on its circumference, and let it be fastened to the center of the line *a b* at the point *d*. Denote the center of the second small circle by *f*, and any point chosen at random on its circumference by *h*. Place the point *h* of the second small circle upon *a*, the end of the given line; and *f* upon *c*. Let *h* describe in one direction, about *f* as center, an angle twice as great as that described in equal time by *f* about *d* in the opposite direction. Clearly, then, in one revolution of the first small circle the point *h* twice describes and traverses the line *a b*, and the second small circle revolves twice.[160]

While thus describing a straight line through the combination of two circular motions, the point *h* moves most slowly near the ends *a* and *b*, and more rapidly near the center *d*. It has therefore pleased my teacher to name this motion of the point *h* along the line *a b* a "libration," because it resembles the motion of objects hanging in the air.[161] It is also called

---

[159] Prowe altered "ab" (Th 471.30) to *a b* (PII, 335.6), thereby making a difficult passage hopeless.

[160] For a detailed explanation of this device see above, p. 88, n. 100.

[161] In the corresponding passage of *De rev.* Copernicus wrote *motus . . . pendentibus similes librationibus* and *pendentium instar* (Th 163.13,19). Menzzer (p. 136) rendered the former expression by: "Pendelschwingungen ähnliche Bewegungen" (motions like the swinging of a pendulum); and the latter by: "den Pendeln ähnlich" (like a pendulum). Accepting Menzzer's interpretation, Dreyer stated that the librations were so named because they are "like the motion of a pendulum" (*Planetary Systems*, p. 330).

Are we justified in attributing the pendulum to Copernicus? I think not. In the sentence under discussion Rheticus's language is *ad similitudinem pendentium in aere*, "it resembles the motion of objects hanging in the air". This formulation is modeled after a phrase used by Copernicus in a wholly different context, *in aere pendentibus* (Th 22.14). Here Menzzer (p. 21) translated by: "in der Luft Schwebende" (objects suspended in the air). We may safely conclude that Copernicus is not referring to the pendulum, but in general to the kind of motion which is quickest in the middle and slowest at the ends (cf. p. 118, above).

E. Wiedemann corrected a false attribution of the pendulum to the Arabs (*Verhandlungen der deutschen physikalischen Gesellschaft*, XXI[1919], 663-64 and *Zeitschrift für Physik*, X[1922], 267-68); his strictures were overlooked by Edmund Hoppe, *Geschichte der Physik* (Braunschweig, 1926), p. 25.

motion along the diameter; for if you imagine a circle with diameter $a\,b$ and center d, the position on the diameter $a\,b$, to which the point h is brought by the aforesaid combined motion of the small circles, is determined from the doctrine of chords; and by this method the table of prosthaphaereses[162] is constructed.

My teacher calls the motion of the first small circle about d the anomaly, since the prosthaphaeresis is derived from this motion.[163] Thus let f, the center of the second small circle, describe an angle by starting from the point c[164] and moving to the left on the circumference of the first small circle; let the angle $c\,d\,f$ be 30°. The line $d\,f\,g$, drawn from the center d, will cut off, on the circumference of the circle $a\,b$, an arc $a\,g$ of the same number of degrees as the arc $c\,f$ of the first small circle. Since the point h of the second small circle moves from g to the right at twice the speed of f, a straight line drawn from the point g to the point h clearly subtends half of double the arc $a\,g$, and $h\,d$ half of double the arc which remains when the arc $a\,g$ is subtracted from a quadrant.[165] Therefore $a\,h$, that is, the distance of h from a along the diameter $a\,b$, is 1,340[166] units, of which the radius constitutes 10,000. But if $a\,b$ is di-

[162] The varying differences between an apparent and mean motion. When the mean motion is smaller than the apparent, the difference is added (prosthesis) to the mean motion, in order to get the apparent motion; conversely, when the mean motion is greater than the apparent, the difference is subtracted (aphaeresis). The Latin equivalent for this Greek term is *aequatio* (Th 180.14-19).

[163] Reading *motu* (Th 472.17) instead of *motus* (PII, 336.18).

[164] Copernicus's discussion of this topic (*De rev.* iii.4) is accompanied by a diagram, which Rheticus follows, save that he interchanges c with d, and g with h. Now the first three editions of the *Narratio prima* contained no figures, and Mästlin supplied them from *De rev*. To eliminate disagreement between Copernicus's diagram and Rheticus's lettering, Mästlin adopted the simple expedient of transposing the letters in the diagram. But Prowe, following Th, resolved to adhere faithfully to Copernicus's figure, and therefore to alter the text of the *Narratio prima* wherever necessary. In the present instance d was left unchanged, although it should have been replaced by c (Th 472.18; PII, 336.19).

[165] Cf. Th 167.4-7. If we employ the notation used by Manitius in his *Ptolemäus Handbuch* (I, 47 n), we should write:

$$gh = \tfrac{1}{2}\, s_{2}b\ ag$$
$$hd = \tfrac{1}{2}\, s_{2}b\ (90° - ag).$$

[166] Because $h\,d$, subtending an arc equal to 60° on the circle $a\,b$, is 8,660 (Th 49.37): 10,000 − 8,660 = 1,340.

vided into 60 units, $a\,h$ will be 4,[167] and $h\,b$ 56. Then by taking the proportional part of 24′, we shall know the position of the point h on the given finite straight[168] line in this case.

Now that we understand this argument in a rough way, it will be easy to see how the greatest obliquity of the equator to the plane of the ecliptic varies and how the true precession of the equinoxes becomes unequal. Since short arcs do not differ sensibly from straight lines, let us begin by imagining that the point d is placed upon the north pole of the mean equator and that the line $a\,b$ is an arc of the mean solstitial colure. Lying between the north pole of the mean equator and the nearby pole,[169] which is one of the poles that move at a uniform distance from the poles of the ecliptic, b[170] marks the least distance of the pole of the daily rotation, or pole of the earth, from the aforesaid pole of the ecliptic.[171] And a, lying between the north pole of the mean equator and the plane of the ecliptic, marks the greatest distance of the pole of the earth from the pole of the ecliptic. Then with the two small circles properly fitted into place by means of the line $a\,b$, it may be understood what part of the 24′ of the line $a\,b$ is described at the present time by the north pole of the earth in the point h by reason of the combined motion of the two small circles. Observing the law of opposition, the south pole moves by a similar device, as the shifting universe alters the greatest obliquity.

Assume that the first small circle completes its revolution in 3,434[172] Egyptian years and that the terminus from which

---

[167] Since $a\,b = 60$, the radius $= 30$. And $1,340 : 10,000 = 4.02 : 30$.

[168] Reading *rectae* (Th 472.27) instead of *recte* (PII, 336.30); and *subsistat*.

[169] The true pole of the equator.

[170] Reading *quare et* (Th 472.36) instead of *Quare* b *est* (PII, 337.7).

[171] In a note Prowe cites this passage as it appeared in the first edition and declares it to be corrupt. It is, however, entirely sound. But a textual difficulty was introduced by the 1566 edition, which gives *et* instead of *ab* after *terrae* (p. 206r); and the difficulty was aggravated by Th, which keeps *et* and inserts *a* after *dictum* (472.36-37).

[172] Reading *IIIMCCCCXXXIIII* (Th 473.4) instead of *XXXIIIIMCCCCXXX-IIII* (PII, 337.17). From two other passages in the *Narratio prima* (PII, 302. 5-6, 339.5-7), Prowe should have seen that the number he gives here is incorrect (cf. p. 117, above and p. 158, below).

the motion of anomaly begins is the point *a* on the circumference of the circle whose diameter is described by the first libration. If the poles of the earth had no libration other than this one, and did not deviate from the mean solstitial colure, it will be clear at once to anyone that only the angle of inclination of the plane of the true equator to the plane of the ecliptic would vary on account of this motion of the poles of the earth, decreasing when they move from *a* through *d* to *b* and increasing while they complete the opposite movement from *b* through *d* to *a*; and that hence no inequality would appear in the precession of the equinoxes.

However, it is certainly clear from the observations that the true equinoctial points move 70′ to either side of the mean equinoctial points in the greatest prosthaphaeresis and that the change in the obliquity takes twice as long as this motion. My teacher was therefore persuaded to introduce,[173] in addition to the first, a second lesser libration, whereby the poles of the earth deviate from the mean solstitial colure toward the sides of the universe in such a way that the arc or straight line *a d b* of the second libration forms four right angles with the mean solstitial colure. In the north let *a* lie to the right side of the universe, *b* to the left; and in the south *a* to the left, *b* to the right. Through the points *h* of the first libration let *d* of the second libration describe lines of 24′ to either side of *a d b*. Finally, let the poles of the earth be in reality fixed to the points *h* of the second libration, and let them be deflected by the second libration only 28′ to either side of the said colure, with *a* and *b* taken as the outermost points. For when the poles are at these points, the true solstitial colure makes with the mean solstitial colure an angle not perceptibly greater than 70′.

Now the prosthaphaereses of precession must be taken in relation to the mean vernal point. Hence my teacher's analysis of the second libration deals with the relation of the true vernal point to the mean, especially since this method of examining the prosthaphaereses is rather easy. Then the line *a b* will be 140′ long; and it will be so placed that it corresponds to the

---

[173] Reading *ad* before *constituendam* (Th 473.15; PII, 338.6).

north line of the second libration, with d at the mean vernal point, the true vernal point at h, and the radius of either small circle $35'$. Moreover, the terminus from which the motion begins is the mean vernal point, from which the true vernal point moves to the right toward a. But the anomaly is measured from the northernmost point of the circle whose diameter is described by the true vernal point; and the northernmost point is marked on the circumference of the circle by the mean equinoctial colure. And since in one cycle of the obliquity the inequality of the precession is twice completed, the anomaly of the second libration has a period of 1,717 Egyptian years.[174] Therefore the anomaly of the obliquity, as taken from the tables and doubled, equals the anomaly of the precession. The name "simple anomaly" is given to the former, "double anomaly" to the latter.

But if the second libration alone were to be assumed, the angle of inclination of the planes of the true equator and ecliptic clearly would not vary; and this would be a serious fault, for every inequality of the phenomena would be observed only in connection with the inequality of the precession of the equinox.[175] However, since both librations occur together and since, as has been said, their motions interact, the poles of the earth describe about the poles of the mean equator the figure of twisted rings.[176]

When the poles of the earth cross the mean solstitial colure, the true[177] colure lies in the same plane with the mean, and the true vernal point coincides with the mean; however, unless the poles of both equators coincide, the planes of the equators and of the mean and true solstitial and equinoctial colures do not[178] completely coincide. Now, when the north pole lies between d of the second libration and a, the outermost point to the right, the south pole occupying the opposite point, the true equinox

[174] Cf. above, p. 156, n. 172.

[175] Omitting *veri* (Th 474.5-6; PII, 339.14).

[176] That is, the figure 8 (cf. Th 163.30–165.15).

[177] Reading *verus* (Th 474.9) instead of *versus* (PII, 339.19).

[178] Reading with Mästlin (1596 ed., p. 128; 1621 ed., p. 126) *non* before *omnino* (PII, 339.23).

follows the mean, and the sun comes to the mean equator before it comes to the true. But when the poles of the earth cross
over to the opposite sides of the universe, so that the north
pole lies to the left of the mean solstitial colure and the south
to the right, the true equinox precedes the mean, and the sun
meets the true equator before it meets the mean. Besides, when
the poles of the earth move from *a* toward *b*, the tropical year
decreases, because the true equinox advances, as it were, to meet
the sun; but when the poles move from *b* toward *a*, since the
equinox, as it were, flees from the sun, the tropical year increases. And when the poles of the earth are near *d*, for a brief
span of years the increase or decrease in the year is distinctly
perceptible. Moreover, since the apparent motion of the fixed
stars is bound up with the length of the tropical year, in the
same way the motion in precedence of the solstitial and equinoctial points among the fixed stars is observed as swifter and
slower.

So far as the solar apogee is concerned,[179] and the distance of
the vernal equinox from it, the conclusions which in the beginning[180] I drew from the observations in accordance with my
teacher's opinion are clarified by the preceding discussion. The
motion of the apogee in the ecliptic depends on the motion of
the center of the small circle and on the uniform motion of the
center of the great circle in the circumference of the small
circle. The diameter of the great circle or ecliptic that passes
through the centers of the sun and small circle is the mean
apse-line of the sun; but the diameter through the centers of
the sun and great circle is the true apse-line. The center of the
great circle is found between the sun and the point on the
ecliptic where the sun is thought to be in perigee.[181] Similarly,
the center of the small circle is situated between the point of
mean perigee and the sun.

In the time of Ptolemy the true apse-line was at one end,
the point of apparent apogee, 57°50′ from the first star of

---

[179] Reading with Mästlin *ad* instead of the first *ab* (1621 ed., p. 127; PII, 340.13).

[180] Pages 119-26. [181] The point of apparent perigee.

Aries; and at the other end, the perigee, 237°50'.[182] But for
the mean apse-line this distance was 60°16', and in the opposite
point, 240°16'. For, starting from that point on the small circle
which is at the greatest distance from the center of the sun, the
center of the great circle had moved about 21⅓° in precedence;
and the simple anomaly, that is, the anomaly of the obliquity,
had at that time an equal value.[183] But since the center of the
small circle moves uniformly about the center of the sun, and
the center of the great circle moves uniformly on the circumference of the small circle, the higher apse of the sun, at the
time of the observation made by my teacher, was found to be
69°25' from the first star of Aries.[184] Because at that time the
simple anomaly was almost exactly 165°, the prosthaphaeresis
was determined as almost exactly 2°10',[185] and the center of
the small circle fixed the point of mean perigee between the
sun and 251°35'.[186] Furthermore, the eccentricity of the great
circle, or eccentric of the sun if this term is preferred, which
Ptolemy computed as ¹⁄₂₄ of the radius of the great circle, is in
our time about ¹⁄₃₁,[187] as the observations show, and as is readily
deduced if the hypotheses of my teacher are adopted and mathematics applied.

The manner in which the eccentricities of the five planets
vary on account of the motion of the center of the great circle
on the small circle, as I pointed out in the reasons for revising
the hypotheses,[188] can be understood with no great effort. In
the investigation of the five planets two considerations are of
special importance: first, in what manner and to what extent the
center of the earth approaches to or withdraws from the centers
of the deferents of the planets; second, what relation this ap-

---

[182] For the true apogee was 64° 30' from the true equinox (see p. 125, above);
            subtract   6° 40', the true precession (see p. 116, above)
                 $\overline{57° 50'} + 180° = 237° 50'$ for the perigee.

[183] Hence the prosthaphaeresis was about 2° 26' (Th 224.14-15): 57° 50' +
2° 26' = 60° 16' + 180° = 240° 16'.

[184] Th 221.23-28.

[185] Th 221.28-30, 224.32.

[186] 69° 25' + 2° 10' + 180° = 251° 35'.

[187] Cf. above, p. 61, n. 9.   [188] See p. 136.

proach or withdrawal bears to the radius of the deferent of each planet. The causes will not be far to seek.

In the case of Saturn the entire diameter of the small circle has no perceptible ratio whatever to the radius of the deferent, since Saturn is the first planet beneath the starry sphere. Hence observations can reveal no variation in the eccentricity of Saturn. As for Jupiter, its apogee is about a quadrant from the apogee of the sun. Hence the motion of the center of the great circle produces no observable change in the eccentricity at the present time, even though the ratio of the diameter of the small circle to the radius of the deferent is perceptible and measurable. And this is the reason why in the case of Mercury also no change is observed in the eccentricity, since its apogee is at a similar distance from the apogee of the sun.

Because the apogee of Mars is about 50° to the left of the sun's apogee, and the apogee of Venus 42° to the right, the centers of their deferents are suitably placed to reveal the change in the eccentricity;[189] and the diameter of the small circle has a perceptible ratio to the deferent of each. By a trigonometrical analysis of the observations of these two planets, my teacher found that the eccentricity of Mars has decreased by $\frac{1}{42}$, of Venus by $\frac{1}{5}$,[190] on account of the approach of the center of the great circle to the sun.

Lest any of the motions attributed to the earth should seem to be supported by insufficient evidence, our wise Maker expressly provided that they should all be observed equally perceptibly in the apparent motions of all the planets; with so few motions was it feasible to satisfy most of the necessary phenomena of nature. Therefore the motion of the center of the great circle affects not only the sun and the planets revolving about it but also the phenomena of the moon. For Ptolemy

[189] Cf. Th 327.16-18.

[190] Prowe's text (PII, 342.9) omits *partem* before *propter* (Th 475.35). According to Copernicus's findings, the eccentricity of Mars had decreased from 1500 to 1460 (cf. above, p. 77, n. 58), a decrease of $\frac{1}{37}$ or $\frac{1}{38}$ rather than $\frac{1}{42}$, as Rheticus has it. As for the eccentricity of Venus, Copernicus explicitly reports a diminution of somewhat more than $\frac{1}{6}$, not $\frac{1}{5}$ (from 416 to 350; cf. Th 369.8-11).

computed the greatest distance of the sun from the earth to be 1,210 units, of which the radius of the earth is one, and the axis of the earth's shadow 268;[191] and my teacher shows that in our time the greatest distance of the sun from the earth is 1,179 units, and the axis of the cone of shadow 265.[192] But I have decided to reserve the other related topics[193] for a "Second Account" to follow this one, wherein I shall examine the motions and phenomena of the sun and moon by the light of the change in the hypotheses.

THE SECOND PART OF THE HYPOTHESES

The Motions of the Five Planets

When I reflect on this truly admirable structure of new hypotheses wrought by my teacher, I frequently recall, most learned Schöner, that Platonic dialogue which indicates the qualities required in an astronomer and then adds "No nature except an extraordinary one could ever easily formulate a theory."[194]

When I was with you last year and watched your work and that of other learned men in the improvement of the motions of Regiomontanus and his teacher Peurbach, I first began to understand what sort of task and how great a difficulty it was to recall this queen of mathematics, astronomy, to her palace, as she deserved, and to restore the boundaries of her kingdom. But from the time that I became, by God's will, a spectator and witness of the labors which my teacher performs with energetic mind and has in large measure already accomplished, I realized that I had not dreamed of even the shadow of so great a burden of work. And it is so great a labor that it is not any hero who can endure it and finally complete it. For this reason, I suppose,

[191] HI, 425.17-21.

[192] Th 282.25-26.

[193] Omitting *his* (PII, 342.24; Th 476.6).

[194] *Epinomis* 990B; θεωρῆσαι here means "observe" rather than "theorize," as Rheticus interpreted it. The authenticity of the *Epinomis* is disputed; for the view that it is genuine see J. Harward, *The Epinomis of Plato* (Oxford, 1928), pp. 26-58 and, for the opposing view, J. Geffcken, *Griechische Literaturgeschichte* (Heidelberg, 1926-34), II, 174-76.

the ancients related that Hercules, sprung of Jupiter most high, no longer trusting his own shoulders, replaced the heavens upon Atlas, who, being long accustomed to the burden, resumed it with stout heart and undiminished vigor, as he had borne it in former days.

Moreover, divine Plato, master of wisdom as Pliny styles him,[195] affirms not indistinctly in the *Epinomis* that astronomy was discovered under the guidance of God.[196] Others perhaps interpret this opinion of Plato's otherwise. But when I see that my teacher always has before his eyes the observations of all ages together with his own, assembled in order as in catalogues; then when some conclusion must be drawn or contribution made to the science and its principles, he proceeds from the earliest observations to his own, seeking the mutual relationship which harmonizes them all; the results thus obtained by correct inference under the guidance of Urania he then compares with the hypotheses of Ptolemy and the ancients; and having made a most careful examination of these hypotheses, he finds that astronomical proof requires their rejection; he assumes new hypotheses, not indeed without divine inspiration and the favor of the gods; by applying mathematics, he geometrically establishes the conclusions which can be drawn from them by correct inference; he then harmonizes the ancient observations and his own with the hypotheses which he has adopted; and after performing all these operations he finally writes down the laws of astronomy—when, I say, I behold this procedure, I think that Plato must be understood as follows.

The mathematician who studies the motions of the stars is surely like a blind man who, with only a staff to guide him, must make a great, endless, hazardous journey that winds through innumerable desolate places. What will be the result? Proceeding anxiously for a while and groping his way with his staff, he will at some time, leaning upon it, cry out in

---

[195] *Natural History* vii.30(31).110.

[196] This proposition is not expressly formulated anywhere in the *Epinomis*, but is derived from the argument in 989D-990A. For the question of the authenticity of the *Epinomis* see n. 194, above.

despair to heaven, earth, and all the gods to aid him in his misery. God will permit him to try his strength for a period of years, that he may in the end learn that he cannot be rescued from threatening danger by his staff. Then God compassionately stretches forth His hand to the despairing man, and with His hand conducts him to the desired goal.

The staff of the astronomer is mathematics or geometry, by which he ventures at first to test the road and press on. For in the examination from afar of those divine objects so remote from us, of what avail is the strength of the human mind? Of what avail[197] dim-sighted eyes? Accordingly, if God in His kindness had not endowed the astronomer with heroic ambitions and led him by the hand, as it were, along a road otherwise inaccessible to the human intellect, the astronomer would not be, I think, in any respect better circumstanced and more fortunate than the blind man, save that trusting in his reason and offering divine honors to his staff, he will one day rejoice in the recall of Urania from the underworld. When, however, he considers the matter aright, he will perceive that he is not more blessed than Orpheus, who was aware that Eurydice was following him as he danced his way up from Orcus; but when he reached the jaws of Avernus, she whom he dearly longed to possess disappeared from view and descended once more to the infernal regions. Let us then examine, as we set out to do, my teacher's hypotheses for the remaining planets, to see whether with unremitting devotion and under the guidance of God, he has led Urania back to the upper world and restored her to her place of honor.

With regard to the apparent motions of the sun and moon, it is perhaps possible to deny what is said about the motion of the earth, although I do not see how the explanation of precession is to be transferred to the sphere of the stars. But if anyone desires to look either to the principal end of astronomy and the order and harmony of the system of the spheres or to

---

[197] Reading *quid* (Th 477.15) instead of *quam* (PII, 344.24). In a note, Prowe attributes the change from *quam* to *quid* to Mästlin; but the 1566 edition has *quid* (p. 207v).

ease and elegance and a complete explanation of the causes of
the phenomena, by the assumption of no other hypotheses will
he demonstrate the apparent motions of the remaining planets
more neatly and correctly. For all these phenomena appear to
be linked most nobly together, as by a golden chain; and each
of the planets, by its position and order and every inequality of
its motion, bears witness that the earth moves and that we who
dwell upon the globe of the earth, instead of accepting its
changes of position, believe that the planets wander in all sorts
of motions of their own. And if it is possible anywhere else to
see how God has left the universe for our discussion, it surely
is eminently clear in this matter. No one can be affected, I
think, by the argument that God permits Ptolemy and other
famous heroes to dissent on this point. For it is not the sort
of opinion which Socrates in the *Gorgias*[198] declares to be
evil for men; and it does not cause any harm to either the
science itself or the divining art derived therefrom.

The ancients attributed to the epicycles of the three superior
planets the entire inequality of motion which they discovered
that these planets had with respect to the sun. Then they saw
that the remaining apparent inequality in these planets did not
occur simply on the theory of an eccentric. The results obtained
by calculating the motions of these planets in imitation of the
hypotheses for Venus agreed with experience and the observations. Hence they decided to assume for the second apparent
inequality a device like that which their analyses established
for Venus. As in the case of Venus, the center of the epicycle of
each planet was to move at a uniform distance from the center
of the eccentric, but at a uniform rate with respect to the center
of the equant; and this point was to be the center of uniform
motion also for the planet, as it moved on the epicycle with its
own motion, starting from the mean apogee. So long as the
ancients strove to retain the earth in the center of the universe,

[198] 458A; Rheticus is quoting not the original Greek but a Latin translation.
Copernicus used the translation of Marsilio Ficino (*Stromata Copernicana*, pp.
306-7). But *perniciosas* in our text shows that Rheticus used Simon Grynaeus's
revision of Ficino's translation, for Grynaeus replaced Ficino's *malum* by *per-
niciosum* (Basel, 1532, p. 342).

they were compelled by the observations to affirm that, just as
Venus revolved with its own special motion on the epicycle,
but by reason of the eccentric advanced with the mean motion
of the sun, so conversely the superior planets in the epicycle
were related to the sun, but moved with special motions on the
eccentric. But in[199] the theory of Mercury, the ancients thought
that they had to accept, in addition to the devices which they
deemed adequate to save the appearances of Venus, a different
position for the equant, and revolution on a small circle for the
center from which the epicycle was equidistant. All these
arrangements were shrewdly devised, like most of the work of
antiquity, and would agree satisfactorily with the motions and
appearances if we granted that the celestial circles admit an
inequality about their centers—a relation which nature abhors
—and if we regarded the especially notable first inequality of
apparent motion as essential to the five planets, although it is
clearly accidental.

Moreover, in the latitudes of the planets, the ancients seem
to neglect the axiom that all the motions of the heavenly bodies
either are circular or are composed of circular motions; unless
perhaps it is proposed to explain the reflexions and declinations
of Venus and Mercury, the inclinations[200] of the epicycles in
the three superior planets, and the deviations in the inferior
planets by motions in libration, as was done just above for the
earth's motion in declination. We may admit this for the reflexions and declinations of Venus and Mercury, inasmuch as
the angles of inclination of the planes of their eccentrics and
epicycles remain everywhere unchanged. But common calculation shows that the inclinations of the epicycles in the three
superior planets, and the deviations of Venus and Mercury
do not occur through librations. Let me speak only of the
deviations. The proportional minutes, by which we compute
the deviations in relation to the distance of the center of the
epicycle from the nodes and apsides, have been investigated
and determined by the same method by which the declinations

---

[199] Omitting *in* (PII, 346.19; Th 478.21).
[200] Reading *declinationes* (Th 478.33) instead of *declinationis* (PII, 347.3).

of the parts of the ecliptic are examined in the doctrine of the first motion. Therefore, when the center of the epicycle of Venus is 60° from any of the apsides of the eccentric, we infer a deviation of 5', and for Mercury 22½'. But if the deferent were assumed to oscillate by means of librations, true science would require for this position of the epicycle of Venus a deviation not greater than 2½' and for Mercury 11¼'. For in this position of the center of the epicycle the angle of inclination of the plane of the eccentric to the plane of the ecliptic would be found not greater than 5' for Venus and 22½' for Mercury, on account of the properties of motion in libration. Perhaps for this reason John Regiomontanus thought it advisable to caution his readers that calculation of latitudes is concerned only with the approximate truth.[201]

Finally, as Aristotle points out at length in another connection,[202] men by nature desire to know. Hence it is quite vexing that the causes of phenomena are nowhere else so hidden and wrapped, as it were, in Cimmerian darkness, a feeling which Ptolemy shares with us. Concerning the hypotheses of the ancients for the five planets I shall say no more for the present than is required perhaps by an explanation of the new hypotheses (if I may so term them) and a comparison of them with the ancient hypotheses. I sincerely cherish Ptolemy and his followers equally with my teacher, since I have ever in mind and memory that sacred precept of Aristotle: "We must esteem both parties but follow the more accurate."[203] And yet somehow I feel more inclined to the hypotheses of my teacher.

[201] As Mästlin (1596 ed., p. 136) indicates, the reference is to the *Epitome*, Book XIII, Prop. 21: "But to find the inclinations of this kind for every position of the epicycle on the eccentric is no mean task. Hence attention was necessarily directed toward another means whereby the latitudes for the remaining positions of the epicycle would be readily determined approximately."

[202] *Metaphysics* i.1 980a21. Rheticus is quoting, not the original Greek, but some Latin translation which is neither the *antiqua translatio* nor Bessarion's; cf. above, p. 142, n. 133.

[203] *Metaphysics* xii.8 1073b16-17 (W. D. Ross's translation). The precept perhaps came to the attention of Rheticus because it was quoted in Simplicius's *Commentary on Aristotle's De caelo* (ed. Heiberg, 506.2-3); cf. above, p. 145, n. 139.

This is so perhaps partly because I am persuaded that now at last I have a more accurate understanding of that delightful maxim which on account of its weightiness and truth is attributed to Plato: "God ever geometrizes";[204] but partly because in my teacher's revival of astronomy I see, as the saying is, with both eyes and as though a fog had lifted and the sky were now clear, the force of that wise statement of Socrates in the *Phaedrus:* "If I think any other man is able to see things that can naturally be collected into one and divided into many, him I follow after and 'walk in his footsteps as if he were a god.' "[205]

The Hypotheses for the Motions in Longitude of the Five Planets

What has been said thus far regarding the motion of the earth has been demonstrated by my teacher. Consequently (as I pointed out[206] in the reasons for revising the hypotheses) the entire inequality in the apparent motion of the planets which seems to occur in their positions with respect to the sun[207] is caused by the annual motion of the earth on the great circle. It likewise follows that the planets in reality have a single inequality, which is observed in relation to the parts of the zodiac, and is one of the two recognized heretofore. Hence only those hypotheses are acceptable which can explain both inequalities of motion. Just as my teacher chose to employ an epicycle on an epicycle for the moon,[208] so, for the purpose of demonstrating conveniently the order of the planets and the measurement of their motion, he has selected, for the three superior planets, epicycles on an eccentric, but for Venus and Mercury eccentrics on an eccentric.

---

[204] This celebrated maxim is not found in the Platonic corpus. See Plutarch's *Moralia: Quaestiones convivales* Book viii, Question 2 (ed. Bernardakis, IV [Leipzig, 1892], 307.11–308.2). The first edition of this work (Venice, 1509) was available to Rheticus; the passage cited appears on p. 882.

[205] *Phaedrus* 266B (H. N. Fowler's translation, Loeb Classical Library, 1913).

[206] Page 138.

[207] A Greek phrase borrowed from Ptolemy; cf., e.g., HII, 250.14-15.

[208] Cf. above, pp. 68-69, 134.

Now since we look up at the motions of the three superior planets as from the center of the earth, but regard the revolutions of the inferior planets as below us, the centers of the deferents of the planets may properly be brought into relation with the center of the great circle; and from this point we may then quite correctly transfer all the motions and phenomena to the center of the earth. Therefore there must be understood for the five planets an eccentric, the center of which lies outside the center of the great circle.

But to gain a better understanding of the method of establishing the new hypotheses, in short to place everything in an increasingly clearer light, let us suppose first that the planes of the eccentrics of the five planets are in the plane of the ecliptic, and that the centers of the deferents and equants are related to the center of the great circle, as with the ancients they were related to the center of the earth. Then let us divide into four equal parts the distances between the center of the great circle and the points or centers of the equants. Next let us place the center of the eccentric of each of the three superior planets at the third dividing point, as you move upward from the center of the great circle toward the apogee. With the remaining fourth part as radius, let us describe an epicycle with its center on the circumference of the eccentric, and the scheme of real motion in longitude will become apparent for each of these planets.[209]

Then, in the opinion of my teacher, as the epicycle revolves, the planet moves in its upper circumference in consequence, in its lower in precedence, so that when the center of the epicycle is in the apogee of the eccentric, the planet is found in the perigee of the epicycle; and conversely, when the center of the epicycle is in the perigee of the eccentric, the planet is in the apogee of the epicycle. By this similarity of motions, the planet completes its periods on the epicycle in equal time with the center of the epicycle on the eccentric. If the equants are removed, the inequality in the motion of the superior planets with respect to the center of the great circle is clearly regular

[209] Cf. above, p. 74, n. 50.

and composed of uniform motions. For the epicycle assumed in this theory succeeds to the function of the equant; and the eccentric describes equal angles about its own center in equal times, while the planet, moving on the epicycle to which it is attached, likewise describes equal angles about the center of the epicycle in equal times.

But the motion of Venus will be established as follows. Rejecting the deferent, which is replaced by the great circle, describe a small circle about the third dividing point, with the remaining quarter of the line as radius. Then let the center of the epicycle of Venus, which will here be called eccentric on the eccentric, second eccentric, and movable eccentric, move on the circumference of the said small circle[210] according to this law, that whenever the center of the earth crosses the apse-line, the center of the eccentric is in the point of the small circle that is nearest to the center of the great circle; and whenever the earth is midway on its circle between the two apsides, the center of the eccentric of Venus is in the point of the small circle that is most remote from the center of the great circle. The center of the eccentric moves in the same direction as the earth, that is, in the order of the signs; but, as follows from the foregoing, it revolves twice in each period of the earth.

While the scheme of motions for Mercury agrees in general with the theory of Venus, on account of the remaining inequality, there is an additional epicycle,[211] whose diameter Mercury describes by a libration. To put the scheme in terms of the earth's motion, the length of the radius of the movable deferent is 3,573,[212] the eccentricity of the first deferent 736, the length of the radius of the small circle, which carries the movable center of the deferent, 211, and the diameter of the said epicycle 380 units, of which 10,000 constitute the line from the center of the great circle to the center of the earth. But in the motion of Mercury the following law is observed:

[210] Cf. above, p. 81, n. 69.

[211] Cf. above, p. 88, n. 96.

[212] Rheticus has chosen to give the minimum value (cf. Th 382.27–383.2, and above, p. 86, n. 90).

the center of the movable eccentric, in contrast with the case of Venus, is most remote from the center of the great circle whenever the earth is in the line of the planet's apsides; and nearest, whenever the earth is at a quadrant's distance from the apsides of the planet. Mercury will have, as is apparent, a fixed epicycle. The diameter of this epicycle is directed to the center of the movable deferent and is described by a motion in libration of the planet moving along it in a straight line according to the following law. Whenever the center of the movable eccentric is most remote from the center of the great circle, the planet is in the perigee of the epicycle, which is the lower limit of the diameter described by the planet. Conversely, Mercury is at the other limit, which may be called the apogee, whenever the center of the movable eccentric is nearest to the center of the great circle. But the motions of the apsides of the planets, like certain other topics, are reserved for the "Second Account."

The foregoing is very nearly the whole system of hypotheses for saving the entire real inequality of the motion in longitude of the planets. Therefore, if our eye were at the center of the great circle, lines of sight drawn from it through the planets to the sphere of the stars would, as the lines of the true motions, be rotated in the ecliptic by the planets exactly as the schemes of the aforementioned circles and motions require, so that they would reveal the real inequalities of these motions in the zodiac. But we, as dwellers upon the earth, observe the apparent motions in the heavens from the earth. Hence we refer all the motions and phenomena to the center of the earth as the foundation and inmost part of our abode, by drawing lines from it through the planets, as though our eye had moved from the center of the great circle to the center of the earth. Clearly it is from this latter point that the inequalities of all the phenomena, as they are seen by us, must be calculated. But if it is our purpose to deduce the true and real inequalities in the motion of the planets, we must use the lines drawn from the center of the great circle, as has been explained. To smooth our way through the topics in planetary phenomena which remain to be discussed and to make the whole treatise easier and more agreeable, let us imagine not only the lines of true apparent motion drawn from the center of the earth through the planets to the ecliptic but also those drawn from the center of the great circle and therefore properly called the lines of the inequality of motion.

When, as the earth advances with the motion of the great circle, it reaches a position where it is on a straight line between the sun and one of the three superior planets, the planet will be seen at its evening rising; and because the earth, when so situated, is at its nearest to the planet, the ancients said that the planet was at its nearest to the earth and in the perigee of its epicycle. But when the sun approaches the line of the true and apparent place of the planet—this occurs when the earth reaches the point opposite the above-mentioned position—the planet begins to disappear by setting in the evening and to attain its greatest distance from the earth, until the line of the true place of the planet passes also through the center of the sun. Then the sun lies between the planet and the earth, and the planet is occulted. After occultation, since the motion of the earth continues uninterrupted and since the line of the true place of the sun withdraws from the line of the true place of the planet, the planet reappears at its morning rising, when it has attained the proper distance from the sun required by the arc of vision.

Moreover, in the hypotheses of the three superior planets, the great circle takes the place of the epicycle attributed to each of the planets by the ancients. Hence the true apogee and perigee of the planet with respect to the great circle will be found on the diameter of the great circle prolonged to meet the planet. But the mean apogee and perigee will be found on the diameter of the great circle that moves[213] parallel to the line drawn from the center of the eccentric to the center of the epicycle. Since in the semicircle closer to the planet the earth approaches the planet, and in the other, opposite semicircle recedes from it, in the former semicircle the ends of the

[213] Reading *movetur* (Th 482.11) instead of *moventur* (PII, 352.11).

diameters of the great circle are the perigees, but in the latter the apogees. For the former semicircle takes the place of the lower part of the epicycle, but the latter, the upper.

Imagine that a conjunction of sun and planet is not far off. Let the center of the earth be in the true place of the apogee of the planet with respect to the great circle; and let the line of the real inequality coincide with the line of the apparent place of the planet. However, as the earth in its motion moves away from this position, the line[214] of the real inequality and the line of the true place of the planet begin to intersect in the planet. The former advances with the regular unequal motion of the planet in the order of the signs; and the latter, as it separates from the former, makes the planet seem to us to move more rapidly in the ecliptic than it really does with its own motion.

But when the earth reaches the part of the great circle that is nearer[215] to the planet, the direction of its motion at once becomes westward, so that the apparent motion of the planet forthwith seems slower to us. Moreover, because the earth mounts toward the planet, the line of the true motion of the sun moves away from the planet, and the planet is thought to approach us, as though it were descending from its upper circumference. However, the motion of the planet seems to be direct, until the center of the earth reaches the point on the great circle with respect to the planet[216] where the angle through which the line of the true place of the planet moves daily in precedence equals the diurnal angle of the real inequality in consequence. For there, since the two motions neutralize each other, the planet appears to remain at its first stationary point for a number of days, depending on the ratio of the great circle to the eccentric of the planet under consideration, the position of the planet on its circle, and the real rate of its motion. Then as the earth moves from this position nearer to the planet, we believe that the planet retrogrades

---

[214] Reading *linea* (Th 482.18) instead of *lineae* (PII, 352.20).

[215] Reading *propiorem* (Th 482.22) instead of *propriorem* (PII, 352.25).

[216] Reading *planetam* (Th 482.27) instead of *planetae* (PII, 352.31).

and moves in precedence, since the regression of the line of the true place of the planet perceptibly exceeds the real motion of the planet. This apparent retrogradation continues until the earth reaches the true perigee of the planet with respect to the great circle, where the planet, at the mid-point of regression, is in opposition to the sun and nearest to the earth. When Mars is found in this position, it has, in addition to the common retrogradation or parallax caused by the great circle, another parallax caused by the sensible ratio of the radius of the earth to the distance of Mars, as careful observation will testify.

Finally, as the earth moves in consequence from this central conjunction with the planet, so to say, the westward regression diminishes exactly as it had previously increased, until when the motions are again equal, the planet reaches its second stationary point. Then as the real motion of the planet exceeds the motion of the line of the true place of the planet, and as the earth advances, the situation is reached where the planet at length appears at the mid-point of its direct motion; and the earth again comes to the true apogee of the planet, whence we started its motion, and produces for us in order all the above-mentioned phenomena of each of the planets.

The foregoing is the first use made of the great circle in the study of the planetary motions; by it we are freed from the three large epicycles in Saturn, Jupiter, and Mars. What the ancients called the argument of the planet, my teacher calls the planet's motion in commutation,[217] for by means of it we explain the phenomena arising from the motion of the earth on the great circle. These phenomena are clearly caused by the great circle, as the parallaxes of the moon are caused by the ratio of the radius of the earth to the lunar circles. The motion of the center of the epicycle of each planet, when subtracted from the uniform motion of the earth, which is also the mean motion of the sun, leaves as a remainder the uniform motion of commutation; and it is computed from the mean apogee, from which the earth also moves uniformly. Hence the true and apparent motion of each planet in the ecliptic is

[217] Th 308.2-8; cf. p. 48, above.

readily obtained from my teacher's tables of the prosthaphaereses of the planets.

Moreover, we shall find the second of the uses[218] of the
great circle, no less important than the first, in the theory of
Venus and Mercury. For since we observe these two planets
from the earth as from a lookout, even if they should remain
fixed like the sun, nevertheless, because we are carried about
them by the motion of the great circle, we would think that
they, like the sun, traverse the zodiac in motions of their own.
Now the observations testify that Venus and Mercury move
on their circles in independent motions of their own. Hence,
in addition to the mean motion of the sun, by which they are
carried in the order of the signs, other accidental phenomena
caused by the great circle are observed in them. For in the
first place we will consider their circles as epicycles which, as
though on their own deferents, traverse the zodiac at an equal
rate with the sun. Thus when the earth is in the perigee of
the first deferents, their entire circles will be thought to be
in the apogee of the eccentric, and conversely in the perigee
with the earth in apogee. Moreover, just as in the superior
planets the apogees and perigees with respect to the planets
are designated on the great circle, so conversely they are
marked on the circles of Venus and Mercury with respect to the
center of the earth, wherever it may be; and, by reason of the
annual motion of the earth, are drawn through all the points
on the deferents. The ends of the diameter of the movable
deferent that moves[219] parallel to the line of the mean motion
of the sun, that is, the line from the center of the great circle
to the center of the earth, are the mean apsides. The apsides
in the part of the movable deferent that is more remote from
the earth are called, not without reason, the higher apsides;
those in the nearer part, the lower.

Venus revolves in nine months, as was stated above,[220] and

---

[218] Reading *utilitatum* (Th 483.20) instead of *utilitatem* (PII, 354.24).

[219] Reading *movetur* (1566 ed., p. 210v; 1596 ed., p. 143; 1621 ed., p. 137)
instead of *moventur* (PII, 355.19).

[220] Page 146.

Mercury in approximately three. Hence, if[221] the annual motion of the earth should cease, each planet would appear to us on the earth to be in each period twice in conjunction with the sun, twice stationary, and twice at the outermost points in the curvature[222] of the deferents, and once morning, evening, retrograde, direct, in apogee, and in perigee. Moreover, if our eye were at the center of the great circle, only the independent unequal motions of Venus and Mercury, as of the other planets, would[223] appear; and as the planets traversed the entire zodiac by their own motions, they would come to be in opposition to the sun and would be seen in the other configurations with respect to it.[224]

But since we do not observe the motions of the planets from the center of the great circle, nor does the annual motion of the earth cease, it will be quite clear why these phenomena appear in such great variety to us who inhabit the earth. In accordance with the size of their circles, Venus and Mercury outrun the earth by their swifter motion, while the earth follows them in its annual motion. Therefore Venus overtakes the earth in about sixteen months,[225] and Mercury in four; with these intervals as their period, the planets show us again and again all the phenomena which God desired to be seen from the earth.

The lines of the real inequalities of motion move[226] regularly, revolving about the center of the great circle in the period allotted to them by God; but the lines of the true places, which are drawn from the center of the earth through Venus and Mercury, move in an altogether different manner, not only because they are drawn from a point outside the orbits, but also because the point is movable. We think that Venus and

---

[221] Reading *Si* (Th 483.39) instead of *Sic* (PII, 355.22).

[222] Reading *curvaturis* (Th 484.3; PII, 355.26).

[223] Reading *offerrent* (Th 484.5-6) instead of *offerent* (PII, 356.3).

[224] Reading *eum* (Th 484.7) instead of *cum* (PII, 356.4).

[225] This estimate of the synodic period of Venus is too low. Mästlin's correction, nineteen months (1596 ed., p. 143; 1621 ed., p. 137), should be unhesitatingly adopted, since it agrees with Th 310.1-7 and the tables (Th 318-19).

[226] Reading *incedunt* (Th 484.15) instead of *incedant* (PII, 356.14).

Mercury move on their circles with the motion with which the ancients said that they moved on the epicycle. But since this motion is merely the difference by which the swifter planet exceeds the mean motion of the earth or sun, my teacher calls this excess the motion in commutation, for exactly the same reasons as in the three superior planets. Consequently all the phenomena of Venus and Mercury which would appear if the earth were fixed recur more slowly on account of the earth's motion; and they occur at all the parts of the deferents and points on the ecliptic where their motions of every sort would be observed. For even without the earth fixed in Cancer, Ptolemy would have found that Mercury has its least elongations from the sun in Libra, and Venus in Taurus.[227] No matter where the earth may be on the great circle, Venus and Mercury seem to us to have their greatest elongation from the sun when they are observed at the sides of the deferent. If both tangents are drawn from the center of the earth to the deferents of Venus and Mercury, the planets will move in the order of the signs in the upper circumference, upper, that is, with reference to the earth; but in the opposite direction in the lower circumference, which is nearer to the earth. For here they appear to the senses to be stationary and retrograde, since the line of the true place of the planet makes about the center of the earth a diurnal angle in precedence equal to the angle of the mean motion, which is also the motion of the earth, in consequence, or a greater angle, etc. It is clear from these considerations why Venus and Mercury are seen to revolve about the sun.

It is also clearer than sunlight that the circle which carries the earth is rightly called the great circle. If generals have received the surname "Great" on account of successful exploits in war or conquests of peoples, surely this circle deserved to have that august name applied to it. For almost alone it makes us share in the laws of the celestial state, corrects all the errors of the motions, and restores to its rank this most beautiful

---

[227] For Ptolemy found the apogee of Mercury in Libra, and of Venus in Taurus (HII, 271.2-4, 300.15-16; cf. above, p. 85, n. 87 and p. 81, n. 71).

part of philosophy. Moreover, it is called the great circle because it has, in comparison with the circles of both the superior and inferior planets, a sensible magnitude which is the explanation of the principal phenomena.

The Apparent Deviation of the Planets from the Ecliptic

In the latitudes of the planets the first point to observe is that the name "great" is correctly assigned to the circle that carries the center of the earth. This circle deserves even higher commendation for the reason that the views of the ancients regarding the latitudes are quite involved and obscure, as is well known. The motions in longitude of the planets offer excellent evidence that the center of the earth describes what we call the great circle; but in the latitudes of the planets, the uses of this circle, as if placed in some well-lighted spot, are more obvious, since the great circle is the principal cause of every inequality of the appearances in latitude, even though it nowhere departs from the plane of the ecliptic. You see, most learned Schöner, that this circle should be honored and embraced with the greatest affection; for when all the causes have been set forth, it puts the whole subject of motion in latitude so briefly and so clearly before our eyes.

First, let the deferents of the three superior planets be inclined to the ecliptic as in Ptolemy's system; let their apogees be found to the north, their perigees to the south; and let the planets revolve on their deferents like the moon on its oblique circle, out of the plane of which it does not move. The lines of the real inequality, the dragons[228] of the planets, as they are commonly called, indicate the inclinations of the deferents

---

[228] This term was employed to designate the deviation of the moon and the planets from the ecliptic. The point on the ecliptic where the moon or planet passes from south latitude to north (ascending node) was called the "dragon's head," *caput Draconis* (Th 261.29); the point where it passes from north latitude to south (descending node) was called the "dragon's tail," *cauda Draconis* (Th 261.30). The usage survives in (*a*) the modern name, draconitic month, for the average time between two successive passages of the moon through the same node, and (*b*) the symbols still used to denote the nodes (for these symbols in a MS of the fourteenth century see Paul Tannery, *Mémoires scientifiques*, Toulouse and Paris, 1912– , IV, 356-57, plate II).

to the plane of the ecliptic and its intersections with the motions of the planets. Intersecting these lines in the centers of the planets are the lines of the true places. The latter, according to the position of the earth's center[229] on the great circle in relation to the planet, and the position of the planet on its oblique circle, mark the true places of the planets as nearer[230] to and remoter from the line through the middle of the signs,[231] in accordance with the size of the angles which the lines of the true places make with the plane of the ecliptic, as mathematical theory requires. Therefore, no matter what part of its deferent and epicycle the planet is in on the oblique circle, when the center of the earth is in the half of the great circle that is more remote from the planet—the half which the ancients called the upper part of the epicycle—the apparent latitudes clearly must be smaller than the angle of inclination of the deferent to the plane of the ecliptic; for in this position of the center of the earth in relation to the planet, the angle of apparent latitude is smaller than the angle of inclination, being an interior angle in comparison with the exterior and opposite. Furthermore, when the center of the earth reaches the half of the great circle that is nearer to the planet, conversely the apparent latitude is seen to be greater than the angle of inclination, obviously for the same reasons; for what was previously the exterior and opposite angle is now the interior angle.

This is the reason why the ancients thought that when the center of the epicycle was outside the nodes, the upper part of the epicycle was always between the planes of the deferent and ecliptic; that the other half of the epicycle was tilted in the same direction as the half of the deferent occupied by the center of the epicycle; that the diameter which passed through the middle longitudes of the epicycle moved parallel to the plane of the ecliptic; and that when the epicycle was in

[229] Reading with Mästlin (1596 ed., p. 145) *centri* instead of *centro* (PII, 358.12); cf. *in tali centri terrae situ* (PII, 358.22).

[230] Reading *propiora* (Th 485.27) instead of *propriora* (PII, 358.14).

[231] A term for the ecliptic; *per signorum medium*, "through the middle of the signs," is a literal translation of the standard Greek term διὰ μέσων τῶν ζῳδίων (cf. HI, 68.17-18).

the nodes, the planet had no latitude wherever it might be on the epicycle. In our hypotheses, the planet has no latitude when it is in one of the nodes, no matter where the earth may be found on the great circle. If the angle between the planes of the epicycle and deferent had been found, in the hypotheses of the ancients, invariably equal to the angle of inclination of the planes of the deferent and ecliptic; that is, if the plane of the epicycle had been found always parallel to the ecliptic, the aforementioned theory of latitudes would be sufficient. But an inequality is implied in the observations geometrically examined, as can be seen in the last book of Ptolemy's *Great Syntaxis*.[232] Therefore, using a motion in libration, my teacher makes the angle of inclination of the deferent to the ecliptic increase and diminish in a definite relation to the mean motion of the planet on its oblique circle, and of the earth on the great circle. This result will be obtained if in each period of the motion in commutation the diameter along which the libration takes place is twice described by the outermost limits of the oblique circle, and if the following condition is observed: that when the planet is at its evening rising the angle of inclination is greatest, and hence the angle of apparent latitude is even greater; but with the planet at its morning rising, minimal, and hence the apparent latitude, as is consistent, even smaller.[233]

But the appearances of Venus and Mercury in latitude, with the single exception of the deviation, are more easily understood than the theories of the superior planets. Let us examine the latitudes of Venus first. Within the great circle the sphere of Venus is the first to occur. According to my teacher, the plane in which Venus moves is inclined to the plane of the ecliptic or great circle along the diameter through the true apsides of the first deferent, so that the eastern half rises northward from the plane surface of the ecliptic by the angle of inclination which would be contained, in Ptolemy's hypotheses, between the planes of the epicycle and deferent; and the western half dips southward. By "eastern half" is to be understood the half that

---

[232] HII, 537.7-542.15.
[233] Cf. pp. 79-80, above.

extends in consequence from the place of the higher apse, etc.
By this simple hypothesis alone we can easily derive all the
rules for the declinations and reflexions, together with their
causes, from the relation of the position of the earth to the
plane of the planet.

For when by the annual motion of the earth we reach the
place opposite the higher apse of the first deferent, where we
think that the circle of Venus is like an epicycle in the apogee
of its deferent, the plane in which Venus moves seems to us to
have a reflexion from the plane of the ecliptic, because in this
position we see the plane of Venus crosswise. And because we
look at this plane from below, the part that rises northward
will be to the left, and the other part, that dips southward, to
the right, for us whose eyes are directed southward. But as the
earth moves upward toward the higher apse of the planet,
the circle of Venus is thought to descend from the apogee of
its eccentric, and we begin to look down as from above upon
the inclined plane of the deferent of Venus. Therefore the
reflexion gradually changes into a declination, so that when the
earth is at a quadrant's distance from its former position, no
matter where the planet may be seen in the part of its path
that tilts upward, it has only a declination from the ecliptic. In
this position, since we on the earth are opposite the half of the
deferent that extends in consequence from the higher apse
and rises northward from the plane of the ecliptic, the ancients
said that the epicycle of Venus was in the descending node and
that the apogee of the epicycle reached its greatest northern
declination, and the perigee its greatest southern.

Then, as the earth in its annual motion carries us upward
toward the place of the higher apse of Venus, its circle, like an
epicycle, seems to approach the lower apse of its deferent;
the plane of the epicycle, which is for us the plane in which
Venus moves, and which previously had a declination to the
plane of the ecliptic, again appears to have a reflexion to us;
and the northern half of the deferent, rising from the plane of
the ecliptic, lies to the right because we see Venus from above.
But when the center of the earth reaches the place of the higher

apse of Venus, no declination and only a reflexion is seen; and
the circle of Venus is believed to be in the lower apse of its
deferent, as the ancients would have said. This is the order of
the phenomena while the center of the earth completes half a
revolution, as it mounts in the order of the signs from the place
of the lower apse of Venus to the place of the higher apse of
Venus.

When the earth descends in the same way, the reflexion,
to our eyes, gradually changes into a declination; and because
the half of the plane of the deferent that extends in precedence from the higher apse becomes, through this motion of
the earth, opposite to us, the apogee of the deferent of Venus
begins to have a southward declination from the plane of the
ecliptic, until when the earth is 90° from the place of the apse
both halves are seen in declination to the plane of the ecliptic
and the circle of Venus like an epicycle is thought to be in the
ascending node at the higher apse. As the earth moves on from
this position, the declination again changes into a reflexion; and
when the earth reaches the place of the lower apse of Venus, it
begins to produce once more the same phenomena of latitude
in Venus. From these considerations it is clear that when the
earth is on the apse-line of Venus, the plane of the deferent
of the planet appears to have a reflexion; when the earth is at
a quadrant's distance from the apsides, a declination; and when
the earth is at the intervening points, mixed latitudes are
seen.[234]

Mingled with these latitudes, which the ancients assigned
to the epicycle of Venus, there is still another, called "deviation" by the ancients, by Ptolemy "tilting of the eccentric
circles,"[235] which they demonstrated by the center of the
deferent of the epicycle of Venus, now eliminated. Hence my
teacher has decided that another theory must be constructed
in better agreement with the observations. To make this theory
of my teacher for saving the deviation easier to understand,
like the other ideas heretofore set forth, let us define the plane

[234] Cf. pp. 83-84, above.
[235] For example, HII, 535.6-7.

discussed above as the mean plane, and therefore fixed; from it the true plane deviates in a definite way, now to one side, now to the other. We comprehend all motions with less effort and expenditure of time by directing our attention to their poles. We should therefore begin with the statement that one of the poles of the mean plane lies north of the plane of the ecliptic by the amount of the angle of inclination; the other pole on the opposite side lies an equal distance to the south; and what we shall prove with regard to the north pole, or the phenomena related to it, must be understood in like manner with regard to the south pole, the law of opposition being, of course, observed.

Accordingly, let us assume that about the north pole of the mean plane there is a movable circle, whose radius equals the greatest inclinations of the mean and true planes. Let the north pole of the true plane describe the diameter of the said circle by a motion in libration. Furthermore, let the movable circle follow the motion of the planet, so that as Venus proceeds with its own motion it observes the following rule: it leaves behind one of the two intersections that follow it, and exactly in a year overtakes the intersection left behind. Draw a great circle through the poles of both planes, mark off 90° on each side of[236] its intersection with the true plane, and the nodes or intersections, as I have called them, are determined when the poles of the true and mean planes do not coincide. While a periodic return of Venus to either one of the nodes is being completed, let the pole of the true plane twice describe the diameter of the said movable circle by a motion in libration. Let these phenomena so occur that the planet appears to have entered into a covenant with the center of the earth whereby, whenever the earth is at the apsides of the deferent, no matter where Venus is on its true deferent, it has its greatest northward deviation from the mean plane, that is, it is at its greatest distance from its mean course; moreover, when the earth is at a quadrant's distance from the apsides of the deferent, the planet, together with its entire true plane, lies in the plane of the mean

[236] Reading *ab* (Th 488.11) instead of *ad* (PII, 362.18).

deferent; and when the earth passes through the intervening
points, the path of the planet likewise has intermediate deviations. That this covenant of earth and planet might be ever-
lasting, God ordained that the first small circle of libration,
to use this term, should revolve once in the time in which one
return of Venus to either of the movable nodes occurs.

Let us make these relations clearer by an example. If at
any beginning of the motion of deviation the north pole of
the true plane is at its greatest southward distance from the
pole of the adjacent mean plane, and if Venus is at the limit
of its deviation, which lies to the north, the center of the earth
being in one of the apsides of Venus, in the fourth part of a
year the earth in its annual motion will come to the mid-point
between the apsides, and in the same time the planet will reach
its movable intersection or node. Because the motion in libration
is commensurable with the periodic return of the planet to its
nodes or intersections, the first small circle of libration will
likewise complete a quadrant; and the second small circle,
which moves at twice the rate of the first, will join the pole
of the true plane to the pole of the mean plane, and therefore
the two planes will coincide. But as the planet moves away
from the node, the earth proceeds toward the other apse of
the first eccentric, and the pole of the true plane moves
northward in libration from the pole of the mean plane. Thus
it happens that even though Venus is in south latitude, as in
our example, the latitude, if south, nevertheless diminishes, if
north, increases. When the earth reaches the other apse, the
pole of the true plane attains the northern limit of its motion
in libration; and the planet, midway between the two intersections in its annual return to the nodes, again has its greatest
northward deviation. It is therefore clear that the motion of the
circle which has been assumed has this advantage, that the
revolution of Venus with respect to the nodes occurs in a year;
and that when the earth is in the apse-line, no matter where the
planet is in its true plane, it always has its greatest deviation
from the mean plane; and that when the earth is midway

between the apsides, the planet is in the nodes. Moreover, by reason of the motion in libration, it happens that when Venus is in one of the nodes, the two planes coincide; and that part of the true plane in which Venus is moving always deviates northward from the mean plane, so that this latitude, as is proper, always remains a north latitude.

The mean plane of Venus, as we have called it, is intersected by the ecliptic in the apse-line of the first eccentric; and the half of this plane that lies in consequence from the higher apse rises northward, and the other half, by the law of opposition, dips southward. In Mercury there is a mean plane of a similar nature. It is inclined to each side of the plane of the ecliptic along the apse-line, as is proper, so that conversely the half of the mean plane that lies in precedence from the higher apse extends northward. Therefore, in the annual revolution of the center of the earth the declinations and reflexions in Mercury will be found interchanged, as compared with those of Venus. To make this contrast more striking, God arranged the deviation of the true plane of Mercury from the mean plane so that the half in which Mercury is moving always deviates southward from the mean plane; and when the earth is at the apsides, the planet lies with its true plane in the mean plane. Consequently Mercury has only the above-mentioned differences in latitude from Venus, except that this deviation is greater in Mercury than in Venus,[237] as the former has also the greater angle of inclination.[238] The other changes of latitude in Mercury will quite easily be found exactly as in Venus.

A part of the task remains, and part is done;
Here let the anchor drop and moor our boat,

to conclude this *First Account* with the words of the poet.[239]

---

[237] The traditional estimates were, respectively, 45' and 10'; although Copernicus departed from them somewhat, he did not alter their relative value (cf. above, p. 90, n. 102).

[238] Cf. pp. 83, 89, above.

[239] Ovid *Ars amatoria* i.771-72. Kepler also used this couplet to close Part III of his work on Mars (*Opera*, ed. Frisch, III, 325).

Just as soon as I have read the entire work of my teacher with sufficient application, I shall begin to fulfill the second part of my promise. I hope that both will be more acceptable to you, because you will see clearly that when the observations of scholars have been set forth, the hypotheses of my teacher agree so well with the phenomena that they can be mutually interchanged, like a good definition and the thing defined.

Most illustrious and most learned Schöner, whom I shall always revere like a father, it now remains for you to receive this work of mine, such as it is, kindly and favorably. For although I am not unaware what burden my[240] shoulders can carry and what burden they refuse to carry, nevertheless your unparalleled and, so to say, paternal affection for me has impelled me to enter this heaven not at all fearfully and to report everything to you to the best of my ability. May Almighty and Most Merciful God, I pray, deem my venture worthy of turning out well, and may He enable me to conduct the work I have undertaken along the right road to the proposed goal. If I have said anything with youthful enthusiasm (we young men are always endowed, as he says, with high, rather than useful, spirit) or inadvertently let fall any remark which may seem directed against venerable and sacred antiquity more boldly than perhaps the importance and dignity of the subject demanded, you surely, I have no doubt, will put a kind construction upon the matter and will bear in mind my feeling toward you rather than my fault.

Furthermore, concerning my learned teacher I should like you to hold the opinion and be fully convinced that for him there is nothing better or more important than walking in the footsteps of Ptolemy and following, as Ptolemy did, the ancients and those who were much earlier than himself. However, when he became aware that the phenomena, which control the astronomer, and mathematics compelled him to make certain assumptions even against his wishes, it was enough, he thought,

[240] Reading *mei* (Th 489.36) instead of *me* (PII, 364.30).

if he aimed his arrows by the same method to the same target as Ptolemy, even though he employed a bow and arrows of far different type of material from Ptolemy's. At this point we should recall the saying "Free in mind must he be who desires to have understanding."[241] But my teacher especially abhors what is alien to the mind of any honest man, particularly to a philosophic nature; for he is far from thinking that he should rashly depart, in a lust for novelty, from the sound opinions of the ancient philosophers, except for good reasons and when the facts themselves coerce him. Such is his time of life, such his seriousness of character and distinction in learning, such, in short, his loftiness of spirit and greatness of mind that no such thought can take hold of him. It is rather the mark of youth or of "those who pride themselves on some trifling speculation," to use Aristotle's words,[242] or of those passionate intellects that are stirred and swayed by any breeze and their own moods, so that, as though their pilot had been washed overboard, they snatch at anything that comes to hand and struggle on bravely. But may truth prevail, may excellence prevail, may the arts ever be honored, may every good worker bring to light useful things in his own art, and may he search in such a manner that he appears to have sought the truth. Never will my teacher avoid the judgment of honest and learned men, to which he plans of his own accord to submit.

[241] This sentence serves as motto for the *Narratio prima* (see p. 108, above) and also for Kepler's *Dissertatio cum nuncio sidereo* (Prague, 1610; see *Kepleri opera omnia*, ed. Frisch, II, 485). It is quoted substantially correctly from the *Didaskalikos* (C. F. Hermann's Teubner edition of Plato, VI [Leipzig, 1892], 152). This elementary textbook of Platonic philosophy was available to Rheticus in the Aldine editions of Iamblichus (1516) and Apuleius (1521); in the latter work the words quoted appear on fol. 1 2r. The *Didaskalikos* was formerly attributed to Alcinous, but now it is held that its author was Albinus, who flourished in Smyrna during the middle of the second century A.D., and was a teacher of Galen (R. E. Witt, *Albinus and the History of Middle Platonism*, Cambridge, 1937, pp. 104-9).

[242] *De mundo* 391a23-24; Rheticus has adapted the original to the structure of his sentence and has shifted the meaning of θεωρία from "spectacle" to "speculation" (cf. above, p. 162, n. 194). The *De mundo* is pseudo-Aristotelian (cf. above, p. 139, n. 121).

In Praise of Prussia[243]

In the ode[244] which is said to be preserved in golden letters in the temple of Minerva and which celebrates the Olympic victory of the boxer Diagoras of Rhodes, Pindar says that Diagoras's native land is the daughter of Venus and the dearly beloved spouse of the sun; that Jupiter, moreover, rained much gold there, inasmuch as the Rhodians worshipped his daughter Minerva; and that in consequence, through Minerva, the Rhodians gained a reputation for wisdom and education, to which they were deeply devoted.

I am not aware that anyone could apply this resounding praise of the Rhodians to any region of our time more suitably than to Prussia, concerning which I propose to say a few words that perhaps you desired to hear. Doubtless the same divinities would be found to be presiding over this region, should some skillful astrologer make careful inquiry about the stars that rule over this most beautiful, most fertile, and most fortunate area. As Pindar says:[245]

But the tale is told in ancient story that, when Zeus and the immortals were dividing the earth among them, the isle of Rhodes was not yet to be seen in the open main, but was hidden in the briny depths of the sea; and that, as the Sun-god was absent, no one put forth a lot on his behalf, and so they left him without any allotment of land, though the god himself was pure from blame. But when that god made mention of it, Zeus was about to order a new casting of the lot, but the Sun-god would not suffer it. For, as he said, he could see a plot of land rising from the bottom of the foaming main, a plot that was destined to prove rich in substance for men, and kindly for pasture.

---

[243] This section was omitted from the Basel edition of 1566, the Warsaw edition of 1854, and Th. It was included in the following editions: Danzig, 1540; Basel, 1541; Tübingen, 1596; Frankfurt, 1621; and PII, 367-77. It was also printed, in incomplete form, in *Acta Borussica*, II (Königsberg and Leipzig, 1731), 413-25; and completely in Hipler, *Spicilegium Copernicanum*, pp. 215-22. It was translated into German by Franz Beckmann (ZE, III[1866], 5-17) and Prowe (PI², 448-63).

[244] Pindar, seventh Olympian ode.

[245] *Ibid.*, 54-63; the translation is by J. E. Sandys in the Loeb Classical Library (London, 1915).

Doubtless the sea once covered Prussia, too. What more definite and more important[246] evidence could anyone produce than that today amber is found inland, at a very great distance from the coast? Therefore, on the principle that it rose from the sea, by an act of the gods Prussia passed into the hands of Apollo, who cherishes it now, as once he cherished Rhodes, his spouse. Cannot the sun reach Prussia as well as Rhodes with vertical rays? I grant that it cannot. But it makes up for this in many other ways; and what it accomplishes in Rhodes by its vertical rays, it performs in Prussia by lingering above the horizon. Moreover, amber is a special gift of God, with which He desired to adorn this region above all others, as I think nobody will deny. Indeed, anyone who considers the nobility of amber and its use in medicine will regard it, not without reason, as sacred to Apollo and as a magnificent gift, an abundance of which he presents, like a most valuable jewel, to his spouse Prussia.

But besides the medical and prophetic arts, which Apollo invented and first practiced, he is filled with a passion for hunting. For this reason he seems to have chosen this land before all others. And since he long foresaw that the savage Turks would despoil Rhodes, he transferred his abode to these parts and migrated hither with his sister Diana, as seems not improbable. For no matter where you turn your eyes, if you look at the woods, you will say that they are game preserves ("paradise" in Greek) and beehives stocked by Apollo; if you look at the orchards and fields, rabbit warrens and birdhouses, lakes, ponds, and springs, you will say that they are the holy places of Diana and the fisheries of the gods. And Apollo appears to have chosen Prussia before other regions, I say, as his paradise. Besides stag, doe, bear, boar, and the kind of wild beast that is commonly known elsewhere, he brought in also urus, elk, bison, etc., species scarcely to be found in other places, to say nothing of the numerous and quite rare types of bird and fish.

The progeny which Apollo received from his spouse Prussia

[246] Reading *propiusque* (1596 ed., p. 153) instead of *propriusque* (PII, 369.13).

is as follows: Königsberg, seat of the illustrious prince, Albrecht, duke of Prussia, margrave of Brandenburg, etc., patron of all the learned and renowned men of our time; Thorn, once quite famous for its market, but now for its foster-son, my teacher; Danzig, metropolis of Prussia, eminent for the wisdom and dignity of its Council, for the wealth and splendor of its renascent literature; Frauenburg, residence of a large body of learned and pious men, famous for its eloquent and wise Bishop, the Most Reverend John Dantiscus; Marien-burg, treasury of His Serene Majesty, the king of Poland; Elbing, ancient settlement in Prussia, where, too, the sacred pursuit of literature is undertaken; Kulm, famous for its literature, where the Law of Kulm had its origin.

You might say that the buildings and the fortifications are palaces and shrines of Apollo; that the gardens, the fields, and the entire region are the delight of Venus, so that it could be called, not undeservedly, Rhodes. What is more, Prussia is the daughter of Venus, as is clear if you examine either the fertility of the soil or the beauty and charm of the whole land.

As Venus is said to have risen from the sea, so Prussia is the daughter of Venus and of the sea. And therefore it is fertile enough to feed Holland and Zealand with its crops and to serve as granary for the neighboring kingdoms and also for England and Portugal. Besides this excellent produce it exports quantities of fish of every sort and other valuable resources, with which it abounds. But Venus is interested in the things that promote culture, dignity, and the good and humane life. These could not grow and develop in this region, for the character of the country forbade it. So she saw to it that with the aid of the sea they could be successfully imported into Prussia from abroad.

But since these facts are so well known to you, most learned Schöner, that there is no need for me to speak of them at greater length; and since they are treated in other books, wholly devoted to this subject, I refrain from further praise. I add only this item, that by the grace of the presiding divinity the Prussians are a numerous people and also possess an unusual talent for culture. Moreover, they worship Minerva with every type of art and for this reason receive the kindness of Jupiter. For, not to speak of the lesser arts attributed to Minerva, like architecture and its allied disciplines, the revival of literature in the world is everywhere welcomed with keenest interest, as befits heroes, by the illustrious duke most of all, and also by all the dignitaries and nobles of Prussia, in whose hands lies the direction of affairs, and by the rulers of states. They strive to encourage and support it, both independently and jointly. Therefore Jupiter forms a yellow cloud and rains much gold. This means, as I interpret it, that because Jupiter is said to preside over kingdoms and states, when the mighty undertake to support studies, learning, and the muses, then God gathers the minds of his subject and neighboring kings, princes, and peoples into a golden cloud; from it he distils peace and all the blessings of peace, like drops of gold; minds in love with tranquillity and public order; cities governed by just laws; wise men; upright and devout education of children; pious and pure spread of religion, etc.

The story is frequently told of the shipwreck of Aristippus, which they say occurred off the island of Rhodes. Upon being washed ashore, he noticed certain geometrical diagrams on the beach; exclaiming that he saw the traces of men, he bade his companions be of good cheer. And his belief did not play him false. For through his great learning he easily obtained for himself and his comrades from educated and humane men the things necessary for sustaining life.

So may the gods love me, most learned Schöner, it has not yet happened to me that I should enter the home of any distinguished man in this region—for the Prussians are a most hospitable people—without immediately seeing geometrical diagrams at the very threshold or finding geometry present in their minds. Hence nearly all of them, being men of good will, bestow upon the students of these arts every possible benefit and service, since true knowledge and learning are never separated from goodness and kindness.

In particular, I am wont to marvel at the kindness of two

distinguished men toward me, since I readily recognize how slight is my scholarly equipment, measuring myself by my own abilities. One of them is the illustrious prelate whom I mentioned at the outset,[247] the Most Reverend Tiedemann Giese, bishop of Kulm. His Reverence mastered with complete devotion the set of virtues and. doctrine, required of a bishop by Paul. He realized that it would be of no small importance to the glory of Christ if there existed a proper calendar of events in the Church and a correct theory and explanation of the motions. He did not cease urging my teacher, whose accomplishments and insight he had known for many years, to take up this problem, until he persuaded him to do so.

Since my teacher was social by nature and saw that the scientific world also stood in need of an improvement of the motions, he readily yielded to the entreaties of his friend, the reverend prelate. He promised that he would draw up astronomical tables with new rules and that if his work had any value he would not keep it from the world, as was done by John Angelus,[248] among others. But he had long been aware that in their own right the observations in a certain way required hypotheses which would overturn the ideas concerning the order of the motions and spheres that had hitherto been discussed and promulgated and that were commonly accepted and believed to be true; moreover, the required hypotheses would contradict our senses.

He therefore decided that he should imitate the Alfonsine Tables rather than Ptolemy and compose tables with accurate rules but no proofs. In that way he would provoke no dispute among philosophers; common mathematicians would have a correct calculus of the motions; but true scholars, upon whom Jupiter had looked with unusually favorable eyes, would easily arrive, from the numbers set forth, at the principles and sources from which everything was deduced. Just as heretofore learned men had to work out the true hypothesis of the motion of the

[247] Page 109.

[248] For Angelus see Gesner, *Bibliotheca universalis*, pp. 382v–383r; and *Allgemeine deutsche Biographie*, I, 457.

starry sphere from the Alfonsine doctrine, so the entire system would be crystal clear to learned men. The ordinary astronomer, nevertheless, would not be deprived of the use of the tables, which he seeks and desires, apart from all theory. And the Pythagorean principle would be observed that philosophy must be pursued in such a way that its inner secrets are reserved for learned men, trained in mathematics, etc.[249]

Then His Reverence pointed out that such a work would be an incomplete gift to the world, unless my teacher set forth the reasons for his tables and also included, in imitation of Ptolemy, the system or theory and the foundations and proofs upon which he relied to investigate the mean motions and prosthaphaereses and to establish epochs as initial points in the computation of time. The bishop further argued that such a procedure had produced great inconvenience and many errors in the Alfonsine Tables, since we were compelled to assume and to approve their ideas on the principle that, as the Pythagoreans used to say, "The Master said so"—a principle which has absolutely no place in mathematics.

Moreover, contended the bishop, since the required principles and hypotheses are diametrically opposed to the hypotheses of the ancients, among scholars there would be scarcely anyone who would hereafter examine the principles of the tables and publish them after the tables had gained recognition as being in agreement with the truth. There was no place in science, he asserted, for the practice frequently adopted in kingdoms, conferences, and public affairs, where for a time plans are kept secret until the subjects see the fruitful results and remove from doubt the hope that they will come to approve the plans.

So far as the philosophers are concerned, he continued, those of keener insight and greater information would carefully study Aristotle's extensive discussion and would note that after convincing himself that he had established the immobility of the earth by many proofs Aristotle finally takes refuge in the following argument:

[249] At this point the text as printed in *Acta Borussica* breaks off.

We have evidence for our view in what the mathematicians say about astronomy. For the phenomena observed as changes take place in the figures by which the arrangement of the stars is marked out occur as they would on the assumption that the earth is situated at the center.[250]

Accordingly the philosophers would then decide:

If this concluding statement by Aristotle cannot be linked with his previous discussion, we shall be compelled, unless we are to waste the time and effort which we have invested, rather to assume the true basis of astronomy. Moreover, we must work out appropriate solutions for the remaining problems under discussion. By returning to the principles with greater care and equal assiduity, we must determine whether it has been proved that the center of the earth is also the center of the universe. If the earth were raised to the lunar sphere, would loose fragments of earth seek, not the center of the earth's globe, but the center of the universe, inasmuch as they all fall at right angles to the surface of the earth's globe? Again, since we see that the magnet by its natural motion turns north, would the motion of the daily rotation or the circular motions attributed to the earth necessarily be violent motions? Further, can the three motions, away from the center, toward the center, and about the center, be in fact separated? We must analyze other views which Aristotle used as fundamental propositions with which to refute the opinions of the *Timaeus* and the Pythagoreans.

They will ponder the foregoing questions and others of the same kind if they desire to look to the principal end of astronomy and to the power and the efficacy of God and nature.

But if it is to be the intention and decision of scholars everywhere to hold fast to their own principles passionately and insistently, His Reverence warned, my teacher should not anticipate a fate more fortunate than that of Ptolemy, the king of this science. Averroes, who was in other respects a philosopher of the first rank, concluded that epicycles and eccentrics could not possibly exist in the realm of nature and that Ptolemy did not know why the ancients had posited motions of rotation. His final judgment is: "The Ptolemaic astronomy is nothing, so far as existence is concerned; but it is convenient

---

[250] *De caelo* ii.14 297a2-6; the translation is from Thomas L. Heath, *Greek Astronomy* (London, 1932), p. 91.

for computing the nonexistent."[251] As for the untutored, whom the Greeks call "those who do not know theory, music, philosophy, and geometry,"[252] their shouting should be ignored, since men of good will do not undertake any labors for their sake.

By these and many other contentions, as I learned from friends familiar with the entire affair, the learned prelate won from my teacher a promise to permit scholars and posterity to pass judgment on his work. For this reason men of good will and students of mathematics will be deeply grateful with me to His Reverence, the bishop of Kulm, for presenting this achievement to the world.

In addition, the benevolent prelate deeply loves these studies and cultivates them earnestly. He owns a bronze armillary sphere for observing equinoxes, like the two somewhat larger ones which Ptolemy says were at Alexandria[253] and which learned men from everywhere in Greece came to see. He has also arranged that a gnomon truly worthy of a prince should be brought to him from England. I have examined this instrument with the greatest pleasure, for it was made by an excellent workman who knew his mathematics.

The second of my patrons is the esteemed and energetic John of Werden, burgrave of Neuenburg, etc., mayor of the famous city of Danzig. When he heard about my studies from certain friends, he did not disdain to greet me, undistinguished though I am, and to invite me to meet him before I left Prussia. When I so informed my teacher, he rejoiced for my sake and drew such a picture of the man that I realized I was being invited by Homer's Achilles, as it were. For besides his distinction in the arts of war and peace, with the favor of the muses he also cultivates music. By its sweet harmony he refreshes and inspires his spirit to undergo and to endure the burdens

---

[251] Averroes, *Commentary on Aristotle's Metaphysics*, Book xii, *summa* ii, *caput* iv, No. 45. A Latin version of Averroes's *Commentary* was printed (Padua, 1473) with the *Metaphysics* in Latin translation (GW, 2,419; see also 2,337-40).

[252] Aulus Gellius *Noctes Atticae* i.9.8. It is at this point that the text as printed in *Acta Borussica* is resumed.

[253] HI, 195.5-7; 197.17-20.

of office. He is worthy of having been made by Almighty and Most Merciful God a "shepherd of the people."[254] Happy the state over which God has appointed such rulers!

In the *Phaedo*[255] Socrates rejects the opinion of those who called the soul a "harmony." And he did so rightly if by harmony they understood nothing but a mixture of the elements in the body. But if they defined the soul as a harmony because in addition to the gods only the human mind understands harmony—just as it alone knows number, wherefore certain thinkers did not fear to call it a number—and also because they knew that souls suffering from the deadliest diseases are sometimes healed by musical harmonies, then their opinion will not seem unfortunate, inasmuch as it is principally the soul of a heroic man that is called a harmony. Hence we might correctly call those states happy whose rulers have harmonious souls, that is, philosophical natures. Surely the Scythian had no such soul who preferred hearing a horse's neighing to a talented musician whom others listened to in amazement. Would that all kings, princes, prelates, and other dignitaries of the realms had souls chosen from the vessel of harmonious souls. Then these excellent studies and those which are chiefly to be pursued for their own sake would doubtless achieve a worthy station.

The foregoing, most distinguished sir, are the things which I thought I should for the present write to you regarding the hypotheses of my teacher, Prussia, and my patrons. Farewell, most learned sir, and do not disdain to guide my studies with your advice. For you know that we young men greatly need the counsel of older and wiser men. And you have not forgotten that charming sentiment of the Greeks, "The opinions of older men are better."[256]

From my library at Frauenburg
September 23, 1539

---

[254] A familiar epithet of kings and chiefs in Homer, e.g., *Iliad* ii.243.

[255] 86B-C, 92A-95A.

[256] This line, from a lost play of Euripedes, is preserved in Stobaeus, *Florilegium* CXV.2.

BIBLIOGRAPHICAL NOTE

THIS NOTE does not offer a complete bibliography of the subject, but only a guide to the basic literature. Books which have been consulted but which are not mentioned here are cited in the notes and indexed by author and title.

The fundamental work on Copernicus is Leopold Prowe, *Nicolaus Coppernicus* (Berlin, 1883–84); for the biography of Copernicus and the social history of his times the book is extremely valuable, but Prowe's judgment in scientific matters was unreliable. The most important work since Prowe has been done by Ludwik Antoni Birkenmajer; his chief publications were, regrettably, in Polish: *Mikołaj Kopernik* (Cracow, 1900); *Stromata Copernicana* (Cracow, 1924); and *Mikołaj Kopernik Wybór pism* (Cracow, 1926). A recent Polish work of merit is Jeremi Wasiutynski's *Kopernik, Twórca nowego nieba* (Warsaw, 1938). For the life of Rheticus the basic study is Adolf Müller's article "Der Astronom und Mathematiker Georg Joachim Rheticus" in the *Vierteljahrsschrift für Geschichte und Landeskunde Vorarlbergs*, neue Folge, II(1918), 5–46.

Sound exposition of Copernicus's astronomical system may be found in Angus Armitage, *Copernicus* (London, 1938), chs. iii–vi; Arthur Berry, *A Short History of Astronomy* (London, 1898), ch. iv; J. L. E. Dreyer, *History of the Planetary Systems from Thales to Kepler* (Cambridge, 1906), ch. xiii; and A. Wolf, *A History of Science, Technology, and Philosophy in the 16th and 17th Centuries* (New York, 1935), ch. ii. Among the older works the most useful treatment is that by J. B. J. Delambre in his *Histoire de l'astronomie moderne* (Paris, 1821), I, 85–142. Students may also consult Rudolf Wolf, *Geschichte der Astronomie* (Munich, 1877; reissued Leipzig, 1933), pp. 222–42, and Ernst Zinner, *Die Geschichte der Sternkunde* (Berlin, 1931), pp. 454–63.

An older guide to the literature of the subject is J. C. Houzeau and A. Lancaster, *Bibliographie générale de l'astronomie* (Brussels, 1882–89), No. 2,503–4 (I, 578–79), No. 6,340–98 (I², 886–90), and Vol. II, columns 109–13, 237. A more recent guide, containing the items published before 1924, is Wilhelm Bruchnalski's *Bibljografja Kopernikowska*, constituting ch. x, pp. 209–46, in *Mikołaj Kopernik* (Lvov, Warsaw, 1924). Useful summaries of subsequent material are provided by Eugen Brachvogel's contributions to the *Zeitschrift für die Geschichte und Altertumskunde Ermlands*, XXIII(1927–29), 193–95, 795–803; XXV(1933–35), 237–45, 548–55, 819–23; XXVI, Heft 1(1936), 249–58. A brief analysis of some recent work appears in *Forschungen und Fortschritte*, XIII(1937), 369–71.

ANNOTATED COPERNICUS BIBLIOGRAPHY
1939—1958

THE LITERATURE concerning Copernicus that has come to my attention since the publication of the first edition of this book is recorded in the following bibliography, which makes no claim to be complete. I shall be grateful to readers who may wish to notify me about additional entries worthy of inclusion in an international scientific bibliography.

For help in procuring rare German material, I desire to thank Professor Willy Hartner, director of the Institut für Geschichte der Naturwissenschaften, Johann Wolfgang Goethe Universität, Frankfurt am Main, and Frieda Henn of the Astronomisches Rechen-Institut, Heidelberg. Professor Aleksander Birkenmajer of Warsaw University and Professor Stanisław Wędkiewicz, director of the Centre Polonais de Recherches Scientifiques in Paris, kindly aided me in coping with Polish authors. Finally, Professor Vasily Pavlovich Zubov, Institute for the History of Science and Technology, Academy of Science, U.S.S.R., patiently assisted me with some Russian items.

Secretarial expenses connected with the preparation of this bibliography for the press were met in part by a grant awarded to me by the Faculty Research Committee, City College of New York.

E. R.

ABETTI, GIORGIO

1. Copernico in Italia. *Sapere*, 1943, *17–18*: 374–375. A review of Copernicus' student days in Italy. If his lecture at Rome in 1500 dealt with his new astronomical ideas, how can he have first "begun to think about his new system" in 1506?

2. *Storia dell' astronomia* (Florence, 1949) xii + 370 p. Translated into English by Betty Burr Abetti as *The History of Astronomy* (New York, 1952) p. 338; (London, 1954) xviii + 345 p., with a foreword by Sir Harold Spencer Jones.

In the Italian ed. Copernicus is discussed mainly at pp. 65–78, and at pp. 67–81 in the English translation. Reviewed **18**.

AERSCHODT, LÉON VAN

3. Copernic et le IIIe Reich. *Ciel et terre*, 1939, *55*: 309–312. After reviewing various contentions that Copernicus was either a German or a Pole, the author asks whether the great astronomer did not rather belong to a select family of geniuses.

AHERN, MICHAEL JOSEPH

4. Copernicus and His Times. *Catholic Mind*, 1943 (September), *41*: no. 969, pp. 48–54. This radio address, broadcast on May 16, 1943, mistakenly asserted that Copernicus' observations "were the basis of the calculations which brought about the Gregorian calendar." Why did Ahern say that Copernicus expounded the heliocentric astronomy at Rome in 1500; that the Pope was "instrumental in the publication" of the *Revolutions*; that he "accepted the dedication"; and that the "work was published at the expense of a Cardinal"?

AITKEN, ROBERT GRANT

5. *De revolutionibus orbium coelestium.* Astronomical Society of the Pacific, Leaflet no. 172 (June, 1943) p. 8. The emeritus director of the Lick Observatory discusses Copernicus with an imaginary man from Mars.

ALEKSEEV, B.

6. Portret Kopernika. *Ogonek*, 1956, *34*: no. 26, p. 32. Brief discussion in Russian of a portrait of Copernicus.

7. Copernicus' Water Supply. *Op. cit.*, 1957, *35*: no. 3, p. 32.

ALTENBERG, K.

8. Kopernikowska sesja akademii nauk ZSRR. *Życie szkoły wyższej*, 1953, *1*: no. 11, pp. 78–81. A report on the Copernicus celebration under the auspices of the Soviet Academy of Science.

AMZALAK, MOSES BENSABAT

9. *As teorias monetárias de Nicolau Copérnico* (Lisbon, 1947) p. 47. Contains a translation into Portuguese of Copernicus' *Monetae cudendae ratio* and his letter to Felix Reich on the same subject.

ANDRISSI, GIOVANNI L.

10. Nessun libro fu condannato perchè copernicano. *L'Osservatore romano*, 1950, *90*: no. 299, p. 3 (December 22). Contends that the Roman Catholic church banned no book simply because it was Copernican; only if a Copernican book ventured into theology was it banned.

11. Ha la terra il terzo moto assegnatole da Copernico? *Op. cit.*, 1951, *91*: no. 85, pp. 3-4 (April 13). Argues that Copernicus was not wrong in attributing to the earth's axis the slow conical motion which he called the earth's third motion (in addition to the daily rotation and the annual revolution).

12. Sulla genesi del sistema copernicano. *Op. cit.*, 1953, *94*: June 8–9, p. 3. Copernicus was dissatisfied with the astronomy accepted in his time, and searched in the writings of his predecessors for a more satisfactory system.

ANTHONY, HERBERT DOUGLAS

13. *Science and its Background* (London, 1948) ix + 303 p.; 2d ed. (London, New York, 1954) 336 p.; 3d ed. (London, 1957) ix + 352 p. In the third ed. Copernicus is discussed at pp. 128-135.

ANNOTATED COPERNICUS BIBLIOGRAPHY

ARAGÓN Y LEÓN, AGUSTÍN

14. Copérnico. *Memorias y revista de la Academia nacional de ciencias* (Mexico), 1935–1944, *55*: 275–280. Commemorating Copernicus' four-hundredth anniversary.

ARBUSOW, LEONID

15. Livländische Beziehungen von Nikolaus Koppernicks Frauenburger Zeitgenossen. *Quellen und Forschungen zur baltischen Geschichte*, 1944, *5*: 3–14. This study of the relations between Livonia and Copernicus' Frombork contemporaries erroneously confers on Copernicus a Master of Arts degree; misdates his receipt of the doctoral degree; falsely asserts that he was designated a priest in 1497; and mistakenly denies that he held more than one church office.

Archeion, 1940, *22*: 137–138

16. L'annexion de Copernic. Under this title the periodical edited by Aldo Mieli reproduced two newspaper items dealing with the Polish-German dispute over the nationality of Copernicus.

ARCHIBALD, RAYMOND CLARE

17. *Outline of the History of Mathematics*, 4th ed. (Oberlin, 1939) p. 66; 5th ed., 1941, p. 76; 6th ed., 1949, p. 114. The sixth ed. of this little work, first published in 1932, discusses Copernicus at pp. 34, 75–76.

ARMITAGE, ANGUS

18. *Sun, Stand Thou Still* (New York, 1947; London, 1948) p. 210. Differs from Armitage's previous work, *Copernicus, the Founder of Modern Astronomy* (London and New York, 1938) principally by an expansion of the earlier Epilogue into a group of excellent chapters on the transformation and triumph of Copernicanism. The best introduction to the subject in the English language. Reprinted in a paperback ed. as a Mentor Book by the New American Library (New York, 1951) under the title *The World of Copernicus*, and translated into Italian by Orazio Nicotra as *Niccolò Copernico e l'astronomia moderna* (Turin, 1956) p. 256. *Copernicus, the Founder*

of Modern Astronomy (New York and London: Yoseloff, 1957) is a further expansion and improvement of the earlier versions.

Sun, Stand Thou Still was reviewed by Paul W. Merrill, *Scientific Monthly*, 1947, *65*: 439–440; by Francis R. Johnson, *Isis*, 1948, *39*: 175–176; by G. J. Whitrow, *Observatory*, 1948, *68*: 195–196; by John W. Streeter, *Sky and Telescope*, 1948, *7*: 157; by Howard Laurence Kelly, *Journal of the British Astronomical Association*, 1948–1949, *59*: 44; by T. M., *Science Progress*, 1949, *37*: 179; by N. H. de V. Heathcote, *Głos Anglii*, 1949, no. 22, p. 135; and by Herbert Dingle, *Annals of Science*, 1948–1950, *6*: 213–214. The Italian translation was reviewed by Giorgio Abetti, *Scientia*, 1957, *92*: 74.

19. *Copernicus and the Reformation of Astronomy* (Historical Association Publications, General Series G 15, London, 1950) p. 23. Reviewed by Edward Rosen, *Archives internationales d'histoire des sciences*, 1950, *3*: 946; by Howard Laurence Kelly, *Journal of the British Astronomical Association*, 1950, *60*: 147; and by A. R. Hill, *History*, 1952, *37*: 259–260.

ASSOCIAZIONE ITALIANA PER I RAPPORTI CULTURALI CON LA POLONIA

20. *Niccolò Copernico* (Rome, 1953) p. 14. The Italian Association for Cultural Relations with Poland salutes Copernicus as a contributor to the freedom of human thought.

Astronomie (*L'*), 1943, *57*: 30

21. Copernic, Galilée et Newton. A brief anonymous quadricentennial commemoration of Copernicus as one of the founders of modern astronomy.

BABINI, JOSÉ

22. *Historia sucinta de la ciencia* (Buenos Aires, 1951) p. 226. The brief discussion of Copernicus at pp. 115–117 misdates the composition of the *Commentariolus*.

See Rey Pastor, Julio (613).

BACHER, RUDOLF

23. *Das astronomische Dreigestirn der Renaissance* (Innsbruck, 1952) p. 62; reprinted

ANNOTATED COPERNICUS BIBLIOGRAPHY

Munich, 1953. The director of the Astrophysical Laboratory of the Vatican Observatory commends this book, published with the permission of the Roman Catholic church, because it demonstrates the possibility of peaceful coexistence between Faith and Science. Both spring from God, says the author. But all God's children were not always recognized as such by God's (self-proclaimed) representative on earth. Copernicus (pp. 9–24) is one of the three stars of the astronomical Renaissance, Galileo and Kepler being the other two. The case of Galileo is "a regrettable chapter in the history of the church. . . . The Congregation [of the Index of Prohibited Books] made a deplorable mistake" (p. 38).

BAEV, KONSTANTIN LVOVICH

24. *Sozdateli novoi astronomii* (Moscow, 1948) p. 116; 2d ed., Moscow, 1955, p. 125. Translated into Slovak (Bratislava, 1949); into Polish by Jerzy Milenband under the title *Twórcy nowej astronomii* (Warsaw, 1950) p. 171, and also by Juliusz Makowski under the title *Twórcy nowożytnej astronomii* (Warsaw, 1950) p. 109; and into Hungarian by Olivér Dessewffy under the title *Az új asztronómia megteremtöi* (Budapest, 1952) p. 99.

This Russian discussion of the founders of modern astronomy deals with Copernicus at pp. 9–40.

25. *Kopernik* (Moscow, 1953) p. 214; a re-issue of a book published originally in 1935.

BAKER, HERSCHEL

26. *The Dignity of Man* (Cambridge, Mass., and London, 1947) p. 365. Emphasizes the religious aspect of Copernicus' thought.

BANFI, ANTONIO

27. *L'Uomo copernicano* (Milan, 1950) p. 407 (Il pensiero critico, vol. 20). Defining a "Copernican man" as one whose mind has been freed from the illusion that he is the center of the universe, Banfi examines the leading contemporary philosophical trends from a Marxist point of view. Reprinted in this volume is "L'Uomo

copernicano," an address delivered to the full assembly of the Tenth International Congress of Philosophy at Amsterdam in August, 1948, and published in *Studi filosofici*, 1949, *10*: 17–35.

See *Sesja kopernikowska*.

BARANOWSKI, HENRYK

28. O początkach teorii Kopernika w Rosji. *Urania* (Kraków), 1955, *26*: 282–284. The introduction of the Copernican theory into Russia, a comment on an entry in Szyc (775), p. 243.

See Copernicus, bibliography (109–110).

BARYCZ, HENRYK

29. *Mikołaj Kopernik—wielki uczony Odrodzenia* (Warsaw, 1953) p. 82. Pp. 30–33, dealing with the neoplatonism of Copernicus, were translated into French by Allan Kosko at pp. 108–111 in Wędkiewicz.

Reviewed by Jan Gadomski, *Urania* (Kraków), 1955, *26*: 152–153.

Copernicus, a great Renaissance scientist, as viewed at the celebration conducted by the Polish Academy of Science on September 15–16, 1953.

30. Mikołaj Kopernik w dziejach narodu i kultury polskiej. *Przegląd zachodni*, 1953, *9*: no. 3, pp. 513–570. The place of Copernicus in the history of the people and culture of Poland. Mainly biographical, with emphasis on Copernicus' education and his relations with Bishop Dantiscus.

31. Polski udział w historii badań nad tekstem De revolutionibus Mikołaja Kopernika. *Kwartalnik historii nauki i techniki*, 1956, *1*: 227–258 (summary in English at pp. 257–258). Polish participation in the effort to establish the text of Copernicus' *Revolutions*.

32. Rozwój nauki w Polsce w dobie Odrodzenia, at pp. 35–153 in *Odrodzenie w Polsce*, vol. II (Warsaw, 1956), a symposium on the Renaissance in Poland. Barycz's contribution, "The Development of Science in Poland in the Age of the Renaissance," deals with Copernicus at pp. 117–121.

BATOWSKI, ZYGMUNT

33. *Coppernicusbildnisse* (Berlin, 1942), 35 mimeographed pp. and 18 plates.

Translated into German by Alfons Triller from the Polish original *Wizerunki Kopernika* (Toruń, 1933). A study of the iconography of Copernicus.

BADOUX, PIERRE
34. Copernic et l'église. *Ciel et terre*, 1954, *70*: 318–320. This brief sketch of the relations between the Copernican astronomy and the Roman Catholic church elicited letters by Cyr. De Bièvre and M. Daisomont at pp. 431–433 of the same volume.

BAUMER, FRANKLIN LEVAN, ed.
35. *Main Currents of Western Thought* (New York, 1952) p. 699. An excerpt from the *Commentariolus* at pp. 271–273.

BEAGLEHOLE, J. C.
See Victoria University College.

BECKER, FRIEDRICH
36. *Geschichte der Astronomie* (Bonn, 1946) p. 83; 2d ed., enlarged (Bonn, 1947) p. 95. The enlarged edition was translated into French by Francis Cusset under the title *Histoire de l'astronomie* (Paris, 1955) p. 174.
The discussion of Copernicus (1st ed., pp. 35–37; 2d ed., pp. 38–40) repeats the traditional misstatement that the Gregorian calendar reform utilized tables based on the Copernican astronomy.

BELL, ERIC TEMPLE
37. *The Development of Mathematics* (New York and London, 1940) xiii + 583 p. Bell erred in saying that in Copernicus' astronomy "the sun itself had a small orbit" (p. 103).

BERG, ALEXANDER
38. Der Arzt Nikolaus Kopernikus und die Medizin des ausgehenden Mittelalters, at pp. 172–201 in 370. A review of Copernicus' medical studies, practice, and books.

BIEDER, THEOBALD
39. Wann hat Kopernikus im Kampfe gegen Ptolemäus und Tycho Brahe gesiegt? *Die Sterne*, 1939, *19*: 193–198. The

definitive triumph of Copernicanism over the Ptolemaic and Tychonic systems in Germany occurred around 1760.

BIEGAŃSKI, PIOTR
40. Pomnik Kopernika w Warszawie. *Ochrona zabytków*, 1953, *6*: 47–54. Traces the history of Thorvaldsen's statue of Copernicus from its erection in Warsaw through the damage of World War II to its reconstruction in 1949.

BIHL, ADOLF
41. Nikolaus Kopernikus zu seinem 400. Todestag am 24. Mai 1943. *Zeitschrift des Vereines deutscher Ingenieure*, 1943, *87*: 288. A salute to Copernicus in commemoration of the four-hundredth anniversary of his death.

BIRKENMAJER, ALEKSANDER
42. Nicolas Copernicus, at pp. 37–50 in *Great Men and Women of Poland*, ed. Stephen P. Mizwa (New York, 1942) xxviii + 397 p. That Aleksander Birkenmajer was the author of this article was not revealed at the time of publication because he was then at the mercy of the Germans who were occupying Poland. He closes his discussion by quoting from the great American scientist Simon Newcomb, who said: "There is no figure in astronomical history, which may more appropriately claim the admiration of mankind through all time than that of Copernicus." But why does Birkenmajer assert that Copernicus "moved to Rome to work in the Pope's chancery"?
43. Mikołaj Kopernik. *Wszechświat* 1953, pp. 1–11. With title unchanged, translated into German in *Kulturprobleme des neuen Polen*, 1953, *5*: no. 5, pp. 1–11, and then reprinted by the Kulturbund zur demokratischen Erneuerung Deutschlands, *Vorträge zur Verbreitung wissenschaftlicher Kenntnisse*, no. 68 (Berlin, 1954) p. 27. Reviewed by Maria Uklejska, *Studia i materiały z dziejów nauki polskiej*, 1956, *4*: 374–376, and by Hans Schmauch, *Zeitschrift für die Geschichte und Altertumskunde Ermlands*, 1956–1957, *29*: 156–158. An

admirable survey of the evolution of Copernicus' thinking about astronomical problems.

44. Z dziejów autografu De revolutionibus. *Nauka polska*, 1953, *1*: no. 3, p. 154. A brief history of Copernicus' holograph manuscript of the *Revolutions* from 1543 to 1953.

45. Uniwersytet krakowski jako ośrodek międzynarodowych studiów astronomicznych na przełomie XV i XVI wieku. This lecture was delivered at the Warsaw Conference on the Renaissance, October 25–30, 1953; was published in *Życie szkoły wyższej*, 1954, *2*: no. 2, pp. 79–83; and was translated into French by Allan Kósko under the title L'Université de Cracovie—centre international d'études astronomiques à la fin du XVe et au début du XVIe siècle at pp. 101–108 in Wędkiewicz.

Examines the teaching of astronomy at the University of Kraków in order to justify the statement made by Albert Caprinus in 1542 that Copernicus "declares he owes everything that he is to the University" of Kraków.

46. Co przyniesie nowe łacińsko-polskie wydanie dzieła "O obrotach sfer niebieskich" Mikołaja Kopernika. *Problemy* 1954, *10*: 98–101. What the new Latin-Polish ed. of Copernicus' *Revolutions* will present.

47. Kawaleryjska osemka Eudoksa a przekręcony wianuszek Kopernika. *Op. cit.*, 308–316. Copernicus used Eudoxus' hippopede.

48. W sprawie studiów kopernikańskich Prof. R. St. Ingardena, in *Odrodzenie w Polsce*, vol. 2, part 2, pp. 85–97 (Warsaw, 1956). A comment on Ingarden's studies of Copernicus.

49. Le commentaire inédit d'Erasmus Reinhold sur le De revolutionibus de Copernic. In a lecture delivered on July 2, 1957 to the International Colloquium on Science in the Sixteenth Century, held at Royaumont, Asnières-sur-Oise, France, Birkenmajer showed that Erasmus Reinhold, who has always been regarded as an adherent of Copernicus, in his unpublished commentary on the *Revolutions* in

fact rejected both the Ptolemaic and Copernican theories, putting forward instead his own view, equivalent to Tycho Brahe's, but formulated a generation earlier. Professor Birkenmajer's lecture will be published in the Actes du Colloque international sur la science au seizième siècle, under the auspices of the Union Internationale d'Histoire et de Philosophie des Sciences.

See Copernicus, editions and translations: *O obrotach sfer niebieskich księga pierwsza* (120), *Mikolaj Kopernik 45 Tablic* (369), and *Sesja kopernikowska*.

BIRKENMAJER, LUDWIK ANTONI

50. *Stromata copernicana*, translated into German by Bassmann (Berlin-Dahlem, 1942) p. 296, mimeographed. The book was originally published in Polish (Kraków, 1924).
See Morstin.

BISHOP, PHILIP W.
See Schwartz, George.

BŁACHUT, WŁADYSŁAW
51. Krakowskie lata Mikołaja Kopernika. *Problemy*, 1953, *9*: 308–311. Copernicus' years at Kraków.

BLATT, FRANZ
52. *Fra Cicero til Copernicus* (Copenhagen, 1940) p. 164. Copernicus' Latin style, his technical vocabulary and understanding of etymology are examined at pp. 152–158.

BLUMENBERG, HANS
53. Der kopernikanische Umsturz und die Weltstellung des Menschen. *Studium generale*, 1955, *8*: 637–648. Copernicus displaced man from a central position in the universe to an eccentric position.

BOLL, MARCEL
54. De Nicolas Copernic à Marie Sklodowska-Curie, at pp. 130–137 in *Pologne—Hommages de M. Neville Chamberlain* etc. (Paris, 1940) p. 174. Mistakenly

ANNOTATED COPERNICUS BIBLIOGRAPHY

asserts that Copernicus acquired a doctor's degree in medicine and that he was urged by Pope Clement VII to publish his astronomical system.

BONDOIS, R.
55. Nicolas Copernic, le père de l'astronomie moderne, at pp. 35–53, vol. I, *Les Conquérants da la science* (Paris, 1945; Les Grands Destins, 4th series).

BORNKAMM, HEINRICH
56. Kopernikus im Urteil der Reformatoren, *Archiv für Reformationsgeschichte*, 1943, *40*: 171–183. Zinner's *Entstehung und Ausbreitung der coppernicanischen Lehre* committed serious errors in discussing the attitude of Luther, Osiander and Melanchthon towards Copernicus.

BOTTO, VINCENZO
See Masotti, Arnaldo (463).

BOUSSER, FRANÇOISE
57. Copernic, at pp. 32–33 in *Les Inventeurs célèbres*, ed. Louis Leprince-Ringuet (Paris, 1950; La Galerie des hommes célèbres, vol. VI). It is deplorable that this little essay should help to perpetuate such grievous errors as the statement that "in 1500 Copernicus was summoned to Rome, where he became acquainted with the famous astronomer Regiomontanus," who had died in 1476.

BOYER, CARL B.
58. Note on Epicycles and the Ellipse from Copernicus to LaHire. *Isis*, 1947–1948, *38*: 54–56. Copernicus knew that an ellipse could be generated by a combination of circular motions.

BOYNTON, HOLMES, ed.
59. *The Beginnings of Modern Science* (New York, 1948) xxi + 634 p. Excerpts from the *Commentariolus* at pp. 5–11.

BRACHVOGEL, EUGEN
60. Nikolaus Koppernikus, at pp. 355–356 in *Altpreussische Biographie*, ed. Christian Krollman (Königsberg, 1936–1944) p. 512; interrupted at the letter "P."

Mistakenly ordains Copernicus a priest before the fall of 1497, and equally mistakenly grants him a Master of Arts degree before June 18, 1499.
See Stein (747).
61. Nikolaus Coppernicus. *Hochland*, 1940–1941, *38*: 55–64. In his eagerness to see in Copernicus a devotee of solar mysticism, Brachvogel mistakenly makes him a priest (pp. 55, 56), grants him an M.A. degree (p. 56), and has him adhere to Aristotle's doctrine of gravitation (p. 60). If Copernicus was famous at Rome in 1500 (p. 56), and if Rome desired his help in reforming the calendar (p. 57), how can Rome have first heard of his astronomy only toward the end of his life (p. 58)?
62. Zur Schreibweise "Coppernicus." *Zeitschrift für die Geschichte und Altertumskunde Ermlands*, 1939–1942, *27*: 260–261. Advocates "Coppernicus" as the spelling of the astronomer's surname in German.
63. Das Copppernicus-Grab im Dom zu Frauenburg. *Op. cit.*, pp. 273–281. The exact spot where Copernicus was buried in the Frauenburg cathedral is not known.
64. Zur kunde der Coppernicus-Bildnisse. *Op. cit.*, pp. 281–286. The portraits of Copernicus collected for various recent exhibitions helped somewhat to answer the question "What did the great astronomer really look like?"
65. Die Sternwarte des Coppernicus in Frauenburg. *Op. cit.*, pp. 338–366. A description and historical sketch of the room in which Copernicus made most of his observations.
66. Des Coppernicus Dienst im Dom zu Frauenburg. *Op. cit.*, pp. 568–591. Copernicus' participation in religious ceremonies.
67. Die Domburg in Frauenburg zur Zeit des Coppernicus. *Op. cit.*, 1943, *28*: 43–46. A historical sketch of the fortified tower in which Copernicus constructed his observatory at Frauenburg (now Frombork).
68. Zur Würdigung des Coppernicus in der 2. Hälfte des 16. Jahrhunderts in Frauenburg. *Op. cit.*, pp. 47–52. This article, left unfinished at the time of its author's death on February 26, 1942,

collects evidence showing that Copernicus' memory was honored in the second half of the 16th century in the town where he had his principal domicile during the last three decades of his life.

A list of Brachvogel's writings about Copernicus from 1912 until his death in 1942 forms the first section of Franz Buchholz's Schriftenverzeichnis von Eugen Brachvogel, *op. cit.*, 1943, *28*: 29–31.

69. Nikolaus Kopernikus in der Entwicklung des deutschen Geisteslebens, at pp. 33–99 in 370. Copernicus arrived at heliocentrism independently, like a good German. As a Nordic, he was markedly affected by the thought of Plato. Modern astronomy is an achievement of German minds (like Galileo, Newton, and Laplace). Reviewed **125.**

BRASCH, FREDERICK E.
70. The First Edition of Copernicus' De revolutionibus. *Library of Congress Quarterly Journal of Current Acquisitions*, 1945–1946, *3*: no. 3, pp. 19–22. Announcing the acquisition by the Library of Congress of a copy of the first ed. of the *Revolutions*. Reviewed **491.**

BRINTON, CRANE et al.
71. A History of Civilization (New York, 1955) 2 vols. *Modern Civilization* (Englewood Cliffs, 1957) p. 868. In the 1955 ed. of this textbook Copernicus is discussed at I, 485–486, and in the 1957 ed. at pp. 68–69.

BRODETSKY, SELIG
See Copernicus, translations: *De revolutionibus*, preface and Book I (126).

BROWN, G. K.
72. Copernicus. *Inquirer*, 1943, July 3, pp. 198–199. Mistakenly keeps Copernicus in Italy from 1506 to 1512. Why does Brown say that Clement VII requested Cardinal Schönberg "to command full publication by Copernicus of his views"?

BROŻEK, LUDWIK
See Copernicus, bibliography (111).

BROŻEK, MIECZYSŁAW
See Copernicus, editions and translations: *O obrotach sfer niebieskich księga pierwsza* (120).

BRÜCHE, ERNST
73. Kopernikus-Jahr in Polen. *Physikalische Blätter*, 1954, *10*: 32–33. A report on the Copernicus celebrations in Poland in 1953, especially those on September 15–16.

BRUGGENCATE, PAUL TEN
74. Zum 400. Todestage von Nikolaus Copppernicus. *Vierteljahrsschrift der astronomischen Gesellschaft*, 1943, *78*: 6–17. Published also in *Göttingische gelehrte Anzeigen*, 1943, *205*: 167–177. This address, delivered on May 24, 1943 at the University of Göttingen, recalls Lichtenberg's famous utterance: "As long as the earth stood still, astronomy stood still." Reviewed **863, 871.**

BRUNNER, WILLIAM
75. Nikolaus Kopernikus als Reformator der Sternkunde. *Vierteljahrsschrift der naturforschenden Gesellschaft in Zürich*, 1943, *88*: 81–98. This expansion of a lecture delivered to the Naturforschende Gesellschaft in Zürich on February 22, 1943 is an excellent survey of Copernicus' reform of astronomy, but unfortunately misdates the composition of the *Commentariolus*.

76. Pioniere der Weltallforschung (Zürich, 1951) p. 296. Copernicus and his astronomy are discussed in chapter 4, pp. 67–84. His final departure from Italy is postponed three years, and from his uncle's residence two years. Why does Brunner say that the Gregorian calendar accepted Copernicus' determination of the length of the year?

BUCHAR, EMIL
77. Zpráva o koperníkových oslavách ve Varšavě. *Věstník československé akademie věd*, 1953, *62*: 292–293. A report by one of the Czechoslovak delegates to the Copernicus celebration at Warsaw on September 15–16, 1953.

BUCHHOLZ, FRANZ
78. Coppernicus als Münzsachverständiger. *Jomsburg*, 1942, *6*: 143–148. A

review of Hans Schmauch's article "Nikolaus Coppernicus und die preussische Münzreform," and of Emil Waschinski's article, "Des Astronomen Nikolaus Coppernicus Denkschrift zur preussischen Münz- und Währungsreform."

79. Kopernikusforscher Eugen Brachvogel. *Jomsburg*, 1942, *6*: 309–311. A eulogy of Eugen Brachvogel, the Copernicus scholar, who died on February 26, 1942.
Reviewed **673, 833**.

Buck, Robert W.
80. Doctors Afield: Nicolaus Coppernic. *New England Journal of Medicine*, 1954, *250*: 954–955. Mistakenly postpones Copernicus' study of medicine until after 1503, the year in which he actually left Italy and its universities.

Büttner, Wolfgang
81. Ein Coppernicus-Institut. *Das Weltall*, 1939, *39*: 61–62. A report of the speech delivered by August Kopff on the occasion when the Recheninstitut, of which he was the director, was officially re-named the "Coppernicus-Institut."

Burmeister, Otto
82. Ueber die Weiterentwicklung des kopernikanischen Weltbildes. *Natur und Kultur*, 1942, *39*: 162–163. To explain why the heavenly bodies move as they do, Burmeister wants to abandon the ideas in vogue since Newton's time, and return to the view, which he thinks he finds in Copernicus, that the universe is a simple organism.

Burns, Edward McNall
83. *Western Civilizations* (New York, 1941) xx + 926 p. Brief treatment of Copernicus at p. 404.

Burtt, Edwin Arthur
84. *The Metaphysical Foundations of Modern Physical Science* (London, 1956) xi + 343 p. (paperback, Garden City, 1954). These re-issues of the revised 1932 ed. of a work first published in 1924 still misdate the composition of the *Commentariolus* in

1530, and alter Heath's baptismal name to Robert. Why does Burtt say that Copernicus had "become convinced that the whole universe was made of numbers"? Copernicus believed "the new hypothesis that the earth is a planet revolving on its axis and circling round the sun . . . to be a true picture of the astronomical universe" (p. 23); "to Copernicus' mind the question was not one of truth or falsity, not, does the earth move?" (p. 39).

Butterfield, Herbert
85. *The Origins of Modern Science* (London, 1949) x + 217 p.; reprinted New York, 1951, p. 187; revised ed., London and New York, 1957, p. 242. At pp. 22–31 of the earlier ed. (pp. 24–33 of the revised ed.) Butterfield discusses Copernicus' modifications of Aristotelian physics and Ptolemaic astronomy.
Reviewed **402**.

Cardús, J. O.
86. En el centenario de Copernico. *Razón y Fe*, 1944, *129*: 118–132. Mistakenly installs Copernicus as a professor at the University of Rome, and brings him home from Italy two years too late.

Carnegie Magazine, 1943–1944, *17*: 9–11
87. The Founder of Modern Astronomy. This unsigned article mistakenly asserts that Copernicus "took a degree in medicine at Cracow," "accepted the chair of astronomy at the University of Rome, remaining there for four years," and "entered the priesthood."

Caspar, Max
88. Nikolaus Kopernikus. *Kant-Studien*, 1943, *43*: 450–467. Opposes various attempts to find supposed metaphysical influences on Copernicus' thinking, which was essentially mathematical, astronomy in his time being a branch of mathematics. Copernicus' universe was immense, not infinite.
89. *Kopernikus und Kepler* (Munich and Berlin, 1943) p. 77. Reviewed by Richard Sommer, *Das Weltall*, 1943, *43*: 160; by Bernhard Sticker, *Die Himmelswelt*, 1943, *53*: 72; by August Kopff, *Vierteljahrsschrift*

ANNOTATED COPERNICUS BIBLIOGRAPHY

der astronomischen Gesellschaft, 1943, *78*: 207–208; and by Maximilian Krafft, *Zentralblatt für Mathematik und ihre Grenzgebiete*, 1943–1944, *28*: 99. This brochure takes its title from the first (pp. 9–45) of the two lectures printed in it. The first lecture was delivered on December 10, 1941 to the Kaiser-Wilhelm-Gesellschaft in Berlin, and was published at pp. 160–204 in the *Jahrbuch 1942 der Kaiser-Wilhelm Gesellschaft zur Forderung der Wissenschaften* (Leipzig, s.a.) 295 p. In it Caspar maintained that the progress from Ptolemy to Copernicus was less than the progress from Copernicus to Kepler. An appendix (pp. 42–45) compares the Ptolemaic and Copernican theories of the outer planets.

90. Kopernikus. *Forschungen und Fortschritte*, 1943, *19*: 166–167. Translated into Spanish as "Copérnico," *Investigación y progreso*, 1943, *14*: 87–90. An appreciation of Copernicus on the four-hundredth anniversary of his death. Reviewed **250, 287, 357, 370, 393, 555, 863 871.**

CECCHINO, GINO
91. *Il cielo* (Turin, 1952) xx + 1147 p. Correctly emphasizes (pp. 14–18) that in Copernicus' view the motion of the earth was a physical reality.

CHAŁASIŃSKI, JÓZEF
92a. Pierwsze polskie wydanie kopernika "O obrotach ciał niebieskich." *Nauka polska*, 1953, *1*: no. 2, pp. 175–198. A study of the first Polish ed. (Warsaw, 1854) of the *Revolutions*.
92b. Problem renesansu i humanizmu w Polsce. *Przegląd nauk historycznych i społecznych*, 1953, *3*: 28–78. This discussion of the Renaissance and humanism in Poland deals with Copernicus at pp. 28–52.

CHALUPCZYNSKI, MIECZYSŁAW
93. In a lecture delivered on May 24, 1943 and published at pp. 7–10 in 108, the Polish ambassador to Colombia mistakenly referred to manuscripts in which Copernicus described himself as a Pole, and to the removal of the *Revolutions* from the Index before Copernicus' death, which

occurred before the *Revolutions* was put on the Index.

CHAMCÓWNA, MIROSŁAWA
94. Studia Jana Śniadeckiego nad życiem i dziełem Mikołaja Kopernika. *Studia i materiały z dziejów nauki polskiej*, 1953, *1*: 80–101. Jan Śniadecki's study of the life and work of Copernicus. See Śniadecki (729).

CHANT, CLARENCE AUGUSTUS
95. The Ptolemaic and Copernican Theories. *Journal of the Royal Astronomical Society of Canada*, 1939, *33*: 166–167. The father of Canadian astronomy points out that the replacement of the Ptolemaic by the Copernican system took a long time.
96. The 400th Anniversary of the Death of Copernicus. *Op. cit.*, 1943, *37*: 74. Reviewed **126, 493.**

CICHOWICZ, LUDOSŁAW
97. Instrumentarium astronomiczne Kopernika. *Przegląd geodezyjny*, 1954, *10*: 215–217. A description of the Copernicus exhibition at the Polytechnic Institute of Warsaw, with special reference to the astronomer's instruments.
98. Jeszcze raz w sprawie "warszawskich poprawek" do rekonstrukcji instrumentów kopernikowskich. *Urania* (Kraków), 1954, *25*: 190–191. A comment on Przypkowski's letter, *op. cit.*, 1954, *25*: 128–129.
99. Po raz ostatni w sprawie uwag dra T. Przypkowskiego o wystawie kopernikańskiej na Politechnice Warszawskiej. *Urania* (Kraków), 1955, *26*: 154–155. A final comment on Przypkowski.

CLARK, JOSEPH T.
100. The Philosophy of Science and the History of Science. A lecture which was delivered on September 3, 1957 to the Institute of the History of Science at the University of Wisconsin and which will be published in the proceedings of that Institute. Section II finds a contradiction in Copernicus between his assertions that a sphere moves circularly and that the sphere of the fixed stars does not move.

CLEMINSHAW, CLARENCE H.
101. Copernicus Moves the Earth. *Griffith Observer*, 1943, 7: 54–61. An admirable little essay on Copernicus' astronomical achievement, particularly in explaining the motion of the planets and the precession of the equinoxes.

CLOUGH, SHEPARD B.
102. *The Rise and Fall of Civilization* (New York, 1951) xiii + 291 p.; 2d ed., New York, 1957. In his brief discussion of Copernicus (p. 210) Clough commits three errors: Copernicus was born in Toruń, not Kraków; he placed the sun at the center of the universe, not of the solar system; and he did not get "much of his training in Nuremberg."

CODINO, G.
103. *Le Monde*, 1954, February 26, p. 8. A letter contending that Copernicus was a German, and using the argument that his surname was "purely German."

COLEMAN, MARION MOORE
104. Copernicana at Columbia University. *Polish Review*, 1943, 3: January 11, pp. 4, 14.

Commemoration of Chu Yuan, Nicolaus Copernicus, François Rabelais, José Martí (Peking, 1953) p. 41
105. "Commemoration of Nicolaus Copernicus" by Chu Ke-chen (pp. 17–21); Kuo Mo-jo (pp. 4–5) and Wojciech Żukrowski also emphasize the revolutionary character of Copernicus' thought.

COMMINS, SAXE and LINSCOTT, ROBERT N., edd.
106. *The World's Great Thinkers*, vol. 4. *Man and the Universe: The Philosophers of Science* (New York, 1947). A portion of the *Revolutions* in Wallis' translation is reprinted at pp. 43–69.

CONWAY, PIERRE
107. Aristotle, Copernicus, Galileo. *New Scholasticism*, 1949, 23: 38–61, 129–146. Calls Thomas Heath "Keith" (p. 39). Transfers Copernicus' observatory to the cathedral (p. 40). Changes Rheticus' given name to Johannes (pp. 42, 44, 56, 57).

Misstates the relation between Copernicanism and the Gregorian calendar (p. 46). Alters the place where the *Narratio prima* was printed to Nuremberg (p. 56). Misdates Tycho Brahe's birth by seventy years, and mislabels him Copernicus' "continuator in the gradual establishment of the Copernican system" (p. 59). Mistakenly asserts that a corrected ed. of the *Revolutions* appeared in 1621 (p. 131). Why does Conway say that the *Revolutions* was received by Pope Paul III "with pleasure" (p. 56), and that · Calvin was anti-Copernican (p. 58)?

108. *Copérnico, Nicolás* (Bogotá, 1943) p. 27. In commemoration of the Copernicus quadricentennial, five lectures were delivered on May 24, 1943 by Rafael Parga Cortés, Mieczysław Chalupczynski, Belisario Ruiz Wilches, Guillermo Hernandez de Alba, and Luis de Zulueta, *q.v.*

COPERNICUS,
BIBLIOGRAPHY:

BARANOWSKI, HENRYK
109. Mikołaj Kopernik—ekonomista. *Ekonomista*, 1953, no. 4, pp. 251–255. Bibliography of Copernicus as an economist.

110. *Bibliografia kopernikowska 1509–1955* (Warsaw, 1958) p. 449. Reviewed by Edward Rosen, *Isis*, 1958, 49: 458–459. The most comprehensive bibliography of Copernicus.

BROŻEK, LUDWIK
111. *Bibliografia kopernikowska, 1923–1948* (Poznań, 1949) p. 30; supplement of *Przegląd zachodni*, 1949, 5: no. 7–8. Reviewed by Alfons Triller, *Zeitschrift für die Geschichte und Altertumskunde Ermlands*, 1956–1957, 29: 158–159. A useful collection of over 300 entries.

DEIKOVA, N. N.
112. Vistavka Nikolai Kopernik, at pp. 91–111 in Kukarkin, ed.

KOSSMANN, EUGEN OSKAR and ZIEMSEN, I.
113. *Uebersicht über das deutsche Kopernikusschrifttum* (Berlin, 1943) 18 mimeographed pp. Lists 208 publications

ANNOTATED COPERNICUS BIBLIOGRAPHY

concerning Copernicus, most of which are in German, while some are in French or Italian.

KUBACH, FRITZ
114. Kleine Kopernikus-Bibliographie, at pp. 286–304, 374–375 in Kubach, ed. (393). A brief bibliography of the secondary literature in German about Copernicus.

LEŚNODORSKI, BOGUSŁAW
115. Mikołaj Kopernik, at pp. 184–191 in *Bibliografia literatury polskiej okresu Odrodzenia* (Warsaw, 1954). The section on Copernicus in a general bibliography of Polish literature of the Renaissance period.

WOŹNOWSKI, MIECZYSŁAW
116. Literatura kopernikowska. *Wszechświat*, 1953, pp. 209–212, 252.

See Diergart (153); Hoff (287); Poznański; and Zeller, Mary Claudia (865).

COPERNICUS, EDITIONS:

117. A facsimile reprint of the first ed. of the *Revolutions* was produced by Vincenzo Bona at Turin in 1943; 200 copies were released through Chiantore in Turin, and 100 copies through Roskam in Amsterdam.
118. *Nikolaus Kopernikus Gesamtausgabe.* Vol. I (Munich and Berlin, 1944) xiv + 212 + xxiv p. A photographic facsimile of the original manuscript of the *Revolutions* as written down by Copernicus himself. The editor, Fritz Kubach, provides a brief introduction (pp. ix–xiv). In an appendix (pp. iii–xxiv) Karl Zeller discusses the physical characteristics of the original manuscript, its history from Copernicus' times to our own, and the process by which it was reproduced in photographic facsimile.
119. Vol II, edd. Franz Zeller and Karl Zeller (Munich, 1949), p. 470. A critical text of the *Revolutions* (pp. 3–404); a discussion of the earlier edd. and of Copernicus' style (pp. 405–430); notes on the text of the *Revolutions* (pp. 431–454); indices (pp. 457–470). The editors' remarks are written in Latin, the quality of which is sharply condemned in K. G. Hagstroem's review, *Lychnos*, 1952, pp.

395–397 (in Swedish). Vol. II was reviewed also by August Kopff, *Astronomischer Jahresbericht*, 1949, *49*: 26; by Heribert Schneller, *Astronomische Nachrichten*, 1950–1951, *279*: 143; and by Hans Schimank, *Physikalische Blätter*, 1952, *8*: 43–44. Both volumes were reviewed rather severely by Ernst Zinner, *Die Sterne*, 1950, *26*: 95–96, and by Franz Hammer, *Deutsche Literaturzeitung*, 1951, *72*: 370–376. See Schmeidler.

COPERNICUS, EDITIONS AND TRANSLATIONS:

120. *O obrotach sfer niebieskich księga pierwsza* (Warsaw, 1953) p. 118 (On the Revolutions of the Heavenly Spheres, Book I). The Latin text of the *Revolutions*, Book I, chapters 1–11 (pp. 15–42), edited by Ryszard Gansiniec; translation into Polish (pp. 43–75) by Mieczysław Brożek; footnotes (pp. 81–119) by Aleksander Birkenmajer, who also supervised the volume.
Reviewed by Jan Zygmunt Jakubowski, *Życie szkoły wyższej*, 1953, *1*: no. 11, pp. 145–146, and by Feliks Rapf, *Urania* (Kraków), 1954, *25*: 31–33. Brief notice by George Sarton, *Isis*, 1955, *46*: 279. Pp. 5–12 of Birkenmajer's preface were translated into French by Allan Kosko at pp. 134–140 in Wędkiewicz.

ROSSMANN, FRITZ
121. *Nikolaus Kopernikus, Erster Entwurf seines Weltsystems* (Munich, 1948) p. 100. In parallel columns Rossmann reprints the Latin text of the *Commentariolus*, accompanied by his own German translation.
Reviewed by Johann Wempe, *Astronomische Nachrichten*, 1949, *277*: 273; by Edward Rosen, *Archives internationales d'histoire des sciences*, 1950, *3*: 700–703; by Nikolaus B. Richter, *Naturwissenschaftliche Rundschau*, 1950, *3*: 232; by Felix Schmeidler, *Sternenwelt*, 1950, *2*: 120; and by Joseph Ehrenfried Hofmann, *Zentralblatt für Mathematik und ihre Grenzgebiete*, 1951, *38*: 146.
122. Rossmann published extracts from 121 in "Der Commentariolus von Nikolaus Kopernikus," *Die Naturwissenschaften*, 1947, *34*: 65–69.

123. Teofilakt Symokatta, *Listy* (Warsaw, 1953) xix + 173 p. Facsimile ed. of Copernicus' Latin translation of Theophylactus Simocatta's *Letters*, together with the Greek text and a Polish translation. The Latin and Greek texts were edited by the late professor of classical philology at the University of Kraków, Ryszard Gansiniec, who also wrote a discussion of Theophylactus Simocatta by way of introduction. The Polish translation was done by Jan Parandowski, who has translated the *Odyssey* as well as *Daphnis and Chloe* into Polish. Ludwik Hieronim Morstin rendered into Polish the complimentary poem by Corvinus.

Gansiniec opposes the conjecture (most recently supported by Ivan Ivanovich Tolstoi; see Mikhailov, ed.) that Copernicus did not use the Greek text published by Aldus Manutius at Venice in 1499. Gansiniec's analysis of Copernicus' Latin style was translated into French by Allan Kosko at pp. 208–209 in Wędkiewicz.

Reviewed by Jan Zygmunt Jakubowski, *Życie szkoły wyższej*, 1953, *1*: no. 11, p. 146, and by Jan Gadomski, *Urania* (Kraków), 1956, *27*: 26. Brief notice by George Sarton, *Isis*, 1955, *46*: 279.

COPERNICUS, TRANSLATIONS:

124. Nicolaus Copernicus, *On the Revolutions of the Heavenly Spheres*, translated by Charles Glenn Wallis in *Great Books of the Western World* (Chicago, 1952), vol. XVI, pp. 479–838. In reviewing this work, Otto Neugebauer, the distinguished historian of astronomy, said: "In general Mr. Wallis' translation is often not much more than a simple replacement of Latin words by English words with very little regard for the sense. . . . The translation frequently requires retranslation to be intelligible. Often one may wonder whether the translator grasped the meaning of the sentences. . . . a truly dilettante attempt" (*Isis*, 1955, *46*: 69–71).

MENZZER, CARL LUDOLF
125. *Nicolaus Coppernicus, Ueber die Kreisbewegungen der Weltkörper* (Leipzig, 1939) xvi + 363 pages of text + 66 pages of notes. A photographic facsimile of the German translation of the *Revolutions* first published at Thorn in 1879. A new preface (pp. v–x) was provided by Josef Hopmann, director of the University Observatory at Leipzig. Many errors in Hopmann's preface as well as in Menzzer's translation were corrected in *Jomsburg*, 1940, *4*: 233–239, by Eugen Brachvogel, who wrote another review in *Zeitschrift für die Geschichte und Altertumskunde Ermlands*, 1939–1942, *27*: 462–463. Other reviews were published in *Coelum*, 1939, *9*: 140 (unsigned); by Hans Ludendorff, *Die Naturwissenschaften*, 1939, *27*: 808; by Bernhard Sticker, *Die Himmelswelt*, 1940, *50*: 60–61; by Heribert Schneller, *Die Sterne*, 1940, *20*: 23; by August Kopff, *Vierteljahrsschrift der astronomischen Gesellschaft*, 1940, *75*: 7; and by Fritz Kubach, *Zeitschrift für die gesamte Naturwissenschaft*, 1940, *6*: 88–89.

Two excerpts from Menzzer's translation were reprinted in *Physikalische Blätter*, 1953, *9*: 145–148.

126. *De revolutionibus*, preface and Book I, translated by John F. Dobson, assisted by Selig Brodetsky, with some notes and a brief biographical sketch. *Royal Astronomical Society, Occasional Notes*, vol. 2, no. 10, pp. 1–32 (London, 1947; reprinted, 1955).

Reviewed by C. A. Chant, *Journal of the Royal Astronomical Society of Canada*, 1948, *42*: 239, and by G. E. Pettengill, *Journal of the Franklin Institute*, 1948, *246*: 178.

127. Zeller, Karl in Kubach, ed. (393). Excerpts from the *Commentariolus* (pp. 27–30), the Latin version of the paper on coinage (p. 32), and the *Revolutions* (pp. 33–60).

ZINNER, ERNST
128. *Astronomie: Geschichte ihrer Probleme* (Freiburg and Munich, 1951) p. 404. Contains a German translation of the *Commentariolus* (pp. 60–73), and of excerpts from the *Revolutions* (dedication, preface, chapters 1–10 of Book I, and chapter 1 of Book V; pp. 73–103).

PETROVSKY, FYODOR ALEXANDROVICH
129. Ob obrashcheniyach nebesnich sfer, at pp. 187–213 in Mikhailov, ed. (483).

TAYLOR, JACK
130. Copernicus on the Evils of Infla-tion. *Journal of the History of Ideas*, 1955, *16*: 540–547. A translation into English of the main passages in Copernicus' Latin trea-tise *Monetae cudendae ratio*, the first purely empirical and pragmatic treatment of the problem caused by the debasement of a metal currency. See Amzalak.

CORRADI, SILVIO
131. *La terra non gira intorno al sole* (Cremona, 1952) p. 91. Enlarged and translated from Italian into French by the author himself under the title *La terre ne tourne pas autour du soleil* (Cremona, 1954) p. 113. A self-taught successful farmer con-tends that the earth rotates on its axis, but does not revolve around the sun.

CORRINGTON, JULIAN D.
132. Four Hundred Years of Research. *Nature Magazine*, 1943, *36*: 555–556. A quadricentennial salute to Copernicus and Vesalius.

COUDERC, PAUL
133. *Les étapes de l'astronomie* (Paris, 1948; Que sais-je?, no. 165); 3d ed. (Paris, 1955) p. 128. Translated by Armando Silvestri into Italian under the title *Le tappe dell' astronomia* (Milan, 1952) p. 119.
Couderc, an astronomer at the Paris Observatory, devotes to Copernicus pp. 76–87 (pp. 70–80 in the Italian translation). With admirable clarity Couderc demon-strates the superiority of Copernicus' planetary theory to Ptolemy's, especially in opening the way to an accurate de-termination of the planetary distances.

COX, JACQUES F.
134. L'oeuvre de Copernic. *Ciel et terre*, 1954, *70*: 316–317. Copernicus ended the anthropocentric view of the universe.

CROMBIE, ALISTAIR CAMERON
135. *Augustine to Galileo* (London, 1952; Harvard University Press, 1953; 2d ed.,

London, 1957). Copernicus is discussed at pp. 308–313.

CROWTHER, JAMES GERALD
136. *Six Great Scientists* (London, 1955) p. 269. The first of Crowther's six great scientists is Copernicus (pp. 13–47), who is discussed in a lively and spirited manner.

CSORBA, HELENA
137. Wystawy kopernikowskie. *Nauka polska*, 1953, *1*: no. 3, pp. 151–153. Copernican exhibitions in Poland.
138. Z wydawnictw roku kopernikow-skiego i roku Odrodzenia. *Op. cit.*, no. 4, pp. 172–178. Exhibitions in the Coper-nican and Renaissance Year, 1953.

CUMME, HAIMAR
139. Der Kampf um das heliozen-trische Weltsystem. *Wissenschaftliche Zeit-schrift der Universität Greifswald*, Gesell-schafts- und sprachwissenschaftliche Reihe, 1953–1954, *3*: 75–81; published also in *Junge Universität*, Monatschrift der Uni-versität Greifswald, 1953–1954, pp. 21–27. A lecture delivered in commemoration of Copernicus.

DAISOMONT, MAURICE
140. *Copernicus: astronomische Sprokke-lingen* (Bruges, 1943) p. 87. A reprint in book form of popular discussions of astro-nomy, which originally appeared in the Flemish newspaper *Nieuws van den Dag*. Copernicus is mistakenly described as a priest.

DAMPIER, WILLIAM CECIL
141. *A History of Science and its Relations with Philosophy and Religion*, 3d ed. (Cam-bridge, 1942) xxiii + 574 p.; 4th ed. (Cambridge, 1948) xxvii + 527 p. A re-vised and enlarged version of a work first printed in 1929. The 4th ed. was translated into French by René Sudre under the title *Histoire de la science et de ses rapports avec la philosophie et la religion* (Paris, 1951) p. 601; into German by Felizitas Ortner under the title *Geschichte der Naturwissenschaft in ihrer Beziehung zu Philosophie und Weltanschauung* (Vienna and Stuttgart, 1952) p. 615; and

into Italian by Luigi A. Radicati di Bròzolo under the title *Storia della scienza* (Turin, 1953) p. 750.

Two errors in the discussion of Copernicus (4th ed., pp. 109–113) were retained from the earlier version: the *Commentariolus* was circulated long before 1530; it was not Pope Clement VII, but Cardinal Schönberg, who requested Copernicus to publish the *Revolutions*. A new blunder was added: since Copernicus regarded space as enclosed by a sphere, how can his space be infinite?

142. *A Shorter History of Science* (Cambridge and New York, 1944) x + 189 p. Paperback reprint (New York, 1957). Translated into German by Ludwig von der Pahlen under the title *Kurze Geschichte der Wissenschaft* (Zürich, 1946) xi + 288 p.

In the English version the discussion of Copernicus occurs at pp. 49–52.

DAVIES, JOHN D. GRIFFITH
143. Pathfinders of Science. *The Listener*, 1943, *30*: 18–19. A lecture on the Home Service program of the British Broadcasting Corporation.

DERMUL, AMÉDÉE
144. Catalogues d'étoiles. *Gazette astronomique*, 1950, *32*: 53–61, 73–78; 1951, *33*: 1–9, 25–33, 45–50, 81–92, 105–114; 1953, *35*: 21–29; 1954, *36*: 17–22, 37–52, 61–71. Copernicus' star catalogue is discussed in 1951, *33*: 3–5.

DIANNI, JADWIGA
145. Pobyt J. J. Retyka w Krakowie. *Studia i materiały z dziejów nauki polskiej*, 1953, *1*: 64–79. Rheticus' sojourn at Kraków, where he served the king of Poland as royal physician.

146. Matematyka Kopernika. *Wszechświat*, 1954, no. 2, pp. 31–35. Copernicus as a mathematician.

DIANNI, JADWIGA and WACHUŁKA, ADAM
147. *Z dziejów polskiej myśli matematycznej* (Warsaw, 1957) p. 140. Chapter IV of this history of Polish mathematics deals with Copernicus' trigonometry.

DICK, JULIUS
148. Zwei unbekannte Entwürfe Gottfried Schadows zu einem Kopernikus-Denkmal. *Die Sterne*, 1952, *28*: 177–181. Two models for Copernicus monuments by the nineteenth-century sculptor Johann Gottfried Schadow, and a watercolor, done in 1847 by Eduard Gärtner, of the room in which Copernicus was then believed to have been born.

149. Nikolaus Kopernikus' De revolutionibus. *Wissenschaftliche Annalen*, 1953, *2*: 450–458. Copernicus as the intermediary between Ptolemy and Newton.

DIERGART, PAUL
150. Koppernick. *Mitteilungen zur Geschichte der Medizin, der Naturwissenschaften und der Technik*, 1939, *38*: 201. Erroneously describes some pious verses, which were first linked with Copernicus some four decades after his death, as being "his epitaph, chosen by himself."

151. Ed. *Proteus der rheinischen Gesellschaft für Geschichte der Naturwissenschaft, Medizin und Technik*, 1940–1943, *3*: no. 4, pp. 93–151.

Fritz Kübach, "Das Werk von Nikolaus Kopernikus' (pp. 95–96); Viktor Stegemann, "Der griechische Philosoph und Astronom Hiketas von Syrakus als Nicetas (-us) bei Kopernikus und Giordano Bruno" (pp. 97–99); B. L. van der Waerden, "Die Vorgänger des Kopernikus im Altertum" (pp. 100–104); Karl Zeller, "Der Forschungsweg des Nikolaus Kopernikus" (pp. 104–110); Joseph Schumacher, "Die griechischen und deutschen Elemente im kopernikanischen Denken" (pp. 110–115); Hans Schmauch, "Neues über die ärztliche Tätigkeit des Astronomen Kopernikus" (pp. 115–119); Josef Hopmann, "Die Lehre des Kopernikus (1543) bis zum Abschluss durch F. W. Bessel (1838)" (pp. 119–120); Bernhard Sticker, "Ursprung und Vollzug der kopernikanischen Wende" (pp. 121–126); Leopold von Wiese, "Das Selbstbewusstsein des Menschen und das kopernikanische Weltbild" (pp. 126–130).

152. Paul Diergart, "Stichworte zur Linie 'Philolaos' aus Kroton über Aristarch von Samos zu Kopernikus und Friedrich Wilhelm Bessel" (pp. 130–135).

153. Paul Diergart, "Gedanken über neueres Kopernikus-Schrifttum" (pp. 135–146).

Reviewed by Bernhard Sticker, *Die Himmelswelt*, 1944, *54*: 11. Each article listed above is annotated under its author. Diergart's "Stichworte" outlined the key developments in astronomy from the ancient Babylonians through Copernicus to Bessel in a sketch which Diergart himself recognized to be in need of correction. His "Gedanken" reviewed recent literature about Copernicus.

DIETRICH, STANISŁAW

154. Gdzie znajduje się grób Mikołaja Kopernika? *Problemy*, 1954, *10*: 65. Where is Copernicus' grave?

DIETZ, DAVID

155. *The Story of Science*, 4th ed. (New York, 1942) xv + 387 p. Copernicus is discussed briefly at pp. 6–7 in this fourth ed. of a work published originally in 1931. The fourth ed. was translated into Italian by Giannetto Barrera under the title *La storia della scienza* (Rome, 1946) p. 377.

DIJKSTERHUIS, EDUARD JAN

156. Van Coppernicus tot Newton, at pp. 111–152 in *Antieke en moderne Kosmologie* (Arnhem, 1941). From the kinematic astronomy of Copernicus to the celestial mechanics of Newton.

157. Coppernicus en zijn Boek. *De Gids*, 1943, *107*: no. 5, pp. 61–78. Finds contradictory elements in Copernicus' thought.

158. *De Mechanisering van het Wereldbeeld* (Amsterdam, 1950) xiii + 590 p. Translated into German by Helga Habicht-Van der Waerden under the title *Die Mechanisierung des Weltbildes* (Berlin, 1956) vii + 594 p.

Part IV, chapter 1, section A, discusses Copernicus, whose diametrical opposition to Osiander's fictionalism is strangely turned into an agreement.

159. *Het Wereldbeeld vernieuwd* (Arnhem, 1951) p. 65. Copernicus is discussed at pp. 11–22.

Reviewed **812.**

DINGLE, HERBERT

160. Copernicus. *Spectator*, 1943, *170*: 471–472.

161. The Work of Copernicus. *Nature*, 1943, *151*: 576–577, 613. Report of an address delivered to the Royal Astronomical Society on May 14, 1943. The address itself was printed as Copernicus' Work, a Landmark in Scientific History,

162. in *Polish Science and Learning*, 1943 (June), no. 3, pp. 24–39, and also

163. in *Observatory*, 1943, *65*: 38–57; it was reprinted as Chapter III (pp.

164. 58–83) in Dingle's *The Scientific Adventure* (London, 1952; New York, 1953) p. 372.

165. Nicolaus Copernicus. *Endeavour*, 1943, *2*: 136–141. An excellent brief account, marred by a few minor slips.

166. Copernicus and the Planets, at pp. 35–44 in *The History of Science*: *Origins and Results of the Scientific Revolution* (London, 1951) p. 184. A lecture delivered on the B.B.C.

At pp. 46–47 in *The Scientific Adventure* Dingle emphasizes Copernicus' conservatism in a lecture delivered on February 12, 1944

167. under the title "Astronomy in the Sixteenth and Seventeenth Centuries," and reprinted in *Science, Medicine and History: Essays . . . in Honour of Charles Singer* (Oxford University Press, 1953), I, 455–468.

Why does Dingle say that "the Pope, Paul III, accepted the dedication of the work [*Revolutions*] to himself"? Reviewed **18.**

DOBRZYCKI, JERZY

168. Kształtowanie się założeń systemu kopernikowskiego. *Przegląd zachodni*, 1953, *9*: no. 3, pp. 571–587. The construction of the foundation of Copernicus' system.

DOBSON, JOHN F.

See Copernicus, translations: *De revolutionibus*, preface and Book I (126).

DOIG, PETER

169. *A Concise History of Astronomy* (London, 1950) xi + 320 p. Copernicus is discussed at pp. 49–53. Some of this book's defects were pointed out in a review by Edward Rosen, *Archives internationales d'histoire des sciences*, 1950, *3*: 941–942.

DOUGLAS, A. VIBERT

170. Copernicus. *Queen's Quarterly*, 1943, *50*: 146–154. Toruń is displaced to the mouth of the Vistula; Copernicus is assigned private pupils at Rome; he is said to have published accounts of astronomical instruments before describing them in the *Revolutions*; and poor Giordano Bruno's death by burning is advanced forty years to 1560, in the published text of a lecture delivered at McGill University in Montreal on March 26, 1943, and at Ottawa on April 9, 1943 (accounts of these lectures

171. are given in the *Journal of the Royal Astronomical Society of Canada*, *37*: 221–224).

DRAPER, ARTHUR L. and LOCKWOOD, MARIAN

172. *The Story of Astronomy* (New York, 1939) xi + 394 p. Copernicus is discussed mainly at pp. 96–99.

DREYER, JOHN LOUIS EMIL

173. *A History of Astronomy from Thales to Kepler* (New York, 1953) x + 438 p. A reprint of Dreyer's *History of the Planetary Systems from Thales to Kepler*, originally published at Cambridge, England, in

174. 1906. The foreword by William H. Stahl discusses the reasons for the Greek failure to continue along the direction taken by Aristarchus in anticipation of Copernicus. The latter is treated in Dreyer's Chapter XIII, pp. 305–344.

DUGAS, RENÉ

175. *Histoire de la mécanique* (Paris and Neuchatel, 1950) p. 649. Translated by

J. R. Maddox as *A History of Mechanics* (Neuchatel and New York, 1955) p. 671. Copernicus' geophysical ideas are discussed briefly at pp. 84–86, where there are two errors concerning his biography.

176. *La mécanique au XVIIe siècle* (Paris and Neuchatel, 1954) p. 620. Copernicus' arguments for the physical motion of the earth are discussed at pp. 37–38.

DUNAJEWSKI, HENRYK

177. W sprawie poglądów ekonomicznych Mikołaja Kopernika. *Ekonomista*, 1952, no. 4, pp. 226–229. Copernicus' economic views.

178. Kilka uwag w związku z artykułem "Wkład Kopernika w postępową myśl ekonomiczną." *Życie szkoły wyższej*, 1953, *1*: no. 12, pp. 76–80. Some comments on Hoszowski's article (*op. cit.*, no. 10, pp. 107–115).

179. Poglądy ekonomiczne Mikołaja Kopernika. *Kwartalnik historyczny*, 1953, *60*: no. 3, pp. 57–80 (summary in French at pp. 9–10). As an economist, Copernicus differed from his predecessors by examining monetary problems from the business point of view, rather than from a moral point of view.

180. *Mikołaj Kopernik—studia nad myślą społeczno-ekonomiczną i działalnością gospodarczą* (Warsaw, 1957) p. 467. An analysis of Copernicus' social and economic thought (summary in English at pp. 453–457).
See *Sesja kopernikowska*.
Reviewed **436**.

DUNGEN, FRANS H. VAN DEN

181. Copernic et son temps. *Ciel et terre*, 1954, *70*: 313–315. Intending to correct a statement in this article, a letter by M. Daisomont falsely asserts (*op. cit.*, p. 433) that Copernicus was a priest.

DURANT, WILL

182. *The Renaissance* (New York, 1953; The Story of Civilization, Part V) xvi + 776 p. Mistakenly asserts (p. 529) that Copernicus taught mathematics and astronomy in the University of Rome.

183. The Ten Greatest Thinkers. *Rotarian*, 1955, *86*: part 2, pp. 38–39, 90–93. Copernicus is No. 5 in the all-time hit parade of Will Durant, who at p. 90 mis-attributes to Copernicus the idea that the entire solar system is in motion.

184. *The Reformation* (New York, 1957; The Story of Civilization, Part VI) xviii + 1025 p. The discussion of Copernicus (pp. 855–863) contains, besides many minor errors, two major mistakes: Calvin never "answered Copernicus" nor did Melanchthon call him "that Prussian astronomer." Why does Durant say that Copernicus went to the University of Kraków "to prepare for the priesthood," and that "Pope Leo X . . . asked a cardinal to write to Copernicus"?

Eagle (yearbook of the students of SS. Cyril and Methodius Seminary, Orchard Lake, Michigan)

185. *Mikolaj Kopernik*, 38 p. (reprinted from the 1943 *Eagle*). A bilingual brochure in Polish and English which does not attempt to make any original contribution to the understanding of Copernicus.

EASTON, STEWART C.
186. *The Heritage of the Past from the Earliest Times to 1715* (New York, 1957) p. 845. Copernicus and Copernicanism are discussed at pp. 781–786.

EBERHARD, OTTO, ed.
187. *Zeugnisse deutscher Frömmigkeit von der Frühzeit bis heute* (Leipzig, 1940) xii + 458 p.; a re-issue of a work first published in 1938. Copernicus' epitaph, quoted at pp. 182–183 to exemplify his piety, is printed in such a way as to give the impression that these four Latin lines were written by the astronomer himself. Eberhard fails to explain that the quatrain was taken from a poet and first attached to Copernicus' grave long after his death.

EIS, GERHARD
188. Zu den medizinischen Aufzeichnungen des Nicolaus Copernicus. *Lychnos*, 1952, pp. 186–209. Copernicus' medical prescriptions were based on traditional opinion, not on experience.

ENGEL, LEONARD
189. Copernicus, Maker of the "New Astronomy." *Science Digest*, 1953, *33*: no. 3, pp. 86–90. Replete with ludicrous errors and internal inconsistencies, this essay was

190. reprinted in *Science Milestones* (Chicago and New York, 1954), pp. 32–36.

ENGELGARDT, MIKHAIL ALEKSANDROVICH
191. *Nikolaj Kopernik* (Sofia, 1946) p. 96. A Bulgarian translation by Georgi Kovačev of a Russian work originally published in 1892.

Engineering, 1943, May 21, pp. 413–414
192. Nicolaus Copernicus. This unsigned article mistakenly asserts that Copernicus "studied at Rome," where he lectured.

ENRIQUES, FEDERIGO and SANTILLANA, GIORGIO DE
193. *Compendio di storia del pensiero scientifico dall' antichità fino ai tempi moderni* (Bologna, 1946) vi + 481 p. A re-issue of the original ed. (Bologna, 1937). Copernicus is discussed mainly at pp. 315–318. Why do the authors say that Copernicus studied at Rome (p. 315) and that he believed an animal force stopped the planets from falling into the sun?

ENSERING, M.
194. Na vier Eeuwen van Rede. *Nederlandsch Tijdschrift voor de Psychologie*, 1949, *4*: 157–162. The Ptolemaic view of the universe was dominant until 1543, with religion as the basis of life; after 1543, the Copernican view prevailed, with science replacing religion; since 1943 the idealistic view has come to the fore, with wisdom supplanting science.

ERDMANN, FRANZ
195. Verwandler der Welt. *Der Deutsche im Osten*, 1943, *6*: 165–175. Brings Copernicus home from Italy two years too late, and also has him leave his uncle's residence two years too late. Transfers to Venice the place where Giordano Bruno was burned,

and has him utter the words "But it still moves," which an unhistorical legend usually assigns to Galileo.

ERHARDT, RUDOLF VON AND ERHARDT-SIEBOLD, ERIKA VON
196. Archimedes' Sand-Reckoner, Aristarchos and Copernicus. *Isis*, 1941–1942, *33*: 578–602. Copernicus may have been familiar with the reference in Archimedes' *Sand-Reckoner* to Aristarchus' heliocentric astronomy. The Erhardts also question Archimedes' authorship of the *Sand-Reckoner* by a series of arguments which were demolished by Otto Neugebauer, *Isis*, 1942–1943, *34*: 4–6.

ERLER, OTTO
197. Die Blutsfreunde. *Die Mittelstelle*, 1943 (May), *2*: no. 19, pp. 43–47. Copernicus near the end of his life, in a scene taken from a drama and first published here.

ESCALANTE, FRANCISCO
198. Nicolas Copérnico; su vida, su obra. *Memorias y revista de la Academia nacional de ciencias* (Mexico), 1935–1944, *55*: 281–302. Attacks Flammarion's assertion that Copernicus was ordained a priest by the bishop of Kraków.

EVERSHED, MRS. JOHN (Orr, Mary Acworth)
199. *Journal of the British Astronomical Association*, 1943, *53*: 159. Rheticus did for Copernicus what Halley did for Newton. Reviewed **491**.

FABER, MILLY
200. Nikolaus Kopernikus. *Die Messtechnik*, 1943, *19*: 97–98. This brief article contains the misstatements that Copernicus studied at Vienna, and that Cardinal Schönberg paid for the publication of the *Revolutions*.

FALK, MARYLA
201. Mikołaj Kopernik, at pp. 3–18 in Indo-Polish Association, *Quadricentennial Celebration of Nicholas Copernicus* (Calcutta, 1944).

FARRAR, STEWART
202. Poland's Greatest Scientist. *New Poland*, 1953, *8*: no. 7, pp. 4–5. A brief commemorative popular article.

FAUST, AUGUST
203. Die philosophiegeschichtliche Stellung des Kopernikus, at pp. 96–211, 318–370, in Kubach, ed. (**393**). A passionate dissertation·intended to prove that the history of German philosophy (including Copernicus') flowered in National Socialism.
204. Nikolaus Kopernikus. *Kant-Studien*, 1943, *43*: 1–52. Copernicus, like all true German philosophers, did not merely think, he also believed. Yet he cannot be expected to have acted in a way which only the Führer, Adolf Hitler, made possible later. He did not open a path to West European and North American positivism, pragmatism or liberalism. He moved the earth away from the center of the universe in order to bring it nearer to God. "Without German philosophy there would be no modern natural science" (p. 38). German philosophy must defend the coming European common culture from Bolshevism and Americanism with their equalitarian tendencies and mechanization of life.

FEDERAU, WOLFGANG
205. *Nikolaus Kopernikus* (Nuremberg, 1949) p. 177. Reviewed by A. Robl, *Sternenwelt*, 1952, *4*: 156. The military occupation authorized an ed. of 5000 copies of this well-written biography of Copernicus for teen-agers.

FESENKOV, VASILII GRIGOREVICH
206. Nikolai Kopernik i geliotsentricheskaia sistema mira. *Vestnik akademii nauk SSSR*, 1943, *13*: no. 6, pp. 17–24. Copernicus and the heliocentric system.
207. Nicolas Copernic, fondateur de l'astronomie moderne. *Etudes soviétiques*, 1953 (August), *6*: no. 65, pp. 57–61. Opposes that interpretation of the theory of relativity according to which Copernicus merely proposed an alternative origin of coordinates. Copernicus' helio-

centric astronomy is not an arbitrary convention, but mirrors physical reality.

FINDEISEN, OTTO
208. War Nikolaus Kopernikus Deutscher oder Pole? *Archiv für Wanderungswesen und Auslandkunde,* 1940, *11*: 20–22. In his eagerness to make Copernicus a German, Findeisen transfers to Copernicus the words "But it still moves" (which an unhistorical legend says were uttered by Galileo in a low voice immediately after his abjuration).

FITZMYER, JOSEPH A.
209. Copernicus. *America,* 1943, *69*: 148–150. Erroneously asserts that Osiander was "the printer engaged to publish" the *Revolutions,* and that he "inserted the word *hypothesis* in the title."

FLIS, STANISŁAW
210. W sprawie epitafium Kopernika. *Problemy,* 1955, *11*: 492–493. The epitaph on Copernicus' tomb.

FLORENTIIS, GIUSEPPE DE
211. Nicola Copernico. *Sapere,* 1943, *17–18*: 365–367. A few serious errors mar this otherwise excellent essay in popularization.

FLUKOWSKI, STEFAN
212. Powieść o Koperniku. *Wszechświat,* 1953, p. 209. Concerning a novel about Copernicus.
See Szancer.

FOK, VLADIMIR ALEXANDROVICH
213. Sistema Kopernika i systema Ptolomeia v svete obshchei teorii otnositelnosti, at pp. 180–186 in Mikhailov, ed. (483). This lecture, delivered at Moscow in 1943 at the Copernicus quadricentennial ceremony organized by the Academy of Science of the USSR, was translated into French at pp. 147–154 in *Questions scientifiques: physique* (Paris, 1952), and by F. Bartels into German under the title Das kopernikanische und das ptolemäische System im Lichte der allgemeinen Relativitätstheorie at pp. 805–809 in *Sowjetwissenschaft,* naturwissenschaftliche Abteilung, 1953.

The rivalry between the Copernican and Ptolemaic systems, which had been settled in favor of the former within the framework of Newtonian mechanics, was renewed when Einstein enunciated his theory of general relativity. This, according to Fok, does not in the least weaken Copernicus' heliocentric theory of the solar system.
214. Sistema Kopernika i sistema Ptolomeia v svete sovremennoi teorii tiagoteniia, at pp. 57–72 in Kukarkin, ed. For this lecture, which was delivered at Warsaw on September 15, 1953 at the Copernicus celebration held under the auspices of the Polish Academy of Science, see *Sesja kopernikowska.*
See Schatzman (661).

FOLKIERSKI, WŁADYSŁAW
215. Voltaire contre Fontenelle ou la présence de Copernic, at pp. 174–184 in *Literature and Science,* Proceedings of the Sixth Triennial Congress, International Federation for Modern Languages and Literatures (Oxford, 1955) xiii + 330 p. The delayed influence of Copernicanism on literature is examined principally at its entrance point, Fontenelle.

FORSTREUTER, KURT
216. Fabian von Lossainen und der Deutsche Orden, at pp. 220–233 in 370. Prints four previously unpublished letters, one of which refers to an unsuccessful search for a map in all the rooms of a Doctor Nicholas (Copernicus?).

FRACASTORO, MARIO GIROLAMO
217. Come il mondo conobbe le teorie copernicane. *L'Universo,* 1940, *21*: 800–801. Commemorates the first publication of the Copernican theory by Rheticus in 1540. The author errs in saying that Copernicus disapproved of Osiander's Preface. That document was probably never seen by Copernicus. He did, however, disagree with the ideas contained in it.

FRANK, PHILIPP
218. The Philosophical Meaning of the Copernican Revolution. *Proceedings of the*

ANNOTATED COPERNICUS BIBLIOGRAPHY

American Philosophical Society, 1944, *87*: 381–386. In replacing the earth by the sun as the center of the universe, Copernicus showed that the earth was not the only legitimate body of reference, and thereby cleared the way for "the great new truth that we have complete freedom in our choice of a system of reference" (p. 386).

FRANZ, GÜNTHER

219. Kopernikus, at p. 478 in Hellmuth Rössler and Günther Franz, *Biographisches Wörterbuch zur deutschen Geschichte* (Munich, 1952) xlviii + 968 p. Utterly misrepresents the reaction of Protestant leaders to Copernicus' teachings by saying that Luther found them acceptable, Melanchthon regarded them as merely hypothetical, and Calvin rejected them.

FRIESEN, H.

220. Die Kopernikus-Gedenkstätte in Frauenburg. *Die Himmelswelt*, 1943, *53*: 67–70. Plans for a Copernicus memorial in Frauenburg (now Frombork).

GABBA, LUIGI

221. I precursori di Copernico. *Istituto lombardo di scienze e lettere*, Rendiconti, classe di scienze, 1943–1944, *77*: 321–327. A lecture delivered on May 8, 1943 at the Copernicus celebration at Ferrara.

GADOMSKI, JAN

222. *Zarys historii astronomii polskiej* (Kraków, 1948, p. 45; Polska akademia umiejętności, Historia nauki polskiej w monografiach, II). Copernicus is discussed at pp. 6–11 in this outline of the history of astronomy in Poland, which constitutes vol. II in the series "History of Polish Science in Monographs," published under the auspices of the Polish Academy.

223. Technika pracy Mikołaja Kopernika. *Urania* (Kraków), 1953, *24*: 129–136. Copernicus' working technique.

224. Kopernik o kometach. *Op. cit.*, pp. 221–223. Copernicus on the subject of comets.

225. Jedyna metryka Kopernika. *Op. cit.*, pp. 253–254. The only evidence concerning the date of Copernicus' birth.

226. Krzywa Lissajous-Kopernika. *Op. cit.*, pp. 285–287. The Lissajous curve of Copernicus (photograph facing p. 300).

227. Rękopis De revolutionibus. *Op. cit.*, pp. 317–322. The manuscript of the *Revolutions*.

228. Katalog gwiazd Kopernika. *Op. cit.* pp. 349–353. Copernicus' star catalogue.

229. Jubileusz Kopernika. *Życie słowiańskie*, 1953, *8*: no. 2, pp. 35–38. The Copernicus jubilee.

230. Kopernik przy pracy. *Problemy*, 1953, *9*: 312–317. Copernicus at work.

231. Kopernik obserwował kometę Halleya. *Op. cit.*, pp. 596–597. Copernicus observed Halley's comet.

232. Dzieje manuskryptów Mikołaja Kopernika. *Op. cit.*, pp. 748–754. Translated into Dutch under the title De Lotgevallen van de Manuscripten van Mikołaj Kopernik, at pp. 22–32 in Vereniging Nederland-Polen, *Mikołaj Kopernik*. The history of Copernicus' autograph manuscript.

233. Ekspertyza astronomiczna we Fromborku. *Urania* (Kraków), 1954, *25*: 1–7. Astronomical test observations at Frombork for the purpose of reconstructing Copernicus' observatory.

234. Obserwatorium Kopernika we Fromborku. *Op. cit.*, pp. 45–47. Copernicus' observatory at Frombork.

235. Grób Kopernika. *Op. cit.*, pp. 81–84. Copernicus' grave.

236. Jedyny uczén Kopernika. *Op. cit.*, pp. 101–105. Copernicus' only pupil, Rheticus.

237. Jeszcze o fromborkskim nagrobku Kopernika. *Op. cit.*, pp. 320–321. Further remarks on the Frombork epitaph of Copernicus.

238. W poszukiwaniu obserwatorium Kopernika we Fromborku. *Problemy*, 1954, *10*: 34–39 (written jointly with Michał Kamieński and Janusz Pagaczewski, a foreword being provided by Stanisław Szymański). In quest of Copernicus' observatory at Frombork.

239. W poszukiwaniu grobu Kopernika. *Op. cit.*, pp. 561–563. The search for the exact place where Copernicus was buried.

ANNOTATED COPERNICUS BIBLIOGRAPHY

240. Poznajmy Kopernika gruntownie. *Urania* (Kraków), 1956, *27*: 155. A comment on Penconek's article bearing the same title (552).
Reviewed **29, 123, 262, 381, 533, 585, 653, 728.**
See Stenz.

GAJEWSKI, MARIAN
241. Czy Kopernik budował wodociągi. *Ochrona zabytków*, 1953, *6*: 67–68. Did Copernicus build waterworks?

GANSINIEC, RYSZARD
242. Rheticus jako wydawca Kopernika. *Polska akademia umiejętności, Sprawozdania z czynności i posiedzeń*, 1952, *53*: 134–137. Translated into French by Allan Kosko at pp. 129–133 in Wędkiewicz. Rheticus' part in the printing of the first ed. of the *Revolutions*.

243. Rzymska profesura Kopernika. *Kwartalnik historii nauki i techniki*, 1957, *2*: 471–484 (summary in English at pp. 482–484); summary in French by Allan Kosko at pp. 285–286 in Wędkiewicz. Copernicus was not a professor of astronomy at the University of Rome in 1500.

244. Tytuł dzieła astronomicznego Mikołaja Kopernika. *Op. cit.*, 1958, *3*: 195–222 (summary in English at pp. 220–222). A lecture delivered on March 21, 1957 to the Kraków Academy of Science and Literature; summary in French translation by Allan Kosko at pp. 259–260 in Wędkiewicz.
Maintains that the authentic title of Copernicus' major work was "Nicolai Copernici Revolutionum libri VI."
See Copernicus, editions and translations: *O obrotach sfer niebieskich księga pierwsza* (120), and Teofilakt Symokatta, *Listy* (123).
An obituary notice of Gansiniec (born March 6, 1888; died March 8, 1958) by Jerzy Łanowski appeared in *Kwartalnik historii nauki i techniki*, 1958, *3*: 629–637, with a photograph of Gansiniec facing p. 629.

GAWEŁ, ANTONI
245. Mikołaj Kopernicki. *Problemy*, 1953, *9*: 708–709. Antoni Skorulski, an

eighteenth-century Jesuit writer, maltreated Copernicus' surname.

Gazette astronomique, 1939, *26*: 17–18
246. La bataille pour Copernic. Reports an announcement in the Paris newspaper *Le Temps*, January 19, 1939, of a legal contest regarding the German-Polish dispute over the nationality of Copernicus.
247. Un institut Copernic à Berlin. Reports the ceremony of February 19, 1939 at which the Astronomisches Recheninstitut of Berlin-Dahlem was renamed the Coppernicus-Institut.

GEHRMANN, KARLHEINZ
248. Der Beweger der Erde, at pp. 198–203 in *Heimat im Herzen, wir von der Weichsel und Warthe*, ed. Erhard Wittek (Salzburg, 1950) p. 408 + xxxii. Repeats the erroneous statement that Copernicus was a priest.

GENGLER, THOMAS
249. *Nikolaus Kopernikus* (Göttingen, 1944, p. 37; Göttinger Universitäts-Reden, no. 14). Erroneously states that Copernicus was ordained a priest by his uncle, the bishop of Ermland, in the autumn of 1496 (p. 12), and that the Gregorian calendar was based on the Copernican astronomy (p. 33). Mistakenly grants Copernicus a Master of Arts degree (p. 13). Mistranslates "Sarmatian" by "Prussian" in Melanchthon's denunciation of Copernicus (p. 29). To prove how thoroughly German Copernicus was, Gengler denies (p. 34) that Copernicus was influenced by ancient Greek authors (who are quoted in their own language by Copernicus).

GEORGENS, AUGUST, ed.
250. *Nikolaus Kopernikus, Persönlichkeit und Werk* (Danzig, 1943) p. 132; Kulturpolitische Schriftenreihe für den Reichsgau Danzig-Westpreussen, *4*
Wilhelm Löbsack, "Nikolaus Kopernikus—ein deutscher Revolutionär" (pp. 5–14); Hans Schmauch, "Leben und Wirken des Nikolaus Kopernikus" (pp. 15–56); J. Sommer, "Kopernikus und die Weltsysteme" (pp. 57–109); Friedrich

ANNOTATED COPERNICUS BIBLIOGRAPHY

Schwarz, "Wie sah Kopernikus aus?" (pp. 110–132).
Each article listed above is annotated under its author.
Reviewed by Max Caspar, *Kant-Studien*, 1943, *43*: 476.

GERASIMENKO, MIKHAIL PETROVICH
251. *Nikolai Kopernik—vydaiushchiisia ekonomist epokhi rannego kapitalizma* (Kiev, 1953) p. 123. Nicholas Copernicus as an outstanding economist in the period of early capitalism.

GERLO, ALOIS
252. Copernic et Simon Stevin. *Ciel et terre*, 1953, *69*: 277–288. Translated into Polish under the title Kopernick i Szymon Stevin, *Problemy*, 1954, *10*: 237–242.
In a book published in 1605–1608 Stevin accepted the Copernican astronomy as physically true.

GIACON, CARLO
253. Copernico, la filosofia e la teologia. *Civiltà cattolica*, 1943, *94*: part 4, pp. 281–290, 367–374. In this lecture, delivered on May 9, 1943 at a quadricentennial commemoration of Copernicus at Ferrara, Cardinal Schönberg is mistakenly made president of the papal calendar commission, the *Commentariolus* is erroneously described as a summary of the *Revolutions* after that book was finished, and Rheticus' belief in astrology is misattributed to Copernicus.
254. Intorno alla condanna di Copernico. *Vita e pensiero*, 1943, anno 29, vol. 34, pp. 182–187. In the main, a condensation of 253. If Copernicus had presented his astronomy merely as a mathematical device (whereas in fact he regarded it as physically true), and if Bruno and Galileo had not insisted on its conformity with reality, the Roman Catholic church would not have condemned it.
255. Copernico e il realismo della sua "ipotesi," at pp. 77–98 in Giacon's *Scienze e filosofia* (Como, 1946) p. 209. The Roman Catholic church took no action against Copernicanism as long as that doctrine was misunderstood, as a result of Osian-

der's false preface, to be merely a mathematical hypothesis and not a blueprint of the real universe.

GINGRICH, CURVIN H.
256. Copernicus, the Founder of Modern Astronomy. *Popular Astronomy*, 1943, *51*: 297–307. Copernicus viewed as one of the few great astronomers of all time. Reviewed **493, 620.**

GLADBACH, WALTER
257. Eppur si muove! *Die Weltbühne*, 1953, *8*: 648–652. The astronomical system devised by Copernicus, who was not especially interested in theology, was condemned by the theologians.

GÓRSKI, JANUSZ
258. Teoria ekonomiczna Mikołaja Kopernika. *Ekonomista*, 1953, no. 4, pp. 89–109; 1954, no. 4, pp. 301–304. Translated into Czech, *Politická ekonomie*, 1954, *2*: 266–279, 340–347. Copernicus' economic theory.
259. Rok kopernikowski a polska nauka ekonomiczna. *Ekonomista*, 1954, no. 1–2, pp. 236–246. The Copernican year (1953) and Polish economic studies.
260. Teoria pieniądza Decjusza i Kopernika. *Roczniki dziejów społecznych i gospodarczych*, 1955, *17*: 9–50 (summary in French at pp. 51–52). Compares Copernicus' monetary theory with the views expressed in *De monete cussione ratio* by a contemporary of Copernicus, Jodocus Ludovicus Decius (or Dietz).

GÓRSKI, KAROL
261. Dom Kopernika w Toruniu. *Ochrona zabytków*, 1953, *6*: 6–8. Copernicus' house in Toruń.
262. *Domostwa Mikołaja Kopernika w Toruniu* (Toruń, 1955) p. 35. Reviewed by Jan Gadomski, *Urania* (Kraków), 1955, *26*: 342–343.

Great Books of the Western World (Chicago, 1952)
263. The Copernican revolution is discussed in vol. II, pp. 89–97. In a work professedly devoted to revealing the unity of western thought, it is distressing to find

three different and mutually contradictory positions regarding Copernicus' attitude toward Osiander's fictionalist interpretation of astronomical hypotheses.

See Copernicus, translations: *On the Revolutions of the Heavenly Spheres* (124).

GREENWOOD, THOMAS
264. Les hypothèses de Copernic. *Revue trimestrielle canadienne*, 1944, *30*: 240–249. Copernicus' hypotheses, understood as first principles on which a science is based and not as conventional or fictional propositions, have permanent value.

GUREV, GRIGORII ABRAMOVICH
265. Uchenie Kopernika. *Nauka i zhizn*, 1948, no. 12, pp. 2–6. An explanation of Copernicus' theory.

266. *Sistemy mira* (Moscow, 1950), p. 393. This Russian history of the various conceptions of the universe deals with Copernicus at pp. 131–160.

GUYOT, EDMOND
267. Le système du monde de Ptolémée à Einstein à propos du quatrième centenaire de la mort de Copernic. *Scientia*, 1946, *79*: 77–82. Copernicus' part in the development of astronomy from the ancient Greeks to Einstein.

GYÖRGI, NÁDOR
268. A Kopernikuszi tan és hatása a tudományos gondolkodásra. *A Magyar tudományos akadémia, Matematikai és fizikai tudományok osztályának, Közleményei*, 1956, *6*: 93–105. The Copernican theory and its influence on scientific thought.

HALL, ALFRED RUPERT
269. *The Scientific Revolution* (London, 1954; reprinted as a paperback, Boston, 1956) p. 390. At pp. 35–36, 51–68, Hall discusses what was new and what was old in Copernicus' astronomy; at pp. 370–371 he asserts the geometrical equivalence of Ptolemy's and Copernicus' planetary theory.

HARDER, ROBERT LINCOLN
270. Copernicus, Galileo and Ideal Conditions. A Columbia University doctoral dissertation, 1956; 177 typewritten pp., available in microfilm as Publication no. 19243, University Microfilms, Ann Arbor, Michigan. Chapter III, pp. 35–79, distinguishes Copernicus' astronomy from Ptolemy's, and challenges E. A. Burtt's view of Copernicus as a Pythagorean (84).

HARTLEB, KAZIMIERZ
271. *Mikołaj Kopernik* (Toruń, 1946) p. 51; 2d ed. (Toruń, 1948) p. 55. A reply to Wasiutyński's pro-German arguments.

HARTNER, WILLY
272. Nicolaus Copernicus, at vol. I, pp. 386–400, in *Die grossen Deutschen*, edd. H. Heimpel, T. Heuss, and B. Reifenberg (new ed., Berlin, 1956) 4 vols. After an admirable survey of the development of planetary theory since Greek antiquity as the background for a discussion of Copernicus, Hartner pleads that the famous astronomer should no longer be regarded as a prize to be fought over by Germans and Poles, but should be honored as a peaceful link between these two great neighboring nations.

HEINZ, RUDOLF
273. Goethe, die deutsche Geologie und Kopernikus. *Beiträge zur Geologie von Thüringen*, 1943, *7*: 269–275. Just as Copernicus showed how to break the chains of scholasticism, so German geology must learn to throw off the shackles of British empiricism.

HEJNOSZ, WOJCIECH
274. Mikołaj Kopernik "decretorum doctor." *Problemy*, 1954, *10*: 477–479. Two passages were translated into French by Allan Kosko at pp. 114–115 in Wędkiewicz.

HENSELING, ROBERT
275. Genius im Reiche der Gestirne. Der kopernikanische Gedanke und seine geisteswissenschaftliche Bedeutung. *Koralle*, 1943, *11*: 211–212. On both occasions when the heliocentric astronomy was advocated (by Aristarchus and by Copernicus), it encountered religious opposition.

HERCZEG, TIBOR
276. *Kopernikusz* (Budapest, 1954) p.91.

HERNANDEZ DE ALBA, GUILLERMO
277. Copernico y los origenes de nuestra independencia, at pp. 19–23 in 108. José Celestino Mutis, the first public adherent to Copernicanism in the American colonies of Spain.

278. *Heroen des Geistes im deutschen Osten* (Königsberg, 1939), p. 54. Contains Erich Przybyllok, "Das Weltbild des Coppernicus" (pp. 7–16); Hans Joachim Schoenborn, "Coppernicus der Deutsche" (pp. 17–23); and Theodor Schieder, "Deutsches Geistesleben Altpreussens von Coppernicus bis Kant" (pp. 24–30).
Reviewed by Hans Schmauch, *Zeitschrift für die Geschichte und Altertumskunde Ermlands*, 1939–1942, *27*: 298–299.

HILDEBRANDT, KURT
279. *Kopernikus und Kepler in der deutschen Geistesgeschichte* (Halle, 1944) p. 13; offprinted from *Die Gestalt*, Abhandlungen zu einer allgemeinen Morphologie, Heft 14. In this printed version of a speech delivered at the Copernicus celebration held at the University of Kiel, Hildebrandt assures us that the astronomical achievements of both Copernicus and Kepler were directed toward Plato's Idea of the Good, the attainment of which was the German goal in World War II.

HILPERT, GERDA
280. Unter dem Zeichen des Skorpion. Kulturpsychologische und astrobiographische Essays: Nicolaus Coppernicus. *Sterne und Mensch*, 1941, *17*: 8–12. A psychoanalytic astrologer looks at Copernicus with unintentionally comical results.

Himmelswelt, 1942, *52*: 58–59
281. Kopernikus-Gesamtausgabe in Vorbereitung. An announcement of the forthcoming publication of the *Nikolaus Kopernikus Gesamtausgabe*.

1943, *53*: 60
282. Die 400jährige Wiederkehr des Todestages von Nikolaus Kopernikus. A report on the Copernicus celebrations in the territory held by Germans in 1943.

1943, *53*: 71
283. Zur Erinnerung an den 400. Todestag von Nikolaus Kopernikus. An additional report on Copernicus celebrations.

HIRSCH, FELIX E.
284. Copernicus after 400 Years. *Saturday Review of Literature*, 1943, *26*: no. 22, pp. 11–12. Reflections after a visit to Frombork in 1933.
Reviewed **354.**

HÖGBERG, PAUL
285. Copernicus-minnen i Uppsala. *Populär astronomisk tidskrift*, 1943, *24*: 31–42. Books now at Uppsala that may once have belonged to Copernicus.

HOF, S. P. VAN'T
286. Nicolaus Coppernicus. *De Natuur*, 1943, *63*: 43–55. Has Copernicus return home from Italy two years too late, and leave Heilsberg also two years too late. Misdates his receipt of a canonry in 1500. Approves the traditional misstatement that he was a priest. Erroneously asserts that the pope accepted the dedication of the *Revolutions*, that the *Commentariolus* is an extract from the *Revolutions*, that Gassend's biography of Copernicus is the earliest, and that Copernicus indirectly provided the basis for the Gregorian calendar reform.

HOFF, ERWIN
287. Zur geistesgeschichtlichen Beurteilung und Bedeutung des kopernikanischen Gedankens in Vergangenheit und Gegenwart. *Die Burg*, 1943, *4*: 86–138. Reviewed by Max Caspar, *Kant-Studien*, 1943, *43*: 477. Analyzes the supporters and opponents of Copernicanism from the 16th to the 20th century. Makes Platonism and Neoplatonism, rather than Aristotelianism, the decisive influence in Copernicus' thought, which did not tend toward a purely mechanical interpretation of nature. Appends a Copernicus bibliography (pp. 134–138), in which an author is labeled a Jew.

288. Die Ursprünglichkeit des koperni-kanischen Gedankens. *Das Vorfeld*, 1943, *3*: 95–98. Repeats Brachvogel's contention that neither Aristarchus nor Nicholas of Cusa started Copernicus on the road to-ward heliocentrism.

289. Nikolaus Kopernikus—Abriss zu Leben und Werk des grossen deutschen Astronomen. *Das Generalgouvernement*, 1943, *3*: no. 2, pp. 1–8. What makes Hoff so sure that Osiander inserted the words "orbium coelestium" in the title of the *Revolutions*?

HOFFLEIT, DORRIT
290. Copernican Manuscript Returns to Poland. *Sky and Telescope*, 1954, *13*: 112. Czechoslovakia's presentation of Coper-nicus' holograph manuscript of the *Revolutions* to Poland. Reviewed **491.**

HOFFMANN, KARL FRANZ
291. Nikolaus Coppernikus als Arzt. *Hippokrates*, 1943, *14*: 444. Enrolls Coper-nicus as a student in 1596, instead of 1496, and mistakenly grants him a Master of Arts degree.

HOFMANN, JOSEPH EHRENFRIED
292. *Geschichte der Mathematik* (Berlin, 1953), Part I, p. 200. At pp. 110–111 Copernicus' *Commentariolus* is erroneously identified with Rheticus' *Narratio prima*. Reviewed **121, 871.**

HOGG, HELEN SAWYER
293. The Introduction of the Coper-nican System to England. *Journal of the Royal Astronomical Society of Canada*, 1952, *46*. Robert Recorde is discussed at pp. 113–117, John Dee at pp. 158–164, Thomas Digges at pp. 195–201, and Thomas Harriot at pp. 239–244 (completed in 1953, *47*: 15–20).

HOLLITSCHER, WALTER
294. Kopernikus heute. *Blick nach Polen*, 1950, no. 7, pp. 22–23. Copernicus, one of Poland's greatest sons, serves as an inspiration to the youth of his country.

HOLTON, GERALD and ROLLER, DUANE H. D.
295. *Foundations of Modern Physical Science* (Reading, Mass.: Addison-Wesley, 1958) p. 782. Copernicus is discussed at pp. 118–127.

HOPMANN, JOSEF
296. Die Lehre des Kopernikus (1543) bis zum Abschluss durch F. W. Bessel (1838), at pp. 119–120 in Diergart, ed. (151). An outline of the steps by which Copernicus' astronomy was extended and confirmed.

See Copernicus, translations: Menzzer (125).

HORBACKI, WŁADYSŁAW
297. Uwagi o roli Kopernika w dziejach myśli naukowej. *Urania* (Kra-ków), 1953, *24*: 161–165. Observations on Copernicus' place in the development of scientific thought.

HORN–D'ARTURO, GUIDO
298. Il sistema Copernicano. *Coelum*, 1948, *16*: 33–34. Compares Copernicus' planetary theory with ours.

299. Letterato francese del cinquecento critico di Copernico. *Coelum*, 1948, *16*: 34–35. Montaigne doubted the Copernican astronomy.

300. Atti notarili del secolo XVI con-tenenti il nome di Copernico rinvenuti nell' Archivio storico capitolino. *Coelum*, 1951, *19*: 40–43. Publishes two notarial docu-ments, dated 1510 and 1519, that name Copernicus as a legal agent in Ermland matters.

301. Onoranze polacche a Nicolò Copernico. *Coelum*, 1953, *21*: 190. The Polish commemoration of Copernicus at Warsaw on September 15–16, 1953.

302. Nicolò Copernico. *Coelum*, 1954, *22*: 33–38. An address delivered on October 4, 1953 at the University of Ferrara, and describing Copernicus' system as a Polish seed that germinated in Italian soil.

HORSKÝ, ZD.
303. Mikuláš Koperník. *Říše hvězd*, 1953, *34*: 103–107.

HOSZOWSKI

304. Mikuláš Koperník. *Časopis československých ústavů astronomických,* 1953, *3*: 61–63. In commemoration of the 480th year of the birth, and 410th year of the death, of Copernicus.

HOSZOWSKI, STANISŁAW
305. Wkład Kopernika w postępową myśl ekonomiczną. *Życie szkoły wyższej,* 1953, *1*: no. 10, pp. 107–115. Copernicus' contribution to progressive economic thought.

HUGONNOT, JEAN
306. Un grand polonais: Copernic. *Démocratie nouvelle,* 1953, *7*: 466–469. Copernicus in his Renaissance setting

HUMBERT, PIERRE
307. Histoire des découvertes astronomiques (Paris, 1948) p. 272. Copernicus is treated at pp. 44–54 of this brief work, intended for young people.

HUMIĘCKA, WANDA
308. Nowe wydanie dzieł Kopernika. *Nauka Polska,* 1953, *1*: no. 2, pp. 205–207. An announcement of a new Polish ed. of Copernicus.

HURWIC, JÓZEF
309. Gród rodzinny Kopernika składa hołd jego pamięci. *Problemy,* 1953, *9*: 562. Copernicus' native city pays homage to his memory.

IDELSON, NAHUM ILYICH
310. Zhizn i tvorchestvo Kopernika, at pp. 5–42 in Mikhailov, ed. (483).
311. Etudii po istorii planetnich teorii, at pp. 84–179, *op. cit.*

INFELD, LEOPOLD
312. Teoria Kopernika a zagadnienie grawitacji w fizyce współczesnej. *Problemy,* 1953, *9*: 442–448. The Copernican theory and the problem of gravitation in contemporary physics.
313. Znaczenie prac Kopernika dla rozwoju fizyki. *Studia i materiały z dziejów nauki polskiej,* 1954, *2*: 33–53. Infeld's conclusions (pp. 52–53) were translated into French by Allan Kosko at pp. 143–144 in Wędkiewicz. In the mathematical for-

mulation of the theory of relativity, the distinction between the Ptolemaic and the Copernican astronomies vanishes; but when the relativity theory is viewed as an instrument for understanding the physical universe, it was Copernicus who set astronomy on the right road.

INGARDEN, ROMAN STANISŁAW
314. Buridan i Kopernik: dwie koncepcje nauki. *Studia i materiały z dziejów nauki polskiej,* 1953, *1*: 51–63. Translated into French under the title Deux conceptions de la science, Buridan et Copernic. *La Pensée,* 1954, no. 53, pp. 17–28. Another partial translation into French by Allan Kosko at pp. 120–129 in Wędkiewicz. Against Duhem's thesis that modern science originated in late medieval scholastic philosophy, Ingarden argues that Copernicus' concepts were essentially different from Oresme's, that the latter's writings were not known to Copernicus, and that Copernicus opposed the fictionalist interpretation of astronomical theory.
315. *Mikołaj Kopernik i zagadnienie obiektywności praw naukowych* (Warsaw, 1953) p. 83. This lecture on Nicholas Copernicus and the problem of the objective character of scientific laws was delivered on October 25, 1953 at the Polish Academy of Science's discussion of the Renaissance, and was published also in *Odrodzenie w Polsce,*
316. vol. 2, part 2, pp. 7–53 (Warsaw, 1956), where Ingarden replied to Aleksander Birkenmajer's criticism (48) in Odpowiedź na wystąpienie Prof. A. Birkenma-
317. jera, *op. cit.*, pp. 97–109.
See *Sesja kopernikowska.*

Isis, 1944, *35*: 30
318. Copernicus Celebrations. A list of memorial meetings in ten cities.

ITAKURA, KIYONORI
319. What We Learn from Copernicus? *Journal of History of Science, Japan,* 1953 (November), no. 27, pp. 14–22 (in Japanese).

JADWINOWSKI, LUDWIK
320. Poglądy monetarne M. Kopernika, at pp. 37–84 in 367, 2d ed. Copernicus' monetary theory.

JANIKOWSKI, STANISLAW
321. Appunti su Niccolò Copernico. *Ecclesia*, 1953, *12*: 610–613. Replete with errors, some long since refuted, and others awaiting refutation.

JEANS, JAMES
322. *The Growth of Physical Science* (Cambridge, 1947; New York, 1948; 2d ed., Cambridge, 1951) x + 364 p. Translated into French by René Sudre under the title *L'Evolution des sciences physiques* (Paris, 1950) p. 318, and into Italian by Gioietta Bompiani under the title *Il cammino della scienza* (Milan, 1953) p. 488. Copernicus' astronomy is discussed at pp. 124–134, and his economics at p. 186. A number of errors mar the discussion. In particular, with regard to the assertion that Copernicus "had been given permission to publish his *Narratio* to a wider circle in 1540," we may well ask who gave him this permission.

Jenseits der Oder, 1953, *4*: no. 5, pp. 8–12
323. Mikołaj Kopernik—oder der Umsturz in der Weltbildvorstellung. An author who gives his name only as "—er" corrects an error committed by Peter Siehm (*op. cit.*, 1952, *3*: no. 3, p. 12), but himself mistakenly attributes elliptical planetary orbits to Copernicus.

JOHNSON, FRANCIS R.
324. Astronomical Text-books in the Sixteenth Century, at Vol. I, pp. 285–302, in *Science, Medicine and History: Essays . . . in Honour of Charles Singer* (London, 1953) 2 vols. The treatment of the Copernican theory in the textbooks of the sixteenth century. Reviewed **18, 354.**

JOLIOT-CURIE, FRÉDÉRIC
325. L'acte révolutionnaire de Nicolas Copernic. *Les lettres françaises*, 1953, May 28 —June 4, no. 467, p. 1. The major portion of an address delivered at a Copernicus commemoration held under the auspices of the periodical *La Pensée*.

JONES, HAROLD SPENCER
326.
327. *Copernicus* (University College of South Wales and Monmouthshire, 1943) p. 32. See *Nature*, 1943, *152*: 408–409. The Astronomer Royal swiftly surveys the technical superiority of Copernicus' astronomy over Ptolemy's as the pathway leading to the later improvements in the Copernican system. This Selby Lecture, delivered by Sir Harold at Cardiff on May 27, 1943, was reprinted under the title "Copernicus and
328. the De revolutionibus" in *Polish Science and Learning*, 1943 (June), no. 3, pp. 11–24, with the addition of two figures and two footnotes.

329. Copernicus and the Heliocentric Theory. *Nature*, 1943, *151*: 573–576. Reviewed by Hermann von Schelling, *Zentralblatt für Mathematik und ihre Grenzgebiete*, 1944, *28*: 385.
See Abetti (2).

JORDAN, PASCUAL
330. Kopernikus und die Entwicklung des abendländischen Denkens. *Aus Politik und Zeitgeschichte*, Beilage zur Wochenzeitung *Das Parlament*, 1954, *19*: 221–224. A speech delivered at a Copernicus celebration held at Aachen on March 28, 1954.

Journal of the British Astronomical Association, 1948–1949, *59*: 83–84
331. The Home of Copernicus. Announcement of the Polish government's decision to reconstruct the memorials of Copernicus' life in Frombork.

JUNKER, ERNEST
332. Kopernikus und die Sonne. *Die Mittelstelle*, 1943 (May), *2*: no. 19, pp. 3–8. A translation into German of Scene 4 of the dialogue "Il Copernico," one of Giacomo Leopardi's *Operette morali*.

KĄCZKOWSKA, ALICJA
333. Epitafium Mikołaja Kopernika we Fromborku. *Ochrona zabytków*, 1953, *6*: 55–56. The epitaph and portrait of Copernicus at Frombork.

KAEMPFFERT, WALDEMAR

334. Copernicus Day. *New York Times*, 1943, May 23, p. 17E. Reflections on the 400th anniversary of Copernicus' death. Reviewed **354**.

KAHRSTEDT, ALBRECHT

335. Kopernikus als Mensch und Wissenschaftler. *Wissenschaftliche Annalen*, 1954, *3*: 311–316. This appreciation of Copernicus as a man and as a scientist maintains that his priesthood played only a minor role in his life. It would have been more accurate to say that it played no role at all, since he never became a priest.

KAISER, EDWIN G.

336. Before Copernicus: Nicolaus of Oresme. *America*, 1943, *69*: 178–180. Repeats Duhem's absurd statement that Copernicus and Galileo "contributed scarcely anything to what had already been taught by Buridan, Oresme and Nicolaus of Cusa."

KALINOWSKI, STANISŁAW

337. Sesja naukowa poświęcona poglądom społecznym Mikołaja Kopernika. *Życie szkoły wyższej*, 1954, *2*: no. 10, pp. 131–135. Succinct summary of a discussion on June 12, 1954, under the auspices of the Polish Society of Economists, of Copernicus' writings on money and social questions.

KAMIEŃSKI, MICHAŁ

338. O właściwy tytuł dzieła Kopernika. *Życie nauki*, 1949, 7: 603–604. The authentic title of Copernicus' major work. See Gadomski (238).

KAMP, PETER VAN DE

339. Copernicus and the Present World Picture. *American-German Review*, 1943–1944, *10*: no. 2, pp. 10–13. Copernicus' heliocentric system provided the model for the recent investigation of the motion of the stars.

KARPINSKI, LOUIS C.

340. Copernicus, First Citizen of a New World Order. *Bulletin of the Polish Institute*

of Arts and Sciences in America, 1942–1943, *1*: 690–694.

341. The Progress of the Copernican Theory. *Scripta Mathematica*, 1943, *9*: 139–154.

342. Copernicus Celebration at the Polish Institute of Arts and Sciences in America. *Science*, 1943, *97*: 549.

343. Copernicus, Representative of Polish Science and Learning. *National Mathematics Magazine*, 1945, *19*: 343–348 (previously published in Polish in *Polonia Almanac*, Detroit, 1945).

The author errs in saying that Schöner wrote a preface to the second ed. of the *Revolutions*. Reviewed **620**.

KARRASCH, ALFRED

344. *Kopernikus* (Brest-Litovsk, 1944) p. 419; reissued, Düsseldorf, 1948, p. 410. An attempt at historical fiction.

KARSTÄDT, O.

345. Koppernick (Kopernikus) war kein Pole. *Die Praxis der Landschule*, 1939–1940, *48*: 253–256. In his eagerness to make Copernicus a German, Karstädt says that the Warsaw ed. of the astronomer's works "translated them into Polish without indicating that a translation was involved, through a desire to give the world the impression that Copernicus wrote his revolutionary teachings in the Polish language." Had Karstädt so much as glanced at this 1854 ed., he would have seen that it printed Copernicus' original Latin and a Polish translation side by side in parallel columns. But have we the right to demand such painstaking research from a man who states that Copernicus dedicated the *Revolutions* "to Pope Pius VIII," who reigned from 1829 to 1830?

KASIANOVA, E. V.

346. Uchenie Kopernika i tserkov. *Nauka i zhizn*, 1956, *23*: no. 2, pp. 41–44. Copernicus' theory and the Church.

KATTSOFF, LOUIS O.

347. Ptolemy and Scientific Method. *Isis*, 1947–1948, *38*: 18–22. The author charges that "Many books speak as if scientific method came into being with Copernicus," but he does not support his charge by citing specific books.

KAUFFELDT, ALFONS

348. Nikolaus Copernicus. *Wissenschaft und Fortschritt*, 1953, *3*: 39–41. The chronology of Copernicus' career is given somewhat inaccurately. Why does Kauffeldt say that Copernicus regarded the use of epicycles as a defect in his system? The assertion that Tycho Brahe rejected Copernicanism because he was fanatically devoted to precision overlooks his fanatical devotion to the Bible, with which he found Copernicanism in conflict.

349. *Nikolaus Kopernikus: der Umsturz des mittelalterlichen Weltbildes* (Berlin, 1954) p. 140. Reviewed by Vasily Pavlovich Zubov, *Voprosy istorii estestvoznaniia i tekhniki*, 1956, no. 1, p. 302.
See *Sesja kopernikowska*.

KELLY, HOWARD LAURENCE

350. Copernicus. *Journal of the British Astronomical Association*, 1943, *53*: 146, 155–159. Despite Copernicus' own shortcomings, his heliocentric theory initiated an enormous advance in science.
Reviewed **18, 19.**

KĘPIŃSKI, FELICJAN

351. Hołd narodów dla Mikołaja Kopernika. *Urania* (Kraków), 1946, *18*: 43–48. A tribute of the nations to Copernicus.

352. Dzieło Mikołaja Kopernika jako astronoma. *Przegląd geodezyjnj*, 1953, *9*: 338–344. Copernicus' work as an astronomer.

353. O polskim przekładzie (1953 r.) głownego dzieła Mikołaja Kopernika. *Nauka polska*, 1954, *2*: no. 1, pp. 202–203. A discussion of the Polish translation of Copernicus' chief work, with special reference to the meaning of the Latin word "orbis."

KESTEN, HERMANN

354. *Copernicus and His World* (New York, 1945; London, 1945, 1946) x + 408 p.

Translated by E. B. Ashton and Norbert Guterman from Kesten's German text, which was issued under the title *Copernicus und seine Welt* (Amsterdam, 1948, p. 511; Frankfurt am Main, 1950; Vienna, Munich and Basel, 1953, p. 332).

Translated from German into French by Eugène Bestaux under the title *Copernic et son temps* (Paris, 1951) p. 427, and from German into Serbian by M. Mezulić under the title *Kopernik i niegov sviet* (Zagreb, 1956) p. 433.

355. An excerpt was published in the *Polish Review*, 1943 (May 24), *3*: no. 19, pp. 3–4.

The American ed. was reviewed by Waldemar Kaempffert, *Saturday Review of Literature*, 1945, *28*: no. 11, p. 26; by Orville Prescott, *New York Times*, 1945, March 23, p. 17; in *Newsweek*, 1945 (March 26), *25*: no. 13, p. 90; by Bart J. Bok, *New York Times*, 1945, April 8, Book Review Section, p. 18; by Felix E. Hirsch, *Library Journal*, 1945, *70*: 117; by Karl F. Herzfeld, *Commonweal*, 1945, *42*: 293; by James B. Macelwane, *Catholic Historical Review*, 1945–1946, *31*: 368–370; and by Francis R. Johnson, *Isis*, 1947, *37*: 82.

The first German ed. was reviewed by "Fy," *Die Weltwoche*, 1949, *17*: April 22, no. 806, p. 9.

This picture of life in Copernicus' time holds the reader's interest despite its loose organization and light-hearted errors. Some mistakes made by Kesten and by his French translator are pointed out by Wędkiewicz, pp. 193–195.

KIENLE, HANS

356. An den Grenzen von Theorie und Beobachtung. *Die Naturwissenschaften*, 1939, *27*: 601–607. A lecture delivered at the Copernicus celebration at the University of Koenigsberg, February 18, 1939.

357. Das Weltsystem des Kopernikus und das Weltbild unserer Zeit. *Die Burg*, 1943, *4*: 63–85. Reviewed by Max Caspar, *Kant-Studien*, 1943, *43*: 476–477. An enlargement of a lecture delivered to the Institut für deutsche Ostarbeit on June 5, 1942, and

358. published originally in *Die Natur-wissenschaften*, 1943, *31*: 1–12. A swift review of the astronomical and other scientific developments that altered Copernicus' conception of the universe to our own. Reprinted in part in *Natur und Kultur*, 1943, *40*: 65–67. Reviewed **370**.

KING, HENRY C.
359. *The History of the Telescope* (London and Cambridge, Mass., 1955) xvi + 456 p. Copernicus is discussed at pp. 15–16.
360. *The Background of Astronomy* (London, 1957) p. 254. Copernicus is discussed at pp. 6–7, 201–208.

KLAUS, GEORG
361. Nikolaus Kopernikus—ein grosser Sohn des polnischen Volkes. *Urania* (Jena), 1953, *16*: 161–168. Translated into Polish under the title Mikołaj Kopernik—wielki syn narodu polskiego. *Myśl filozoficzna*, 1953, no. 1, pp. 191–208. Klaus mistakenly has Copernicus study at Rome, obtain a doctorate in theology, and return to Frombork only after his uncle's death.
362. Bemerkungen über das Verhältnis von Kopernikus und Rheticus. *Wissenschaftliche Zeitschrift der Humboldt-Universität zu Berlin*, Gesellschafts- und sprachwissenschaftliche Reihe, 1953–1954, *3*: 5–11; published also in *Urania* (Jena), 1954, *17*: 161–165. In this article, based on a lecture before the Polish Academy of Science on September 16, 1953, Klaus opposes the tendency to magnify Copernicus' links with the past and to minimize the significance of his break with it. In particular, he finds the importance of observation in astronomy evaluated utterly differently by Plato and by Copernicus. In the scientific cooperation between the Polish Catholic Copernicus and the German Protestant Rheticus, he sees a model for the future conduct of the two nations. See *Sesja kopernikowska.*

KLIBANSKY, RAYMOND
363. Copernic et Nicolas de Cues, at pp. 225–235 in *Léonard de Vinci et l'expérience*

scientifique au XVIe siècle (Paris, 1953; Colloques internationaux du Centre national de la recherche scientifique, sciences humaines). On the basis of a marginal note (which is assumed to be in Copernicus' handwriting) Klibansky argues that Copernicus was familiar with the writings of Nicholas of Cusa. On the basis of two supposed similarities in their thought, Klibansky maintains that Cusa influenced Copernicus.

KLINE, MORRIS
364. *Mathematics in Western Culture* (New York, 1953) xv + 484 p. The discussion of Copernicus at pp. 110–112 is marred by the erroneous statement that the pope requested the publication of Copernicus' *Revolutions*.

KNEDLER, JOHN WARREN, Jr., ed.
365. *Masterworks of Science* (Garden City, 1947) ix + 637 p. Chapters 1–11 of Book I of the *Revolutions* in English translation at pp. 53–72. Knedler's introductory remarks (pp. 49–52) mistakenly reduce Copernicus' stay in Italy to three years.

KOŁACZEK, B.
366. Wystawa kopernikowska w Politechnice Warszawskiej. *Urania* (Kraków), 1954, *25*: 49–50. The Copernicus exhibition at the Polytechnical Institute of Warsaw.

KOŁO PRZYRODNIKÓW IM M. KOPERNIKA W PALESTYNIE
367. *Mikołaj Kopernik* (Tel Aviv, 1942) p. 35; 2d ed. (Tel Aviv, 1943) p. 84 = *Kosmos*, no. 1 (Jerusalem, 1943). The Copernicus Society of Natural Scientists in Israel commemorates its eponymous hero with the following articles in the 2d ed.: Kazimierz Rouppert, "Z życia Mikołaja Kopernika" (pp. 5–9); "Kopernik a księżyc" (pp. 18–21); "Kopernik o złych sąsiadach-niemcach" (pp. 34–37); Bronisław Żelazowski, "O wielkości Kopernika jako astronoma" (pp. 9–14); Zawilec, "Czym Kopernik dla Polski" (pp. 14–18); L. S. Łukawiecka, "Postać M. Kopernika na scenie teatralnej" (pp. 21–23); Alfred

Laskiewicz, "Działalność Kopernika na polu lecznictwa" (pp. 23–33); and Ludwik Jadwinowski, "Poglądy monetarne M. Kopernika" (pp. 37–84).

368. *Kopernik, Mikolaj* (Warsaw, 1953) p. 12. An unsigned pamphlet, in English, eulogizing Copernicus as a benefactor of mankind. Issued also in French, German, Spanish, and Russian.

369. *Kopernik, Mikolaj—45 Tablic* (Warsaw, 1953). With an introduction by Aleksander Birkenmajer, these forty-five sumptuous photographic plates illustrate the work of Copernicus, his adversaries, and his followers. Reviewed by Alfons Triller, *Zeitschrift für die Geschichte und Altertumskunde Ermlands*, 1956–1957, *29*: 155–156.

370. *Kopernikus-Forschungen*, edd. Johannes Papritz and Hans Schmauch (Leipzig, 1943; Deutschland und der Osten, Quellen und Forschungen zur Geschichte ihrer Beziehungen, *22*) viii + 233 p.

Hans Schmauch, "Nikolaus Kopernikus —ein Deutscher" (pp. 1–32); Eugen Brachvogel, "Nikolaus Kopernikus in der Entwicklung des deutschen Geisteslebens" (pp. 33–99); Hans Schmauch, "Die Jugend des Nikolaus Kopernikus" (pp. 100–131); Johannes Papritz, "Die Nachfahrentafel des Lukas Watzenrode" (pp. 132–142); Friedrich Schwarz, "Kopernikus-Bildnisse" (pp. 143–171); Alexander Berg, "Der Arzt Nikolaus Kopernikus und die Medizin des ausgehenden Mittelalters" (pp. 172–201); Hans Schmauch, "Nikolaus Kopernikus und der Deutsche Ritterorden" (pp. 202–219); Kurt Forstreuter, "Fabian von Lossainen und der Deutsche Orden" (pp. 220–233).

Each article listed above is annotated under its author.

Reviewed by Max Caspar, *Kant-Studien*, 1943, *43*: 476; by Henrik Sandblad, *Lychnos*, 1943, p. 374; by Dietrich Wattenberg, *Das Weltall*, 1943, *43*: 131–132; in *Astronomische Nachrichten*, 1943–1944, *274*: 144; by Hans Kienle, *Die Naturwissen-*

schaften, 1944, *32*: 89; and by Johannes Larink, *Die Himmelswelt*, 1944, *54*: 11.

KOPFF, AUGUST

371. Eine neue dem Coppernicus-Institut gestiftete Büste des Nicolaus Coppernicus. *Die Himmelswelt*, 1939, *49*: 161–164. On February 18, 1939 the Astronomisches Rechen-Institut in Berlin-Dahlem was renamed the Coppernicus-Institut, and it received a bronze bust of its eponymous hero, the sculptor being Kurt Lehmann.

372. Das Kopernikus-Institut in Berlin-Dahlem. *Zeitschrift für die gesamte Naturwissenschaft*, 1943, *9*: 107–110. The Kopernikus-Institut is devoted to the studies pursued by Copernicus, not to the study of Copernicus himself. Reviewed **89, 119, 125, 871.**

KOSKO, ALLAN

373. La prétendue chaire d'astronomie de Copernic à la Sapienza de Rome, at pp. 283–286 in Wędkiewicz. Rheticus did not say that Copernicus had been a professor at the University of Rome.

KOT, STANISŁAW

374. The Cultural Background of Copernicus. *Polish Science and Learning*, 1943 (June), no. 3, pp. 5–11. Nationalité et culture polonaises de Copernic. *Le Monde*, 1954, April 7. This letter was kindly called to my attention by Prof. Alexandre Koyré.

KOWALENKO, WŁADYSŁAW

375. Z kroniki roku kopernikowskiego. *Urania* (Kraków), 1953, *24*: 18–21. The chronicle of the "Copernicus Year" 1953.

376. Bałtyk i Pomorze w historii kartografii (VII–XVI w.). *Przegląd zachodni*, 1954, *10*: no. 2, pp. 353–389. At p. 381, repeats earlier objections to L. A. Birkenmajer's suggestion that Copernicus was associated with Marco Beneventano's publication of Ptolemy's *Geography* (Rome, 1507).

KOWRACH, E. J.

377. Nikolaus Copernicus, a Four-Hundredth Anniversary. *Catholic World*,

1943, *157*: 130–134. Misdates Ptolemy in the pre-Christian era, and mistakenly asserts that an abstract of the *Revolutions* was printed in 1531.

KOYRÉ, ALEXANDRE

378. "Traduttore—traditore" à propos de Copernic et de Galilée. *Isis*, 1942–1943, *34*: 209–210. Menzzer's translation of the *Revolutions* (125) betrayed Copernicus by equating *orbium* in the title with "bodies" instead of "spheres."

379. Nicolas Copernicus. *Bulletin of the Polish Institute of Arts and Sciences in America*, 1942–1943, *1*: 705–730. A valuable review of the philosophical foundations and implications of Copernicus' astronomy.

380. *From the Closed World to the Infinite Universe* (Baltimore, 1957) xii + 313 p.; paperback reprint (New York, 1958) x + 312 p. The eminent French historian of science emphasizes (pp. 28–35) that for Copernicus the universe was still finite and closed; yet by denying the motion of the stars, he made it possible for his followers to assert the infinity of the universe.

KRAJEWSKI, WŁADYSŁAW

381. *Mikolaj Kopernik, twórca nowożytnej astronomii* (Warsaw, 1953) p. 54; 2d ed. (Warsaw, 1954) p. 57. Reviewed by Jan Gadomski, *Urania* (Kraków), 1956, *27*: 25–26. Nicholas Copernicus, the founder of modern astronomy.

382. Z kroniki roku kopernikowskiego. *Urania* (Kraków), 1953, *24*: 18–21. Polish celebrations of the Copernicus Year, 1953.

383. Sesja kopernikowska polskiej akademii nauk. *Myśl filozoficzna*, 1953, no. 4, pp. 347–352. A report on the Copernicus celebration under the auspices of the Polish Academy of Science on September 15–16, 1953. Reviewed **483**.

KRIEGER, ERHARD

384. Nikolaus Kopernikus—Begründer unseres Weltbildes. *Ostdeutsche Monatshefte*, 1955–1956, *22*: 13–16. A hysterical attribution to Copernicus of all sorts of astronomical discoveries which he never made.

KROEBER, ALFRED LOUIS

385. *Configurations of Culture Growth* (Berkeley and Los Angeles, 1944) x + 882 p. Copernicus viewed as a Slavo-German (pp. 155–156).

KRUG, ERICH

386. Coppernicus, der grosse Deutsche. *Die Sterne*, 1939, *19*: 81–83. The name of the Astronomisches Recheninstitut in Berlin-Dahlem was changed to "Coppernicus-Institut" in honor of the greatest German astronomer.

387. Die Coppernicus-Gedenkstätte der Reichshauptstadt. *Op. cit.*, pp. 203–204. The presentation of a bust of Copernicus, carved by the sculptor Kurt Lehmann, to the Coppernicus-Institut, Berlin-Dahlem.

388. Der unbekannte Kopernikus. *Die Sterne*, 1943, *23*: 59–70. An examination of some portraits of Copernicus.

389. Zeittafel zum Leben und Schaffen des Nikolaus Kopernikus. *Op. cit.*, pp. 70–74. A chronological table of some events connected with Copernicus' life and work, including the misstatement that the Gregorian calendar reform was based on Reinhold's *Tabulae prutenicae*.

390. Nikolaus Kopernikus. *Das Himmelsjahr*, 1943, pp. 80–85. Krug says that Copernicus studied at Rome, where in fact he lectured.

KRZESINSKI, ANDRÉ J.

391. Nicolas Copernic, humaniste et savant polonais. *Le Canada français*, 1943, *30*: 772–778. Copernicus viewed as a Polish humanist and scientist.

KUBACH, FRITZ

392. Nikolaus Kopernikus—Das Leben, Schaffen und Weltgebäude des grossen deutschen Naturforschers und die heutige Aufgabe der Kopernikusforschung. *Die Burg*, 1941, *2*: no. 2, pp. 7–23. Copernicus, like all later Aryan German investigators of nature, began his researches with a definite idea, founded it on observations, and checked it before publishing it; Kubach plainly implies that this admirable procedure was not followed by any scientists other than the Aryan Germans (whoever they may have been). He repeats Zinner's deliberate falsification of a letter written by Melanchthon, who called Copernicus a "Sarmatian" (or Pole), consciously mistranslated by Zinner and Kubach as a "Prussian." Kubach is so well informed about such matters that he describes Tycho Brahe, the famous Danish astronomer, as a Swede. He understands history so well that he looked upon the Nazi occupation of Poland as a final settlement.

393. *Nikolaus Kopernikus: Bildnis eines grossen Deutschen*, ed. Fritz Kubach (Munich and Berlin, 1943) x + 378 p.

394. In the Preface (pp. vii–viii) Kubach appeals to his readers to exert themselves to complete the (Nazi) revolution of their time, a vastly more important revolution than that accomplished by Copernicus. He also contributes a short biography of Copernicus

395. (Leben und Schaffen des Nikolaus Kopernikus, pp. 1–26, 313), a bibliography (see Copernicus, bibliography: 114), and a description of a projected nine-volume

396. ed. of Copernicus (Die Kopernikus-Gesamtausgabe, pp. 305–311; reprinted in *Zeitschrift für die gesamte Naturwissenschaft*, 1943, *9*: 111–114). In addition to these three contributions by Kubach, the volume also contains:

Excerpts from Copernicus' works in a new German translation by Karl Zeller, pp. 27–60; Hans Schmauch, "Nikolaus Kopernikus' deutsche Art und Abstammung," pp. 61–95, 314–318; August Faust, "Die philosophiegeschichtliche Stellung des Kopernikus," pp. 96–211, 318–370; Bruno Thüring, "Nikolaus Kopernikus—der grosse deutsche Astronom," pp. 212–232; Hans Schmauch, "Nikolaus Kopernikus und der deutsche Osten," pp. 233–256, 370–373; and Eberhard Schenk, "Kopernikus-Bildnisse," pp. 257–285, 373–374.

Reviewed by Max Caspar, *Kant-Studien,* 1943, *43*: 475–476, and by Eduard May, *Zeitschrift für die gesamte Naturwissenschaft,* 1944, *10*: 33–36.

397. Nikolaus Kopernikus, sein Leben und Schaffen und ihre Bedeutung für unsere Zeit. *Nationalsozialistische Monatshefte,* 1943, *14*: 468–479. Kubach repeats Brachvogel's contention that Copernicus arrived at heliocentrism independently of Aristarchus. He recognizes the "German style of thinking" in Copernicus' search for an orderly universe (a concept which others have found to be a commonplace in ancient Greek writers).

398. Die Kopernikus-Gesamtausgabe. *Mitteilungen der Sternwarte Königstuhl-Heidelberg,* 1943, no. 44. The plan for the ed. of the collected works of Copernicus.

399. Das Werk von Nikolaus Kopernikus, at pp. 95–96 in Diergart, ed. (151). The projected Nikolaus Kopernikus Gesamtausgabe.

See Copernicus, editions: *Nikolaus Kopernikus Gesamtausgabe* (118).

Reviewed **125**.

KUCHARZYK, HENRYK

400. The First Disciples of Copernicus in England. *Polish Science and Learning,* 1943 (June), no. 3, pp. 47–57. From Robert Record's *Castle of Knowledge* (1556) to William Gilbert's *De magnete* (1600).

KÜHLE, LUDWIG

401. *Punkt im All: Nikolaus Kopernikus, Künder eines neuen Weltbildes* (Berlin, 1943) p. 159. A somewhat imaginative presentation of the biography of Copernicus and his accomplishments within a framework of the history of astronomy.

KUHN, THOMAS S.

402. *The Copernican Revolution* (Cambridge, Mass., 1957) xviii + 297 p. Reviewed by C. Doris Hellman, *Renaissance News,* 1957, *10*: 217–220; by Henri Michel, *Ciel et terre,* 1957, *73*: 406–407; by Hugo N. Swenson, *Scientific Monthly,* 1957, *85*: 276–277; by James R. Newman, *Scientific American,* 1957 (October), *197*: no. 4, pp. 155–160; by Herbert Butterfield, *American Historical Review,* 1957–1958, *63*: 656–657; by Charles C. Gillispie, *American Scientist,* 1958, *46*: 64A; by Harry Woolf, *Isis,* 1958, *49*: 366–367; and by Edward Rosen, *Scripta mathematica* (forthcoming).

Copernicus viewed as the heir of an ancient tradition and as the initiator of the modern viewpoint.

KUKARKIN, BORIS V., ed.
403. *Nikolai Kopernik* (Moscow, 1955) p. 112. This symposium, published under the auspices of the Akademiia nauk SSSR, astronomicheskii soviet, contains:
A commemorative address delivered by Aleksandr Nikolaevich Nesmeianov on June 3, 1953 (pp. 5–6); Aleksandr Aleksandrovich Mikhailov, "The Life and Work of Copernicus" (pp. 7–32); Boris Fedorovich Porshnev, "The Age of Copernicus" (pp. 33–56); Vladimir Aleksandrovich Fok, "The Copernican and Ptolemaic Systems in the Light of Present-Day Theories of Gravitation" (pp. 57–72); F. J. Nesteruk, "Copernicus' Waterworks" (pp. 73–89); and a list, prepared by N. N. Deikova, of books and articles displayed at the Copernicus exhibit (pp. 91–111).
Brief notice in *Mathematical Reviews*, 1956, *17*: 1170.

KULCZYŃSKI, STANISŁAW
See *Pressebulletin*.

KULIKOVSKII, P. G.
404. Nikolai Kopernik. *Slaviane*, 1953, no. 5, pp. 51–52. A brief commemorative article in Russian.

Kulturprobleme des neuen Polen
405. 1951, *3*: no. 7, pp. 15–17. Zum Todestag von Mikołaj Kopernik. This article in the magazine published by the Polish Information Bureau in Berlin puts Copernicus' uncle in the wrong diocese, postpones the astronomer's acquisition of his canonry, and misplaces its location. The statement that Copernicus' *Revolutions* remained on the Roman Catholic Index of Prohibited Books only twenty years presumably contains a misprint.
406. 1953,*5*: no. 5, p. 20. Wissenschaftliche Beiträge zum Kopernikusjahr. Enumerates the articles on Copernicus published in Polish journals in 1953, the Copernicus Year.

407. 1953, *5*: no. 6, pp. 15–16. Kopernikus-Gedenkfeiern in aller Welt. An account of Copernicus celebrations in several countries.
408. 1953, *5*: no. 7, pp. 9–10. Das Kopernikus-Museum in Frombork. A description of the Copernicus Museum in Frombork.
409. 1954, *6*: no. 1, pp. 10–11. Die Krakauer Kopernik-Ausstellung. A description of the Copernicus exhibit in Kraków.
410. 1954,*6*: no. 3, pp. 20–21. Mikołaj Kopernik—der Erste in der komplexen Ausnutzung des Wassers. Claims for Copernicus the construction of the first municipal waterworks on a hydrotechnical level surpassing the conventional medieval installation.

KUNITSKY, R. V.
411. *Razvitie vzgliadov na stroenie solnechnoi sistemy*, 5th ed. (Moscow, 1952) p. 80. This brief historical outline of the development of theories concerning the structure of the solar system discusses Copernicus at pp. 35–41.

KURDYBACHA, ŁUKASZ and ZONN, WŁODZIMIERZ
412. *Mikołaj Kopernik* (Warsaw, 1951) p. 33. A popular pamphlet, written jointly by a historian of culture and an astronomer.

LABÉRENNE, PAUL
413. Nicolas Copernic. *La Pensée*, 1953 (September–October), no. 50, pp. 28–40. A lecture delivered on May 21, 1953. Besides his other achievements, Copernicus sought to alleviate the misfortunes of the people among whom he lived.

LANDOVÁ-ŠTYCHOVÁ, LUISA
414. Polské oslavy Mikuláše Koperníka. *Říše hvězd*, 1953, *34*: 149–151. Polish celebrations of Copernicus.

LANG, JOHANNES
415. *Die Widerlegung des kopernikanischen Weltbildes* (Frankfurt am Main, 1938) p. 59. The author of this attempt to refute the Copernican conception of the universe

still believed (pp. 44–45) that the preface to the *Revolutions* was written by Copernicus; yet it was Osiander, not Copernicus, who there labeled the *Revolutions* a merely hypothetical work. Lang denies the motion of the earth and the convexity of its surface.

LANNING, JOHN TATE
416. El sistema de Copérnico en Bogotá. *Revista de historia de América*, 1944, *18*: 279–306. Documents pertaining to the controversy aroused by the defense of Copernicanism at Bogotá in 1770 by José Celestino Mutis, a professor of mathematics.

LARINK, JOHANNES
417. Kopernikanisches System und astronomische Messkunst. *Die Naturwissenschaften*, 1944, *32*: 178–185. If Copernicus was right in attributing an orbit to the earth, then the stars should have shown an annual parallax. This shift is so minute that vastly improved observational instruments were needed to detect it. Reviewed **370.**

LASKIEWICZ, ALFRED
418. Działalność Kopernika na polu lecznictwa, at pp. 23–33 in 367, 2d ed. Copernicus' activities in the field of medicine.

LAUE, MAX VON
419. Von Kopernikus bis Einstein. *Naturwissenschaftliche Rundschau*, 1957, *10*: 83–89; published also in *Jahrbuch 1956 der Max-Planck-Gesellschaft*, pp. 150–172. This printed version of a lecture delivered on June 25, 1956 at Lindau on Lake Constance, at the sixth meeting of recipients of the Nobel Prize, accepts that interpretation of the relativity theory according to which the Ptolemaic and Copernican systems are mathematically equivalent, the choice between them being entirely a matter of emotional preference.

LAZZARI, ALFONSO
420. La cultura scientifica a Ferrara nei tempi di Copernico. *Atti dell' Accademia delle scienze di Ferrara*, 1947–1948, *25*: 115–133. A brief sketch of the scientific atmosphere at the University of Ferrara, in the days when Copernicus took his degree in canon law there.

LENARD, PHILIPP
421. *Grosse Naturforscher*, 4th ed. (Munich, 1941); 5th ed. (Munich, 1942); 6th ed. (Munich, 1943) p. 348. This enlarged and revised ed. of a work first published in 1929 contains a brief biography of Copernicus (pp. 24–27). Why did Lenard say that Copernicus passed through Vienna?

LEOPARDI, GIACOMO
422. *Storia della astronomia dalla sua origine fino all' anno MDCCCXI.* Reprinted in *Tutte le opere di G. Leopardi*, ed. Francesco Flora, Poesie e prose, vol. II (Milan, 1940; 2d ed., 1945; 3d ed., 1949). This history of astronomy from its beginnings to the year 1811 was written by the great Italian poet in 1813, when he was a child prodigy of fifteen, but was first published in 1880. The section dealing with Copernicus occurs in this ed. at pp. 895–898. At vol. I, pp. 989–1000, will be found
423. Leopardi's dialogue, "Il Copernico," written in 1827 and first published in 1845. See Junker, Ernst.

LEŚNODORSKI, BOGUSŁAW
424. *Kopernik—człowiek   Odrodzenia* (Warsaw, 1953) p. 65. Copernicus—man of the Renaissance.
425. Mikołaj Kopernik. *Nowe drogi*, 1953, *7*: no. 6, pp. 60–79. Part of a speech delivered on September 15, 1953 before the Polish Academy of Science.
426. Niektóre elementy założeń poznawczych Kopernika, at pp. 57–84 in *Odrodzenia w Polsce*, vol. 2, part 2 (Warsaw, 1956). Some aspects of Copernicus' fundamental principles.
427. Elementy materializmu w twórczości Kopernika, *op. cit.*, pp. 111–112. The elements of materialism in Copernicus' works.

ANNOTATED COPERNICUS BIBLIOGRAPHY

See *Sesja kopernikowska* and Copernicus, bibliography (115).

LESSER, J.
428. Copernicus. *Contemporary Review*, 1943, *163*: 370–373.

LICHTENBERG, GEORG CHRISTOPH
429. *Nikolaus Kopernikus* (Königsberg, 1943) p. 46. Reviewed by Max Bense, *Europäische Revue*, 1944, *20*: 50.
This essay was originally published in *Pantheon der Deutschen* (Chemnitz, 1794–1800), vol. III, 116 pp. with separate numeration. In a slightly abbreviated form it was republished in 1943, with a postscript (pp. 45–46) by the editor, Götz von Selle. He did not utilize the results attained by the research of the last hundred and fifty years to remove the obsolete portions of Lichtenberg's essay. For example, Selle retained (p. 10) without any comment Lichtenberg's remark that the given name of Copernicus' brother is not known.

LIGOCKI, EDWARD
430. *Galileusz o Koperniku. Problemy*, 1953, *9*: 7–15. The discussion of Copernicus in Galileo's *Dialogue*, by the translator of the *Dialogue* into Polish.
431. *Kopernik na tle epoki. Op. cit.*, pp. 522–531. A popular article on Copernicus against the background of his time.

LIMBERGEN, JOS. VAN
432. *De groote revolutie van Copernicus* (Brussels, 1944) p. 193. A popular account in Flemish of the great Copernican revolution.

LIPIŃSKI, EDWARD
433. *Kopernik jako ekonomista. Wiedza i życie*, 1953, *20*: 248–252. Copernicus as an economist.
434. *Nauka Mikołaja Kopernika o pieniądzu. Polska akademia nauk, Sprawozdania z czynności i prac*, 1953, *1*: 125–139. A summary of discussions concerning Copernicus' monetary theories.
435. *Ekonomiczne poglądy Mikołaja Kopernika. Nauka polska*, 1953, *1*: no. 3, pp. 63–100. The economic views of Copernicus.

436. *Poglądy ekonomiczne Mikołaja Kopernika* (Warsaw, 1955) p. 188. Reviewed by Henryk Dunajewski, *Ekonomista*, 1955, no. 4, pp. 181–190.
Two excerpts, one on bimetallism and the other on Copernicus' conception of the ideal society, were translated into French by Allan Kosko at pp. 111–114 in Wędkiewicz.
An expansion of 435.
437. *Myśl ekonomiczna polskiego Odrodzenia*, at pp. 161–190 in *Odrodzenie w Polsce*, vol. I (Warsaw, 1955), a symposium on the Renaissance in Poland. Lipiński's contribution, "Economic Thought in the Polish Renaissance," discusses Copernicus as an economist at pp. 165–167.
438. *O interpretację myśli ekonomicznej Kopernika. Ekonomista*, 1956, no. 1, pp. 136–153. Interpretation of Copernicus' ideas on economics.
439. *Studia nad historią polskiej myśli ekonomicznej* (Warsaw, 1956) p. 536. Reviewed by Jack Taylor, *American Historical Review*, 1957–1958, *63*: 418–419. These studies in the history of Polish economic thought deal with Copernicus in Chapter II (pp. 28–61).

LIST, HORST FRIEDRICH
440. *Weg zu den Sternen* (Neustadt, 1948) p. 100. Juvenile fiction, in a series called "Gelebtes Leben."

LOCKWOOD, MARIAN
See Draper, Arthur L.

LÖBSACK, WILHELM
441. *Coppernikus und unsere Zeit. Der Deutsche im Osten*, 1941, *4*: 11–20. An expansion of a lecture delivered in December, 1940 at Thorn (now Toruń). Löbsack mistakenly asserts that Poland knew nothing about its supposed national hero Copernicus before 1800. Actually Szymon Starowolski's biography of the astronomer as one of Poland's hundred greatest authors was printed four times before 1800 (twice in Frankfurt am Main). Löbsack concludes that just as Copernicus' mastery of astronomy overthrew the unnatural medieval view of the world, so National Socialist

knowledge of race and blood will overthrow the French Revolution's doctrine of the equality of all men.

442. Nikolaus Kopernikus—ein deutscher Revolutionär, at pp. 5–14 in Georgens, ed. The question whether Copernicus was a German or a Pole was settled by the Nazi conquest of Poland.

LOMONOSOV, MIKHAIL VASILIEVICH
443. O ruchu ziemi. *Problemy*, 1953, *9*: 794. The great eighteenth-century Russian scientist-poet's verses About the Earth's Motion are translated into Polish by Julian Tuwim.

LORENTZ, STANISŁAW
444. *The Renaissance in Poland* (Warsaw, 1955) p. 95. A splendid collection of photographs, one group being devoted to Copernicus and the Copernicus Museum maintained by the Polish government in Frombork.

LOSADA Y PUGA, CRISTÓBAL DE
445. *Copérnico: de la astronomía antigua a la moderna* (Lima, 1943) p. 36. Offprinted from *Revista de la Universidad católica del Perú*, 1943, *11*: 149–178. In a lecture delivered at a quadricentennial commemoration of Copernicus at the Catholic University of Peru on June 30, 1943, the author mistakenly asserted that the astronomical system in the *Commentariolus* was propounded as a hypothesis, not as the truth. He even blundered so badly as to claim Copernicus, rather than Osiander, as the author of the preface to the *Revolutions*.

LÜCK, KURT
446. *Der Mythos vom Deutschen in der polnischen Volksüberlieferung und Literatur* (Posen, 1938; Ostdeutsche Forschungen, 7); 2d ed., Leipzig, 1943, x + 518 p. Pp. 431–439, 506–507, of the first ed., and pp. 451–459, 518, of the second ed., stridently exclaim that Copernicus was German, not Polish.

ŁUKAWIECKA, L. S.
447. Postać M. Kopernika na scenie teatralnej, at pp. 21–23 in 367, 2d ed. Copernicus as a character in the theater.

LUNDMARK, KNUT
448. Nicolaus Kopernikus and His Astronomical Reformation. *Meddelande från Lunds astronomiska observatorium*, series II, no. 112, Historical Notes and Papers, no. 19 (1944) p. 18; published also in *Kungliga Fysiografiska Sällskapets i Lund*, Förhandlingar, 1944, *14*: 22–39. Emphasizes Copernicus' simplification of the Ptolemaic astronomy and revival of the sun's importance.

A. D. M.
449. Nicola Copernico gloria della Università di Cracovia. *L'Osservatore romano*, 1940, July 15–16, *80*: no. 162, p. 3. Mistakenly asserts that Copernicus was ordained a priest about 1495, and that he wrote the letters of 1516 and 1521 denouncing the Teutonic Knights. Why does A. D. M. say that Copernicus composed the *Commentariolus* in 1507; that he started to write the *Revolutions* in 1517 and finished it in 1530; and that Osiander added the words "orbium coelestium" to the title of *De revolutionibus*?

MADWAR, MOHAMED REDA
450. Nicolas Copernic. *Bulletin de l'Institut d'Egypte*, 1942–1943, *25*: 286–276 (in Arabic). This lecture dealing with the biography of the astronomer was delivered at a special meeting in commemoration of Copernicus on May 24, 1943.

MAJEWSKI, ZBIGNIEW
451. Rocznica kopernikowska w naukowej prasie radzieckiej. *Myśl filozoficzna*, 1954, no. 1, pp. 353–357. The Copernicus Year in scientific publications.

MAJOWSKA, YOLANDA
452. The Copernican Quadricentennial in Milwaukee. *Popular Astronomy*, 1943, *51*: 322–323.

MAKEMSON, MAUD WORCESTER
453. Changing Ideas of the Universe. *Popular Astronomy*, 1943, *51*: 307–316. A review of the astronomical beliefs out of which Copernicanism emerged.
454. A Tribute to Copernicus. *Sky and Telescope*, 1943, *2*: no. 7, pp. 11–14.

MAŁACHOWSKI, STANISŁAW
455. Biruni a koncepcje Kopernika. *Problemy*, 1954, *10*: 606–607. Al-Biruni and Copernicus.

MALITA, MIRCEA
456. Nicolai Copernic. *Revista matematica si fizica*, 1953, no. 7, pp. 161–164.

MANCE, EVELYN M.
457. Some Centenaries for 1943. *Journal of the British Astronomical Association*, 1943, *53*: 65–66. Repeats the misstatement that Copernicus "became a priest."

MARCONI, BOHDAN
458. Toruński portret Mikołaja Kopernika. *Biuletyn historii sztuki*, 1953, *15*: no. 2, pp. 3–5. A study of the Copernicus portrait at Toruń.
459. W sprawie współczesności toruńskiego portretu Kopernika. *Op. cit.*, 1954, *16*: no. 2, pp. 277–279. This sequel to 458 maintains that the Toruń portrait was done in Copernicus' lifetime.
Reviewed **799.**

MAROŃ, KAZIMIERZ
460. Nieżnana dzieło Kopernika. *Wzechświat*, 1953, p. 208. Copernicus' *Commentariolus*.

MARSHALL, ROY K.
461. Fels Planetarium Celebration of the Copernican Quadricentennial. *Popular Astronomy*, 1943, *51*: 316–321. Mistakenly dates the composition of the *Commentariolus* "about 1530."

MASON, STEPHEN FINNEY
462. A History of the Sciences: Main Currents of Scientific Thought (London, 1953); Main Currents of Scientific Thought: A History of the Sciences (New York, 1953) viii + 520 p. Somewhat revised for translation into French by Marguerite Vergnaud under the title *Histoire des sciences* (Paris, 1956) p. 476.
Chapter 12 is entitled "The Copernican System of the World," and pp. 99–105 deal with Copernicus himself.

MASOTTI, ARNALDO
463. I "Septem sidera" attribuiti a Niccolò Copernico tradotti in italiano da Mons. Vincenzo Botto. *La scuola cattolica*, 1943, *71*: 430–441. Although Masotti recognizes the validity of the objections to attributing the *Septem sidera* to Copernicus, he prints a translation into Italian by Vincenzo Botto.
464. Niccolò Copernico. *Memorie della Società astronomica italiana*, 1944, *16*: 193–207. A compact little essay, based on extensive acquaintance with the secondary literature.

MATHES, JOH.
465. Luther und Kopernikus. *Evangelisch-Lutherische Kirchenzeitung*, 1951, *5*: 66. Summarizes Norlind's attempt to interpret as an interpolation Luther's characterization of Copernicus as a fool.

MAY, EDUARD
466. Gedanken über die Wirkung und Ausbreitung der kopernikanischen Lehre. *Zeitschrift für die gesamte Naturwissenschaft*, 1943, *9*: 102–107. By destroying the traditional distinction between the heavens and the earth, that is, by making the earth a heavenly body among heavenly bodies, Copernicus paved the way for a truly universal science.
Reviewed **393.**

MCCOLLEY, GRANT
467. Milton Opposed Copernicus. *Sky*, 1938–1939, *3*: no. 5, pp. 6–7, 24. McColley denounces two lines in John Milton's *Paradise Lost* as "cosmological nonsense" since they imply "that the earth was ascribed spheres in Ptolemaic astronomy, which it was not." McColley's comment shows not only that he could not write English, but also that he could not even understand it, since Milton plainly implies that the spheres in question are outside the earth.
468. The Universe of De revolutionibus. *Isis*, 1939, *30*: 452–472. A hopelessly confused contention that Copernicus conceived the universe to be infinite. This mis-

take, accompanied by gratuitous attacks on competent scholars, was later withdrawn by McColley, not in an open and manly manner, but only by implication and without reference to his previous error, in a paper abounding in fresh blunders and written in atrocious

469. English: Humanism and the History of Astronomy, at pp. 321–357 in *Studies and Essays in the History of Science and Learning Offered in Homage to George Sarton* (New York, 1946): "In Copernicus' system, the distance from the sun to the fixed stars was in-de-finite" (p. 351).

470. An Early Poetic Allusion to the Copernican Theory. *Journal of the History of Ideas*, 1942, *3*: 355–357. This "early poetic allusion to the Copernican theory" (p. 355) "may not refer to the Copernican theory" (p. 357).

471. The Eighth Sphere of Copernicus. *Popular Astronomy*, 1942, *50*: 133–137. According to McColley, Copernicus believed that the fixed stars were at unequal distances from the center of the universe. As evidence, McColley uses a familiar *Revolutions* passage, which he translates in his usual muddled way, and from which he draws wholly unwarranted inferences. "The diagram used in the *Revolutions* is precisely that previously employed by Rheticus in the *Narratio prima*," says McColley, citing a nineteenth-century ed. and unaware that both edd. of the *Narratio prima* previous to the *Revolutions* contained no diagrams. This piece of stupidity is only one of the multitude with which McColley polluted the literature of the subject until his death on July 5, 1953.

MEICHNER, FRITZ

472. *In der Mitte steht die Sonne* (Munich, 1943) p. 115. An interesting fictionalized biography of Copernicus, based on the conventional German treatment of the facts, with which it takes further liberties.

MELCHIOR, PAUL J.

473. Sur une observation faite par Copernic et Dominique Maria. *Académie*

royale de Belgique, Bulletin de la classe des sciences, 1954, *40*: 416–417. Copernicus' observation at Bologna on March 9, 1497 of a lunar occultation has been questioned. To uphold the validity of the observation, Melchior maintains that Copernicus saw something which Copernicus himself never claimed to have seen.

METTENLEITER, FRITZ

474. *Nikolaus Kopernikus* (Stuttgart, 1941) p. 344. This (unhistorical) historical novel was printed for the Reichswehr in Paris under German control. By crudely falsifying the documentary sources it turned Copernicus into a German patriot at a time when there was no Germany, and made him sound like Alfred Rosenberg or Joseph Paul Goebbels. How popular such reading matter was then may be judged by the fact that Mettenleiter's novel reached its fourth ed. in 1943.

MEYER, HEINRICH

475. More on Copernicus and Luther. *Isis*, 1954, *45*: 99. Corrects Norlind's misunderstanding of a part of Luther's famous remark about Copernicus.

MEYERHOF, MAX

476. Aristarque de Samos, le Copernic de l'antiquité. *Bulletin de l'Institut d'Egypte*, 1942–1943, *25*: 269–274. The heliocentric astronomy proposed by Aristarchus was unknown to the medieval Muslims.

MIĄCZYŃSKI, JAN ANTONI

477. *Poznaj Muzeum Mikolaja Kopernika we Fromborku* (Warsaw, 1949) p. 35. A guide to the Copernicus Museum at Frombork.

MICHAEL, GEORGE

478. *The Big Five* (London, 1944) p. 24. Translated into Polish under the title *Wielka piątka* (London, 1945) p. 24. Copernicus (pp. 5–7) is one of the big five Poles of history.

MIEDZIŃSKI, FLORIAN

479. Kopernik nie odkrył żadnej nowości. *Problemy*, 1953, *9*: 404–406. Examining the relation between Copernicus and Celio Calcagnini, the author deals with the question whether the former discovered anything new.

MIELI, ALDO

480. *Sumario de un curso de historia de la ciencia* (Santa Fe, 1943) vii + 251 p. Brief discussion of Copernicus at pp. 32–34.

481. *Panorama general de historia de la ciencia*, vol. III (Buenos Aires, 1951): La eclosión del renacimiento. xxii + 400 p. Contains a lively account of Copernicus (pp. 235–258), marred by some unfortunate errors regarding his doctoral degree, friendship with Calcagnini, the date when the latter's discussion of the earth's motion was published, the title of the *Revolutions*, and Copernicus' use of the word "hypothesis."
See **16.**

MIKELEITIS, EDITH

482. *Die Sterne des Kopernikus* (Braunschweig, Berlin and Hamburg, 1943) p. 150. Published also in *Westermanns Monatshefte*, 1942–1943, *87*: 449–456, 489–496, 525–532; 1943–1944, *88*: 13–21. This novel reached its eighth ed. in 1948.

MIKHAILOV, ALEKSANDR ALEKSANDROVICH, ed.

483. *Nikolai Kopernik sbornik statei k chetyrekhsot letiiu so dnia smerti* (Moscow and Leningrad, 1947) p. 220. Reviewed by Z. A. Zeitlin, *Voprosy filosofii*, 1948, no. 1, pp. 306–311; by A. P. Yushkevich, *Vestnik akademii nauk SSSR*, 1948, *18*: no. 7, pp. 114–115; and by Władisław Krajewski, *Nowe drogi*, 1949, *3*: 292–293.

This collection of articles celebrating the fourth centenary of Copernicus' death appeared under the auspices of the Committee for the History of the Mathematical and Physical Sciences, Academy of Science, U.S.S.R. It contains a commemorative address delivered at the Academy on June 6, 1943 by Nahum Ilyich Idelson. In addition to this speech on the life and work of Copernicus (pp. 5–42), Idelson writes on the history of planetary theory from the

Greeks to Copernicus (pp. 84–179). Sergei Danielovich Skazkin sketches the cultural background of Copernicus' time (pp. 43–63: The Epoch of Copernicus). Ivan Ivanovich Tolstoi discusses Copernicus' translation of the letters of Theophylactus Simocatta (pp. 64–83). Vladimir Aleksandrovich Fok examines the Ptolemaic and Copernican systems in the light of the theory of general relativity (pp. 180–186). Fyodor Aleksandrovich Petrovsky translates into Russian the dedication of the *Revolutions* and chapters 1–10 of Book I (pp. 187–213); the accompanying notes by Idelson occupy pp. 214–217. I owe this description to the kindness of Prof. Vasily Pavlovich Zubov of the Institute for the History of Science and Technology, Academy of Science, U.S.S.R., who also informs me that the first complete translation of the *Revolutions* into Russian is in preparation as a part of the series "Classics of Science." The translation is being done by Ivan Nikolaievich Vesselovsky, and will be accompanied by a commentary.

484. Nikolai Kopernik. *Vestnik akademii nauk SSSR*, 1953, no. 6, pp. 32–41.

485. Nikolai Kopernik—revoliutsioner v nauke. *Voprosy filosofii*, 1953, no. 4, pp. 89–98. Translated into German under the title Nikolaus Kopernikus—ein Revolutionär der Wissenschaft. *Forum* (Organ des Zentralrats der FDJ für die deutschen Studenten), 1953 (November 21), 7: Wissenschaftliche Beilage no. 40, p. 8.

Since Copernicus provided the first reliable method of ascertaining planetary distances, his astronomical system is like an architect's plan of a building, whereas Ptolemy's system rather resembles a photograph of the building's exterior.

486. Nikolai Kopernik, ego zizn i tvorchestvo, at pp. 7–32 in Kukarkin, ed.

MILES, F. F.
See Victoria University College.

MILLER, BARBARA ANNE

487. Solem fixit, movit terram. p. 12. An undergraduate essay, awarded the Conant Prize for 1957 by Harvard University.

Minnaert, Marcel G. J.

488. Het levenseinde van Copernicus. *Hemel en Dampkring*, 1955, *53*: 97–98. Discusses the question raised by Dijksterhuis in a review (*op. cit.*, p. 16) whether Copernicus did or did not see some proof sheets of the *Revolutions* before his death. See *Sesja kopernikowska* and Vereniging Nederland-Polen.

Miró Quesada, Oscar

489. *Copernico: su vida y su obra* (Lima, Peru, 1950) p. 191. Although loyal to tradition, Copernicus was an intellectual liberator, freeing the human mind from subjection to mere authority.

Mizwa, Stephen P.

490. Libraries in the Days of Copernicus. *Wilson Library Bulletin*, 1942–1943, *17*: 616, 619.

491. *Nicholas Copernicus* (New York, 1943) p. 88. Reviewed by Frederick E. Brasch, *Science*, 1943, *98*: 40–42; by Mary A. Evershed, *Observatory*, 1943, *65*: 129–130; by Otto Struve, *Astrophysical Journal*, 1943, *97*: 276; by Dorrit Hoffleit, *Sky and Telescope*, 1943, *2*: no 6, p. 18; and by Stefan Oświęcimski, *Życie nauki*, 1947, *4*: 374–375.
Reprinted in part in *Publications of the Astronomical Society of the Pacific*, 1943, *55*, 65–72.
Presents a Polish point of view.

492. Nicholas Copernicus, the Father of Modern Astronomy. *Science*, 1943, *97*: 192–194.

493. Mizwa, ed. *Nicholas Copernicus, a Tribute of Nations* (New York, 1945) xix + 268 p. Reviewed by C. A. Chant, *Journal of the Royal Astronomical Society of Canada*, 1945, *39*: 194–195, and by Curvin H. Gingrich, *Popular Astronomy*, 1945, *53*: 525–526.
A record of the celebrations commemorating Copernicus' four-hundredth anniversary, with an amusing Foreword on Quadricentennials by Harlow Shapley. See Birkenmajer, Aleksander (42).

Moepert, Adolf

494. Kopernikus und sein Abstammungsnachweis. *Zeitschrift des Vereins für* *Geschichte Schlesiens*, 1943, *77*: 56–65. Derives the astronomer's surname from the Czech word "koprník" (meaning "dill") rather than from the Polish word "kopr" (meaning "fennel").

Mohr, J. M.

495. Nicholas Copernicus. *Czechoslovak Journal of Physics*, 1953, *3*: 316–318. Emphasizes Copernicus' belief in the physical truth of his system.

496. *Mikuláš Kopernik* (Prague, 1953) p. 24). A lecture in which Rheticus is praised and Osiander is condemned for their part in the first publication of the *Revolutions*.

Morstin, Ludwik Hieronim

497. *Kłos panny*, 3d ed. (Warsaw, 1947) viii + 269 p. A novel in which Copernicus has many wonderful adventures, including an encounter with a magician, Dr. Faust. The first ed. (1929) was dedicated to the eminent Copernicus scholar, Ludwik Antoni Birkenmajer, and was translated into French by Paul Cazin, professor of Polish literature at the University of Aix-en-Provence, under the title *L'Epi de la vierge* (Paris, 1937), p. 236. The novel was translated into Slovak by Andrej Žarnov under the title *Kopernik* (Trnava, 1948) p. 213.
See Copernicus, editions and translations: Teofilakt Symokatta, *Listy* (123).

Mosharrafa, Ali Mustafa

498. Nicolaus Copernicus and the Evolution of Scientific Thought. *Bulletin de l'Institut d'Egypte*, 1942–1943, *25*: 287–291. Copernicus' achievement is comparable to Einstein's: the former transferred the center of his coordinate axes from the earth to the sun, while the latter "asks us to put our faith in no axes of reference whatever."

Moulton, Forest Ray and Schifferes, Justus J.

499. *The Autobiography of Science* (Garden City, 1945) xxxi + 666 p. Reprints an excerpt from the *Commentariolus* at pp. 59–63.

ANNOTATED COPERNICUS BIBLIOGRAPHY

MUNIER-WROBLEWSKI, MIA
500. *Niklas Koppernigk* (Munich, 1943) p. 397. Reviewed by Bernhard Sticker, *Die Himmelswelt*, 1944, *54*: 48. A novel.

MUNITZ, MILTON K.
501. *Space, Time and Creation* (Glencoe, 1957) p. 182. Copernicus was uncertain whether the universe is finite or infinite.
502. Munitz, ed. *Theories of the Universe* (Glencoe, 1957) x + 437 p. The editor discusses Copernicus at p. 142, and at pp. 149–173 reprints the Dobson-Brodetsky translation of the dedication of the *Revolutions* as well as chapters 1–11 of Book 1.

MURAWA, FELIKS
503. Żywot i dzieła Mikołaja Kopernika, at pp. 5–14 in *Wystawa kopernika w Olsztynie* (Olsztyn, 1946) p. 20, a guide to the Copernicus exhibition at Olsztyn.

Myśl filozoficzna, 1953, no. 1, pp. 137–143
504. Rocznica kopernikańska. The Copernicus anniversary viewed in the long perspective of the history of Polish culture.

NADOLSKI, BRONISŁAW
505. Walka o myśl Kopernika i losy jej w Polsce. *Problemy*, 1954, *10*: 13–20. The struggle over Copernicanism and its outcome in Poland.

Nation und Staat, 1938–1939, *12*: 469–470
506. Ein Gerichtsurteil über die Nationalität von Nikolaus Coppernicus. Cites an account in *Deutsche Rundschau in Polen* (January 21, 1939) of a trial in Bydgoszcz regarding the distribution, by Germans living in Poland, of a postcard describing Copernicus as their greatest son.

Nature, 1943, *151*: 583
507. Copernicus and His Influence on Astronomical Thought. A quadricentennial editorial.

1948, *162*: 445
508. Nicolas Copernicus. The three places in which Copernicus lived and died.

Naturwissenschaftliche Rundschau, 1953, 6: 509
509. Das Kopernikus-Jahr in Polen.

An announcement of Polish plans to commemorate Copernicus in 1953.

Nauka i zhizn, 1953, *20*: no. 2, p. 47
510. Nikolai Kopernik. A brief commemorative article in the Russian journal of popular science.

Nederland Polen
511. 1952, *6*: no. 12, p. 31. 1953 Copernicusjaar.
512. 1953, *7*: no. 1–2, p. 24. 1953 Copernicusjaar. Two announcements that Poland would celebrate 1953 as Copernicus Year.
513. 1953, *7*: no. 3, p. 16. Jubileumjaar van Mikolaj Kopernik.
514. 1953, *7*: no. 6, pp. 12–13. Het Kopernikjaar.
515. 1953, *7*: no. 7–8, p. 12. Mikolaj Kopernik. A Dutch report of the Soviet celebration of Copernicus.
516. 1953, *7*: no. 9, p. 36. Postzegelhoekje. The Polish series of postage stamps commemorating Copernicus.

NEEDHAM, JOSEPH
517. *Thoughts on the 400th Anniversary of Copernicus* (Szechuan Provincial Education Authority, 1943), p. 20 (Chinese) + p. 5 (English summary). In this lecture delivered in May, 1943 to the middle schools at Chengtu, the distinguished historian of western and Chinese science aptly compares what Galileo did for Copernicus with what Huxley did for Darwin.

NESMEIANOV, ALEKSANDR NIKOLAEVICH
See Kukarkin, ed.

NESTERUK, F. J.
518. Nikolai Kopernik kak gidrotekhnik. *Izvestia akademii nauk SSSR, Otdelenie tekhnicheskikh nauk*, 1953, pp. 1341–1349. Translated into Polish by Wojciech Suchorzewski under the title Mikołaj Kopernik — budowniczy wodociągow, *Studia i materiały z dziejów nauki polskiej*, 1955, *3*: 207–219, with a summary in English at pp. 218–219. Nesteruk's conclusions were translated into French by Allan Kosko at p. 231 in Wędkiewicz.

ANNOTATED COPERNICUS BIBLIOGRAPHY

NEUGEBAUER

Nicholas Copernicus as a hydraulic engineer.

519. K 410–letiiu so dnia smerti veliko-go polskogo uchenego Nikolaia Kopernika. *Gidrotekhnicheskoe stroitelstvo*, 1953, *22*: no. 8, p. 47. Translated into Polish under the title Mikołaj Kopernik jako hydrotechnik, *Problemy*, 1954, *10*: 54. In commemoration of the 410th anniversary of Copernicus' death.

520. Ingenernie raboti Nikolaia Kopernika, at pp. 73–89 in Kukarkin, ed.

NEUGEBAUER, OTTO E.

521. *The Exact Sciences in Antiquity*, 2d ed. (Providence, 1957) xvi + 240 p. Copernicus' theories of the moon and Mercury, and his contributions to astronomy (pp. 196–205). Reviewed **124, 620.** See Erhardt.

New York Herald Tribune, 1943, May 25, section II, p. 6

522. Copernicus after 400 Years. A quadricentennial editorial. Reprinted at pp. 205–207 in 493.

New York Times, 1943, May 23, p. 16E

523. Symbolic Copernicus. A quadricentennial editorial. Reprinted at pp. 202–205 in 493.

NEYEL, HANNS

524. De revolutionibus, ein polnischer Beitrag zum Humanismus. *Jenseits der Oder*, 1954, *5*: no. 2, pp. 12–13. Mistakenly asserts that the *Commentariolus* was printed in Copernicus' lifetime. Why does the author say that Cardinal Schönberg wanted the manuscript of the *Revolutions* copied "for the use of the Pope"?

NIWELIŃSKI, JÓZEF

525. Mikołaj Kopernik jako lekarz. *Wszechświat*, 1953, pp. 190–192. Copernicus as a physician.

NOBILE, VITTORIO

526. Il conflitto fra copernicisti e aristotelici nella sua essenza e nel pensiero di Galileo. *Atti della Accademia nazionale dei lincei, Rendiconti, classe di scienze fisiche*, 1950,*9*: 299–306; 1951,*10*: 337–343; 1951, *11*: 311–319. A re-examination, from the vantage point of modern relativity theory, of Galileo's attitude toward the truth or falsity of the Copernican astronomy.

NORDENMARK, N. V. E.

527. Laurentius Paulinus Gothus, Sveriges förste professor i astronomi. *Populär astronomisk Tidskrift*, 1951, *32*: 94–99. Paulinus, Sweden's first professor of astronomy, introduced the Copernican system into that country.

528. Laurentius Paulinus Gothus, Föreläsningar vid Uppsala universitet 1599 över Copernicus hypotes. *Arkiv för astronomi*, 1952, *1*: 261–299. Reviewed by Sten Lindroth, *Lychnos*, 1952, pp. 398–399 (in Swedish). Nordenmark publishes the Latin text, accompanied by a translation into Swedish, of the lectures on the Copernican system which were delivered by Laurentius Paulinus Gothus (1565–1646) at the University of Upsala in 1599. Reviewed **871.**

NORLIND, WILHELM

529. Copernicus and Luther: A Critical Study. *Isis*, 1953, *44*: 273–276. Copernicus och Luther. *Nordisk astronomisk tidsskrift*, 1954, pp. 53–58 (the Swedish equivalent of 529). In one version of Luther's famous remark about Copernicus, the religious reformer did not call the astronomer a fool. See Meyer, Heinrich, and Mathes, Joh.

NOWICKI, ANDRZEJ

530. Kościół w walce z Kopernikiem. *Wiedza i życie*, 1953, *20*: 261–263. The Church in the struggle against Copernicus.

531. Atomistyczne poglądy Kopernika. *Op. cit.*, pp. 814–816. Atomistic ideas in Copernicus.

532. Kościół przeciw Kopernikowi. *Myśl filozoficzna*, 1953, no. 1, pp. 209–229. Although the Roman Catholic church first opposed the materialistic side of Copernicanism, it later tried to depict him as the personification of piety.

ANNOTATED COPERNICUS BIBLIOGRAPHY

533. *Kopernik, człowiek Odrodzenia* (Warsaw, 1953) p. 176. Reviewed by Jan Gadomski, *Urania* (Kraków), 1955, *26*: 376–377. Copernicus, man of the Renaissance.

534. Mikołaj Kopernik, at pp. 5–46 in *Wielcy polacy Odrodzenia* (Warsaw, 1956, p. 176; Great Poles of the Renaissance).

535. Mikołaj Kopernik, at vol. I, pp. 94–124, in *Z dziejów polskiej myśli filozoficznej i społecznej* (Warsaw, 1956–1957, 3 vols.; History of Polish Philosophical and Social Thought).

See Śniadecki (728).

Observatory, 1943–1944, *65*: 204
536. Copernicus Celebrations in New Zealand. A brief notice of the volume published by Victoria University College (816).

OLSZEWICZ, BOLESŁAW
537. I lavori cartografici di Niccolò Copernico, at Vol. I, pp. 425–426, in *Actes du VIIIe Congrès international d'histoire des sciences* (Vinci and Paris, 1958). In executing the map of Poland which he published at Kraków in 1526, Bernard Wapowski received the help of his friend Copernicus, who is not to be identified with the German cartographer Nicolaus Germanus.

O'NEILL, JOHN J.
538. Copernican Quadricentennial. *New York Herald Tribune*, 1943, May 23, section II, p. 8. A series of trite misstatements about Galileo's defense of Copernicanism.

ORESTANO, FRANCESCO
539. Copernico. *Die Mittelstelle*, 1943 (May), *2*: no. 19, pp. 14–29. A speech delivered at Ferrara on May 9, 1943 to celebrate Copernicus' quatercentenary.

540. Copernico, at pp. 157–184 in *La conflagrazione spirituale* (Milan, 1944; vol. XVIII of Orestano's *Opere complete*).

PAGACZEWSKI, JANUSZ
541. Muzeum Mikołaja Kopernika we Fromborku. *Urania* (Kraków), 1948, *19*:

86–90; enlarged in *Wiedza i życie*, 1948, *15*: 1031–1034. The Copernicus Museum at Frombork.

542. Ogólnopolska wystawa Mikołaja Kopernika w Krakowie. *Ochrona zabytków*, 1953, *6*: 69–72. The national Copernicus exhibition at Kraków.

See Gadomski (238).

PANNEKOEK, ANTONIE
543. A Remarkable Place in Copernicus' De revolutionibus. *Bulletin of the Astronomical Institutes of the Netherlands*, 1945, *10*: 68–69, no. 366. A computational error made by Copernicus in determining the orbit of Jupiter.

544. The Planetary Theory of Copernicus. *Popular Astronomy*, 1948, *56*: 2–13. The Dutch astronomer reviews the details of Copernicus' planetary theory, which he holds is too close in spirit to antiquity to be regarded as the beginning of modern science.

545. De groei van ons wereldbeeld (Amsterdam and Antwerp, 1951) p. 440. Section 18 (pp. 152–162) is devoted to Copernicus.

PANOFSKY, ERWIN
546. More on Galileo and the Arts. *Isis*, 1956, *47*: 183–185. A discussion of the Copernicus portrait in two edd. of Galileo's *Dialogue*, a theme with which the distinguished art historian is evidently unfamiliar and in which he goes sadly astray.

PAPRITZ, JOHANNES
547. Die Nachfahrentafel des Lukas Watzenrode. *Jomsburg*, 1937, *1*: 192–197. A genealogical table of Copernicus' relatives. An expanded version was printed at pp. 132–142 in 370.

PARANDOWSKI, JAN
548. *Szkice* (Warsaw, 1953) p. 244. At pp. 96–109 Parandowski discusses the Greek letters of Theophylactus Simocatta and Copernicus' translation of them into Latin.

See Copernicus, editions and translations: Teofilakt Symokatta, *Listy* (123).

ANNOTATED COPERNICUS BIBLIOGRAPHY

PARGA CORTÉS, RAFAEL

549. On May 24, 1943 Colombia's Minister of National Education delivered a lecture which was published at pp. 5–6 in 108.

PASCHINI, PIO

550. Copernico, at Vol. IV, p. 503, in *Enciclopedia cattolica* (Vatican City, 1949–1954), 12 vols. Mistakenly has the University of Padua, instead of Ferrara, grant Copernicus his law degree, and muddles the history of the edd. of the *Revolutions*.

PAYR, BERNHARD

551. Die kopernikanische Revolution. *Bücherkunde*, 1943, *10*: 135–138. Astronomy was founded by the Nordic German spirit (to which the soul of ancient Greece was racially akin). The Copernican revolution was continued by Alfred Rosenberg, who was opposed to universalism.

PENCONEK, ADAM

552. Poznajmy Kopernika gruntownie. *Urania* (Kraków), 1955, *26*: 343–344. An appeal for an edition of Copernicus' memoranda as a sound introduction to his thought.

See Gadomski (240).

PERRIN, FERNAND

553. *Histoire des sciences* (Paris, 1956) p. 613. The brief treatment of Copernicus (p. 82) incorrectly states that he made the empyrean a part of his universe.

PETRI, WINFRIED

554. Nikolaus Kopernikus. *Stimmen der Zeit*, 1952–1953, *152*: 145–148. Misdates Copernicus' doctoral degree in 1493 instead of 1503.

PEUCKERT, WILL ERICH

555. *Nikolaus Kopernikus, der die Erde kreisen liess* (Leipzig, 1943) p. 351. Reviewed by Richard Sommer, *Das Weltall*, 1943, *43*: 147–148; by Max Caspar, *Kant-Studien*, 1943, *43*: 474; by Henrik Sandblad, *Lychnos*, 1943, pp. 372–374; and by Edward Rosen, *Isis*, 1952, *43*: 136–137.

PFAFFE, HERBERT

556. Nikolaus Kopernikus—ein Revolutionär in der Wissenschaft. *Der Bibliothekar*, 1953, *7*: 631–636 The contention that the Copernican and Ptolemaic systems are two equally acceptable alternative astronomies is incompatible with modern cosmogony's ideas concerning the origin of the planets. The *Commentariolus* is mistakenly said to have been published around 1530.

PIOTROWSKI, JAN

557. Sesja polskiej akademii nauk poświęcona Mikołajowi Kopernikowi. *Nauka polska*, 1953, *1*: no. 4, pp. 140–150. An extended report of the lectures and discussions at the Copernicus celebration conducted by the Polish Academy of Science at Warsaw in 1953.

PISKORSKA, HELENA

558. Copernicana w archiwum toruńskim i na wystawie archiwalnej w Toruniu. *Archeion* (organ naczelnej dyrekcji archiwów państwowych), 1955, *24*: 344–346. Documents pertaining to Copernicus in the archives at Toruń.

PLEDGE, HUMPHRY THOMAS

559. *Science since 1500* (London, 1939); reprinted with minor corrections, 1940; New York, 1947, 1949; p. 356. Copernicus is discussed principally at pp. 36–38.

POLIKAROV, A.

560. *Ot Kopernik do Ajnštajn* (Sofia, 1942) p. 463. This Bulgarian history of the evolution of physical concepts deals with Copernicus at pp. 65–72.

Polish Foreign Trade, 1951 (November–December), no. 8, pp. 11–13

561. Ancient Astronomical Instruments in Poland. An account of the successful reconstruction of Copernicus' astronomical instruments.

Polish Review, 1943, May 24, *3*: no. 19

562. Copernicus on Polish Monetary Policy (p. 4).

563. "Criminals . . . !" Copernicus Called Germans (pp. 5–6).

564. Unveiling the Statue of Copernicus in 1900 in the Courtyard of the Jagiellonian University in Cracow (p. 7).

Popular Astronomy, 1943, *51*
565. Nicolas Copernicus Commemoration (pp. 172–173).
566. The Copernican Quadricentennial in Carnegie Hall, New York (pp. 323–329).
567. Copernican Quadricentennial Observed by the Joliet Astronomical Society (p. 466).

1944, *52*: 105–106
568. A Portrait of Nicholas Copernicus. Reporting the presentation of Maxim Kopf's portrait of Copernicus to the Harvard College Observatory.

1945, *53*: 1
569. A New Copernican Era. Modern travel demonstrates the relatively small dimensions of the earth.

PORSHNEV, BORIS FEDOROVICH
570. Epocha Kopernika, at pp. 33–56 in Kukarkin, ed.

POZNAŃSKI, MARCELI
571. Bibliografie kopernikowskie. *Bibliotekarz*, 1953, *20*: 89–92. A survey of the literature about Copernicus, including the bibliographical guides.

PRAAG, SIEGFRIED VAN
572. Mikolaj Kopernik. *Nederland Polen*, 1953, *7*: no. 5, pp. 25–27.
573. Misvattingen omtrent Kopernik. *Op. cit.*, no. 6, pp. 14–15. Among "mistakes about Copernicus," the most vigorously attacked is the claim that he was a German.
574. Kopernik in Hollands gouden Eeuw. *Op. cit.*, 1954, *8*: no. 12, p. 12. The attitude toward Copernicus in Holland's golden age.
See Vereniging Nederland-Polen.

Pressebulletin der diplomatischen Mission der Volksrepublik Polen
575. Sondernummer anlässlich des Kopernik-Jahres in Volkspolen (Berlin, 1953) p. 20. Contains an address by Stanisław Kulczyński on Copernicus, which was

delivered at a commemoration in Frombork on May 24, 1953 (pp. 11–14), and a lecture by Eugeniusz Rybka (pp. 14–20), who emphasizes Copernicus' use of the principle of the relativity of motion and his avoidance of astrology.

PRICE, DEREK J.
576. Precision Instruments to 1500, at pp. 582–619, Vol. III, in *A History of Technology*, edd. Charles Singer *et al.* (Oxford and New York, 1954–1958) 5 vols. Copernicus' parallactic instrument is mentioned briefly at p. 589.
577. Contra Copernicus—a Critical Re-estimation of the Mathematical Planetary Theory of Ptolemy, Copernicus and Kepler. In a lecture which was delivered on September 4, 1957 to the Institute of the History of Science at the University of Wisconsin and which will be published in the proceedings of that Institute, Price maintained that Copernicus was not a great or original mathematician.

PRZYBYLLOK, ERICH H. G.
578. Das Weltbild des Coppernicus, at pp. 7–16 in 278. Errors regarding the biography of Copernicus were pointed out in Hans Schmauch's review, *Zeitschrift für die Geschichte und Altertumskunde Ermlands*, 1939–1942, *27*: 298–299.
579. Nikolaus Kopernikus und der Wandel im Weltbilde. *Veröffentlichungen der Universitäts-Sternwarte zu Königsberg*, 1944, *12*, p. 14. This speech, delivered at the Copernican quadricentennial celebration at the University of Königsberg on May 24, 1943, generously but mistakenly grants Copernicus a Master of Arts degree, a professorship of mathematics at the University of Rome, a determination of the latitude of Bologna, and the honor of providing the Gregorian calendar with its measurement of the length of the year.

PRZYPKOWSKI, TADEUSZ
580. Ze studiów nad instrumentarium astronomicznym Mikołaja Kopernika. *Polska akademia umiejętności, Sprawozdania z czynności i posiedzeń*, 1948, *49*: 309–314. Copernicus' astronomical instruments.

ANNOTATED COPERNICUS BIBLIOGRAPHY

581. Instrumentarium Mikołaja Kopernika odtworzone w Polsce. *Urania* (Kraków), 1948, *19*: 83–86. Reconstruction in Poland of Copernicus' astronomical instruments.

582. Les instruments astronomiques de Nicolas Copernic. *L'Astronomie*, 1951, *65*: 33–36. Three astronomical instruments used by Copernicus are reconstructed: quadrant, armillary sphere, and triquetrum.

583. Z dziejów rozpowszechniania światopoglądu heliocentrycznego w Polsce i w Rosji. *Problemy*, 1951, 7: 329–336. The history of the spread of the heliocentric theory of the universe in Poland and Russia.

584. Z dziejów heliocentryzmu w Polsce. *Myśl filozoficzna*, 1953, no. 1, pp. 176–190. Many Polish scientists supported the heliocentric system, although it was condemned by the Roman Catholic church.

585. O Mikołaju Koperniku (Warsaw, 1953) p. 136. Reviewed by Jan Gadomski, *Urania* (Kraków), 1955, *26*: 218–220. A popular biography.

586. Problemy konserwacji przyrządów naukowych używanych przez Mikołaja Kopernika oraz innych zabytków astronomicznych w Polsce. *Ochrona zabytków*, 1953, *6*: 30–39. Problems in the conservation of the scientific instruments used by Copernicus and other astronomical antiquities in Poland.

587. Les instruments astronomiques de Nicolas Copernic et l'édition d'Amsterdam (1617) de De revolutionibus. *Archives internationales d'histoire des sciences*, 1953, *6*: 220–226; *Actes du VIe Congrès international d'histoire des sciences*, Amsterdam, 1950 (Paris, 1955), pp. 537–543. The instruments used by Copernicus for astronomical observation are lost, but they may be reconstructed from descriptions by Copernicus and his contemporaries. Przypkowski corrects an erroneous reconstruction in the third ed. of the *Revolutions* (Amsterdam, 1617) as well as mistakes by later writers. He also announces a plan to reconstruct the quarters used by Copernicus as his observatory in Frombork, where the Polish government maintains a Copernicus Museum.

588. *Dzieje myśli kopernikowskiej* (Warsaw, 1954) p. 113. The development of Copernicus' thought.

589. W sprawie warszawskich "poprawek" do krakowskiej rekonstrukcji instrumentów Mikołaja Kopernika. *Urania* (Kraków), 1954, *25*: 128–129. The Warsaw correction of the Kraków reconstruction of Copernicus' instruments.

590. W sprawie nagrobka Mikołaja Kopernika. *Op. cit.*, pp. 222–224. Copernicus' epitaph.

591. W sprawie przyrządu niwelacyjnego Mikołaja Kopernika. *Op. cit.*, pp. 224–225 (photograph opposite p. 212). Copernicus' leveling apparatus.

592. Jeszcze o konstrukcji sfery armillarnej Kopernika. *Op. cit.*, pp. 256–257. Further remarks on the construction of Copernicus' armillary sphere.

593. Postęp techniczny między przyrządami astronomicznymi Kopernika, Brahego i Heweliusza. *Postępy astronomii*, 1955, *3*: 24–27. Technical progress in astronomy from Copernicus to Brahe to Hevelius.

594. W związku z dyskusją o instrumentarium Kopernika. *Op. cit.*, p. 92. Concerning the discussion of Copernicus' instruments.

595. Oryginalny egzemplarz chorobates jakiego używał Kopernik, zachowany w Polsce. *Urania* (Kraków), 1955, *26*: 153–154. An original example of the chorobates used by Copernicus is still preserved in Poland.

596. La gnomonique de Nicolas Copernic et de Georges Joachim Rheticus. *Actes du VIIe Congrès international des sciences* (Vinci and Paris, 1958), I, 400–409. Discusses Copernicus' theoretical and practical interest in sundials, and Rheticus' construction of a 45-foot gnomon near Kraków.

Publications of the Astronomical Society of the Pacific, 1943, 55

597. The Copernican Quadricentennial (p. 73);

598. Copernican Quadricentennial Celebrations (pp. 163–164).

ANNOTATED COPERNICUS BIBLIOGRAPHY

Pyramide, 1953, *3*: 141

599. Rhaeticus (1514–1576). This brief sketch of Rheticus' life twice misdates the first publication of his *Narratio prima*.

RAMBERG, JÖRAN M.

600. Nicolaus Copernicus. *Populär astronomisk tidskrift*, 1943, *24*: 81–108. An article commemorating the 400th anniversary of Copernicus' death.

RAMSAUER, REMBERT

601. Neue Ergebnisse zur Coppernicus-forschung aus schwedischen Archiven. *Forschungen und Fortschritte*, 1942, *18*: 316–318. Reviewed by Harald Geppert, *Zentralblatt für Mathematik und ihre Grenzgebiete*, 1943, *27*: 289–290.

See Richard Sommer (735).

Re-examines results reached by previous investigators of the Copernicus material surviving in Swedish libraries, and indicates that their conclusions need revision and further study.

602. *Nicolaus Coppernicus, Wandler des Weltbildes* (Berlin, 1943) p. 77. Reviewed by Richard Sommer, *Das Weltall*, 1943, *43*: 170, and in *Astronomische Nachrichten*, 1943–1944, *274*: 144.

A brief elementary account, with 60 illustrations.

REICHENBACH, HANS

603. From Copernicus to Einstein (New York, 1942) p. 123. This book's treatment of Copernicus and his successors abounds in mistakes. It was translated from the German *Von Kopernikus bis Einstein* (Berlin, 1927; p. 121) by Ralph B. Winn into woefully unidiomatic English.

RENKAWITZ, WALTER

604. Von Kopernikus bis Kepler. *Die Sterne*, 1954, *30*: 98–101. In reviewing the development of astronomy from Copernicus to Kepler, the author mistakenly asserts that the fourth ed. of the *Revolutions* was printed in 1640 at Amsterdam (instead of 1854 at Warsaw).

Research and Progress, 1944, *10*: 143

605. The Complete Works of Copernicus. Announcement of the plans for the *Nikolaus Kopernikus Gesamtausgabe*.

REST, WALTER

606. Nikolaus Coppernikus ein Deutscher. *Die Mittelstelle*, 1942 (July–August), *1*: no. 9–10, pp. 14–16. The earth revolves around the sun; the political world revolves around the Rome–Berlin axis.

607. Il significato della commemorazione copernicana 1943.

Der Sinn der Kopernikus-Feier 1943. *Die Mittelstelle*, 1943 (May), *2*: no. 19, pp. 9–11 (German and Italian in parallel columns).

Today's giant telescopes scrutinize the universe's boundaries in order to find man there.

608. Genealogia e patria di Nicolò Copernico. *Op. cit.*, pp. 48–52. A lecture delivered at Ferrara on March 6, 1943. "In 1454 Thorn . . . became an independent city, which safeguarded its freedom from Polish overlordship . . . until 1793" (p. 49). "Two hundred years after the Second Peace of Thorn [1466], that is, after two hundred years of the Polish protectorate over the 'free city' . . ." (p. 49).

Revista astronomica, 1954, *26*: 163

609. De revolutionibus orbium coelestium libri VI. An editorial announcement that Copernicus' autograph of the *Revolutions*, having been given by Czechoslovakia to Poland, was on exhibit in Warsaw.

Revista de la academia colombiana de ciencias exactas, fisicas y naturales, 1942–1944, *5*: 276–285

610. De Copernico a Laplace. Mistakenly has Copernicus return to Padua until 1505, instead of going home after obtaining his doctoral degree at Ferrara in 1503.

REVZIN, G. I.

611. *Nikolai Kopernik* (Moscow, 1949) p. 432. Translated into Czech as *Mikuláš Kopernik* (Prague, 1952) p. 256. In this popular biography for young people in a series devoted to lives of illustrious persons, Copernicus' heliocentric system is treated as the first of the four steps by which human reason arrived at a correct understanding of planetary motion.

REY PASTOR, JULIO

612. El sistema de Copernico y su influjo en la historia de la cultura. *Revista astronomica,* 1943, *15*: 197 ff. In a lecture delivered on June 17, 1943 to commemorate the quatercentenary of Copernicus, the author mistakenly ascribes to Melanchthon, instead of to Luther, the use of the epithet "fool" as a description of Copernicus.

REY PASTOR, JULIO and BABINI, JOSÉ

613. *Historia de la matemática* (Buenos Aires and México, 1952) xx + 369 p. Copernicus' trigonometry is briefly mentioned at p. 189.

RIGHINI, GIULIO

614. Copernico "Doctor Ferrariensis" e "Magister" a Bologna. *R. Deputazione di storia patria per l'Emilia e la Romagna, sezione di Ferrara, Atti e memorie,* 1942, *1*: 149–160 (written in 1936, not published until 1942). In a Bolognese legal document dated June 18, 1499, Copernicus was called "magister." This title means that while still a law student, he taught Arts courses at Bologna, just as he lectured on mathematics at Rome about 1500, without actually having received the degree Master of Arts.

RIGONI, ERICE

615. Un autografo di Niccolò Copernico. *Archivio veneto,* 1951, anno 81, vol. 48–49, 5th series, no. 83–84, pp. 147–150 (published in 1952). Prints the text of two legal documents which were found in the Archivio Notarile of Padua. In the first document, written in Copernicus' own handwriting, the astronomer designates two proxies to take possession of a benefice recently granted to him in Wrocław. In the second document a notary attests the designation.

See Schmauch (688).

ROBERTS, VICTOR

616. The Solar and Lunar Theory of Ibn ash-Shāṭir. *Isis,* 1957, *48*: 428–432. A Muslim astronomer of the 14th century developed a lunar theory resembling Copernicus'. A forthcoming paper, written jointly with E. S. Kennedy, shows that the planetary theories of the two astronomers were similar. Did Copernicus know about the work of his predecessor from Damascus?

RÖMER, ERNST

617. Nikolaus Kopernikus. *Der Seewart,* 1943, *12*: 62–64. Misdates Copernicus' earliest astronomical observation by fourteen years, and mistakenly makes Copernicus a priest.

RÖRIG, FRITZ

618. Nikolaus Kopernikus und der deutsche Lebenskreis. *Historische Zeitschrift,* 1943, *168*: 263–279. This speech, delivered at the Copernicus celebration held by the University of Berlin on May 24, 1943, emphasizes the astronomer's German cultural environment; his predecessors were Peurbach and Regiomontanus, his successors Kepler and Bessel.

ROLLER, DUANE H. D.

See Holton.

ROMAÑA PUJO, ANTONIO

619. La difusión del sistema de Copérnico. *Euclides,* 1944, *4*: 3–13, 164–174. The supporters and opponents of Copernicanism, from the 16th century to the decisive proofs of the earth's revolution and rotation in the 19th century.

ROSEN, EDWARD

620. *Three Copernican Treatises* (New York: Columbia University Press; London: Oxford University Press; 1939) xi + 211 p. Reviewed by Lawrence W. Friedrich, *Historical Bulletin,* 1939–1940, *18*: 93–94; by Otto Neugebauer, *Mathematical Reviews,* 1940, *1*: 129; by George H. Sabine, *American Mathematical Monthly,* 1940, *47*: 386–387; by Mervyn A. Ellison, *Journal of the British Astronomical Association,* 1940, *50*: 289–291; by A. S. D. Maunder, *Observatory,*

ANNOTATED COPERNICUS BIBLIOGRAPHY

1940, *63*: 161–166; by Henry C. Plummer, *Nature*, 1940, *146*: 343–344; by Daniel Norman, *Isis*, 1940, *32*: 358–359; by Otto Struve, *Journal of Modern History*, 1940, *12*: 424–425; by Ernest Nagel, *Journal of Philosophy*, 1940, *37*: 194; by Henry T. Edge, *Theosophical Forum*, 1940, *16*: 392–393; in *Christian Century*, 1940, *57*: 146; in *Scientia*, 1940, *68*: July–August, p. XI; in *Science News Letter*, 1940, *37*: 224; by Louis C. Karpinski, *National Mathematics Magazine*, 1940–1941, *15*: 104–106; by Alexander Pogo, *Astrophysical Journal*, 1941, *94*: 555–556; by Curvin H. Gingrich, *Popular Astronomy*, 1941, *49*: 398–399; by Florian Znaniecki, *American Historical Review*, 1941, *46*: 624–625; by Ralph Tyler Flewelling, *Personalist*, 1941, *22*: 85; and by Hugh S. Rice, *Scripta mathematica*, 1943, *9*: 185–187.

621. The Ramus-Rheticus Correspondence. *Journal of the History of Ideas*, 1940, *1*: 363–368 (reprinted in *Roots of Scientific Thought*, edd. Philip P. Wiener and Aaron Noland, New York, 1957, pp. 287–292). Ramus erroneously believed Rheticus to be the author of Osiander's unsigned preface to Copernicus' *Revolutions*.

622. The Copernican Theory. *Sky*, 1940, *4*: no. 11, pp. 6, 19 (reprinted in part in *Sky and Telescope*, 1943, *2*: no. 7, pp. 5, 14). Commemorating the 400th anniversary of the publication of Rheticus' *Narratio prima*.

623. The Authentic Title of Copernicus' Major Work. *Journal of the History of Ideas*, 1943, *4*: 457–474. The words "orbium caelestium" in the title of the *Revolutions* are unobjectionable and were not inserted by Osiander.

624. Copernicus and the Discovery of America. *Hispanic American Historical Review*, 1943, *23*: 367–371. When Copernicus said that "America" was named after its discoverer, he was referring only to an area in the southern hemisphere, not to the entire New World.

625. Nicholas Copernicus, the Man and His Work. *Sky and Telescope*, 1943, *2*: no. 7, pp. 3–5. Commemorating the 400th anniversary of the publication of Copernicus' *Revolutions*.

626. Nicholas Copernicus, the Founder of Modern Astronomy, at pp. 29–35 in Mizwa, ed. (493). An address delivered at the Copernicus Quadricentennial Celebration at Carnegie Hall in New York on May 24, 1943.

627. Maurolico's Attitude toward Copernicus. *Proceedings of the American Philosophical Society*, 1957, *101*: 177–194. A refutation of Augustus De Morgan's effort to interpret away Maurolico's recommendation that Copernicus should be whipped.

628. Galileo's Misstatements about Copernicus. *Isis*, 1958, *49*: 319–330. Reprinted in the *Massachusetts Institute of Technology Publications in the Humanities*, no. 32 (1958). Five widely repeated mistakes about Copernicus' life and work were originally made by Galileo.

629. Nicholas Copernicus. *World Book Encyclopaedia* (forthcoming).

Reviewed **19, 110, 121, 169, 402, 555, 837, 871.**

ROSENBERG, ALFRED, ed.

630. *Handbuch der Romfrage* (Munich, 1940); Vol. I, xv + 828 p.; the planned 2d vol. was not published. An unsigned article on "Kopernicus, Nikolaus" (I, 809–810) commits numerous errors and claims Copernicus as a German.

ROSENBERG, BERNHARD-MARIA

631. *Nikolaus Koppernikus* (Lustadt, 1949) p. 44. Taking the form of a series of talks given by an adult to some children, this brochure is so pro-German as to omit Galileo from its account of the development of the telescope, and so pro-Catholic as to deny that the Roman Catholic church condemned the teachings of Copernicus.

ROSENBLATT, ALFRED

632. La posicion de Copérnico en la historia de la ciencia. This lecture on the place of Copernicus in the history of science was delivered at a Copernicus celebration in Lima, Peru, on May 24, 1943, and was published in *Revista de ciencias*, 1943, *45*: 409–442, as well as in *Actas de la Academia nacional de ciencias exactas, físicas y naturales de Lima*, 1943, *6*: 165–198.

ROSS, JULIUSZ

633. Rękopis Kopernika. *Ochrona zabyt-ków*, 1953, *6*: 68–69. Copernicus' autograph manuscript.

ROSSITER, ARTHUR PERCIVAL

634. *The Growth of Science* (Cambridge, England, 1939) p. 372. This history of science, written in Basic English, discusses Copernicus at p. 85.

ROSSMAN, FRITZ

See Copernicus, editions and translations (121).

ROUPPERT, KAZIMIERZ

635. Z życia Mikołaja Kopernika, at pp. 5–9 in 367, 2d ed. The life of Copernicus.

636. Kopernik a księżyc, *op. cit.*, pp. 18–21. Copernicus and the moon.

637. Kopernik o złych sąsiadach-niemcach, *op. cit.*, pp. 34–37. Copernicus on his evil German neighbors.

ROUSSEAU, PIERRE

638. *Histoire de la science* (Paris, 1945) p. 823. Copernicus is discussed at pp. 168–173.

RUDNICKI, JÓZEF

639. *Nicholas Copernicus* (London, 1943) viii + 53 p. Translated from Polish by B. W. A. Massey, with a foreword by Arthur Eddington.

Reviewed by Henry C. Plummer, *Observatory*, 1943, *65*: 80–82, and in *Nature*, 1943, *152*: 33–34.

RUDNICKI, KONRAD

640. Sesja kopernikańska polskiej akademii nauk. *Postępy astronomii*, 1953, *1*: 103–104.

641. Kopernikańska sesja naukowa P. A. N. *Urania* (Kraków), 1953, *24*: 331–335. The Copernicus celebration sponsored by the Polish Academy of Science.

642. W tej samej sprawie. *Urania* (Kraków), 1955, *26*: 155–156. On the same subject (as Cichowicz's comment on pp. 154–155, *op. cit.*, about Przypkowski's reconstruction of Copernicus' instruments). Reviewed **767.**

RUFUS, WILL CARL

643. The Quadricentennial of the "First Account" of the Copernican Theory. *Scientific Monthly*, 1940, *51*: 474–477. A summary of Rheticus' *Narratio prima*.

644. Copernicus and the History of Science. *Scientific Monthly*, 1943, *57*: 181–182. Ptolemy interpreted "nature as she seems, not as she is." Copernicus interpreted not only Ptolemy but nature also.

645. Copernicus, Polish Astronomer. *Journal of the Royal Astronomical Society of Canada*, 1943, *37*: 129–142. Copernicus viewed as one of a valiant band of scientists who strove to free men's minds in the search for truth.

RUIZ WILCHES, BELISARIO

646. La obra de Nicolás Copérnico. A lecture delivered on May 24, 1943 and published at pp. 11–16 in 108.

RUSSELL, BERTRAND

647. *A History of Western Philosophy* (New York, 1945) p. 916. Translated into Italian by Luca Pavolini under the title *Storia della filosofia occidentale* (Milan, 1948) 3 vols. An otherwise excellent treatment of Copernicus (pp. 525–529) is marred by the mistaken assertion that he regarded his astronomy as a hypothesis.

RYBKA, EUGENIUSZ

648. Kopernik we Włoszech. *Urania* (Kraków), 1953, *24*: 33–39. This lecture, which was delivered in translation in 1952 at Rome to the Associazione Italiana per i Rapporti Culturali con la Polonia, deals with Copernicus in Italy.

649. Velikii polskii astronom. *Priroda*, 1953, *42*: no. 5, pp. 3–14. A great Polish astronomer.

650. Kopernik jako astronom. *Wiedza i życie*, 1953, *20*: 241–247. Copernicus as an astronomer.

651. Sesja kopernikowska P.A.N. *Problemy*, 1953, *9*: 779–780. A report of the Copernicus celebration conducted by the Polish Academy of Science on September 15–16, 1953.

ANNOTATED COPERNICUS BIBLIOGRAPHY

652. Światopogląd kopernikański. *Op. cit.*, pp. 809–811. The Copernican view of the world.
See *Pressebulletin* and *Sesja kopernikowska.*

RYBKA, EUGENIUSZ and RYBKA, PRZE-MYSŁAW
653. *Mikołaj Kopernik i jego nauka* (Warsaw, 1953) p. 207. Reviewed by Jan Gadomski, *Urania* (Kraków), 1955, *26*: 246–248. Nicholas Copernicus and his theory.

RYTEL, ALEXANDER
654. Nicolaus Copernicus. *Medical Journal of Australia*, 1955, *42*: 586. This report on Copernicus' medical activities was presented at the Australasian Medical Congress in Sydney, Australia, on August 25, 1955.
655. Nicolaus Copernicus—Physician and Humanitarian. *Polish Medical History and Science Bulletin*, 1956, *1*: 3–11. An expansion of 654.

SANDBLAD, HENRIK
656. Det Copernikanska Världssystemet i Sverige. *Lychnos*, 1943, pp. 149–188; 1944–1945, pp. 79–131 (in Swedish, with English summary at pp. 127–131 in 1944–1945). The spread of Copernicanism in Sweden was delayed by the prevalence of the Aristotelian philosophy and by the acceptance of the Bible, interpreted literally, as the key to the understanding of nature. These two obstacles to clear thinking were not overcome in Sweden until the late 17th and early 18th centuries. Reviewed **370, 555, 871.**

SANDOVAL, ROSENDO OCTAVIO
657. La idea copérnica y sus efectos en el mundo cristiano. *Memorias y revista de la Academia nacional de ciencias* (Mexico), 1935–1944, *55*: 303–313. It is Galileo's fault that the Roman Catholic church condemned Copernicanism.

SANTILLANA, GIORGIO DE, ed.
658. *The Age of Adventure* (New York, 1956) p. 283. Excerpts from the *Revolutions* at pp. 160–166; Santillana's introductory

remarks (pp. 157–160) make Copernicus "a consultant on the Gregorian calendar reform," which was initiated a generation after his death.
See Enriques, Federigo.

SARTON, GEORGE
659. *Six Wings: Men of Science in the Renaissance* (Bloomington, 1957) xiv + 318 p. A valuable treatment of Copernicus at pp. 54–62; the few minor slips would undoubtedly have been corrected had Dr. Sarton lived to see the proofs.
Reviewed **120, 123.**

SCHATZMAN, EVRY
660. La révolution copernicienne. *La Pensée*, 1953 (September–October), no. 50, pp. 41–50. A lecture delivered on May 21, 1953. Copernicus was the first to proclaim the study of nature a science independent of theology.
661. Rencontres scientifiques à Varsovie et à Prague. *La Pensée*, 1953 (December), no. 52, pp. 93–99. A report on the Copernicus celebration in Warsaw on September 15–16, 1953. The section dealing with Fok's lecture and the discussion provoked by it (pp. 94–95) was reprinted at pp. 141–142 in Wędkiewicz.
662. Copernic et la science moderne. *Ciel et terre*, 1954, *70*: 321–323. Interprets Einstein's relativity theory as a new proof of the physical reality of the Copernican system.
See *Sesja kopernikowska.*

SCHELLING, HERMANN VON
663. Paul von Middelburg (1445–1533) und Nicolaus Kopernikus. *Geistige Arbeit*, 1942, *9*: no. 16, pp. 5–6. Paul of Middelburg, who presided over the papal commission on calendar reform, urged Copernicus to investigate the length of the year and month.
Reviewed **329, 863.**

SCHENBERG, MARIO
664. Mikołaj Kopernik. *Postępy astronomii*, 1954, *2*: 165–168.

SCHENK, EBERHARD
665. Kopernikus-Bildnisse, at pp. 257–285, 373–374, in Kubach, ed. (393). There

were two genuine portraits of Copernicus, one a self-portrait done in Italy under Italian influence, and the other a portrait of Copernicus in his old age.

SCHIEDER, THEODOR

666. Deutsches Geistesleben Altpreussens von Coppernicus bis Kant, at pp. 24–30 in 278.

SCHIFFERES, JUSTUS J.

See Moulton, Forest Ray.

SCHIMANK, HANS

667. Nikolaus Kopernikus. *Progressus, Fortschritte der deutschen Technik*, 1943, *8*: 333–336. Brings Copernicus home from Italy two years too late, and also has him leave his uncle's residence two years too late. Postpones the writing of the *Commentariolus* to "about 1530."

668. Zur Nationalität des Wissenschaftlers. *Physikalische Blätter*, 1948, *4*: 382–383. Copernicus was a Prussian patriot of German descent under the feudal sovereignty of the kingdom of Poland. Reviewed **119**.

SCHLESINGER, FRANK

669. Astronomy, at pp. 53–89 in *The Development of the Sciences*, second series (New Haven, 1941). The discussion of Copernicus at pp. 65–66 repeats what was said at pp. 142–143 in the 1923 volume bearing the same title.

SCHMAUCH, HANS

670. Nikolaus Coppernicus—ein Deutscher. *Jomsburg*, 1937, *1*: 164–191. Copernicus is one of the four great German astronomers, together with Regiomontanus, Kepler and Bessel. In an expanded form this **671.** essay was reprinted at pp. 1–32 in 370.

672. Zur neuen polnischen Coppernicusbiographie von J. Wasiutyński. *Jomsburg*, 1938, *2*: 215–230. A review of Jeremi Wasiutyński, *Kopernik twórca nowego nieba* (Warsaw, 1938). Schmauch participates in the controversy aroused by the publication of Wasiutyński's biography of Copernicus.

673. Nikolaus Coppernicus und die preussische Münzreform, 40 pp., separately numbered, in *Staatliche Akademie zu Braunsberg, Personal- und Vorlesungs-Verzeichnis*, 1940, 3. Trimester. Reviewed by Franz Buchholz in *Zeitschrift für die Geschichte und Altertumskunde Ermlands*, 1939–1942, 27: 463–464, and also in *Jomsburg*, 1942, 6: 143–148, as well as by Emil Waschinski, *Zeitschrift des westpreussischen Geschichtvereins*, 1941, 76: 187–189.

Schmauch publishes (pp. 27–34) the first draft, dated August 15, 1517, of Copernicus' treatise on currency. This first draft, in Latin, is accompanied by Copernicus' German version of 1519, and the text of the first Latin draft is compared with the Latin version of a decade later.

674. Nikolaus Kopernikus und der Deutsche Ritterorden. *Jomsburg*, 1941, *5*: 69–80. Reviewed by Harald Geppert, *Zentralblatt für Mathematik und ihre Grenzgebiete*, 1942, 25: 290–291.

In opposition to Prowe and L. A. Birkenmajer, Schmauch argues that Copernicus did not write the complaints made by the Ermland Chapter against the Teutonic Knights on July 22, 1516 and July 25, 1521. In **675.** an expanded form this article was reprinted at pp. 202–219 in 370.

676. Die Gebrüder Coppernicus bestimmen ihre Nachfolger. *Zeitschrift für die Geschichte und Altertumskunde Ermlands*, 1939–1942, 27: 261–273. Like his brother before him, Copernicus obtained a coadjutor in his canonry in anticipation of his death.

677. Der Altar des Nicolaus Coppernicus in der Frauenburger Domkirche. *Op. cit.*, pp. 424–430. In the Frauenburg cathedral Copernicus' altar was the fourth on the south side.

678. Nikolaus Coppernicus und die Wiederbesiedlungsversuche des ermländischen Domkapitels um 1500. *Op. cit.*, pp. 473–541. Copernicus assisted his cathedral chapter in its efforts to resettle abandoned farms.

ANNOTATED COPERNICUS BIBLIOGRAPHY

679. Neue Funde zum Lebenslauf des Coppernicus. *Op cit.*, 1943, *28*: 53–99. Publishes twenty-eight new documents which make it possible to correct errors and fill gaps in the biography of Copernicus.

680. Nikolaus Kopernikus' deutsche Art und Abstammung, at pp. 61–95, 314–318, in Kubach, ed. (393).

681. Nikolaus Kopernikus und der deutsche Osten, at pp. 233–256, 370–373, *op. cit.* Some aspects of Copernicus' activities outside the field of astronomy.

682. Leben und Wirken des Nikolaus Kopernikus, at pp. 15–56 in Georgens, ed. Schmauch errs in saying that Copernicus indirectly contributed to the Gregorian calendar reform by way of Reinhold's *Tabulae prutenicae* (which he misdates). Schmauch denies Greek influence on Copernicus (who admits it) and asserts German influence (about which Copernicus says nothing).

683. Die Jugend des Nikolaus Kopernikus, at pp. 100–131 in 370. Before matriculating at the University of Kraków in 1491, Copernicus probably went to school, not at Włocławek (Leslau), but at Kulm. In an appendix (pp. 113–131) Schmauch assembles every known reference in the sources to the activities and whereabouts of Copernicus' maternal uncle, Lucas Watzenrode, until he became bishop of Ermland in 1489.

684. Neues über die ärztliche Tätigkeit des Astronomen Kopernikus, at pp. 115–119 in Diergart, ed. (151). Some new light on Copernicus' medical activities.

685. Nikolaus Kopernikus in Italien. *Die Mittelstelle*, 1943 (May), *2*: no. 19, pp. 30–37. Copernicus spent seven years in Italy. He entered that country for the first time in the autumn of 1496 and remained there until the autumn of 1503, returning home briefly in the summer of 1501.

686. *Nikolaus Kopernikus* (Kitzingen, 1953, p. 45; Der Göttinger Arbeitskreis, Schriftenreihe no. 34). Translated into English by Helen Taubert under the title *Nicolaus Copernicus* (Goettingen, 1954, p. 63; Goettingen Research Committee, publication no. 95).

687. Nikolaus Kopernikus in Allenstein, at pp. 17–23 in *Südostpreussen und das Ruhrgebiet*, ed. Erwin Nadolny (Leer, 1954) p. 91. Reviews Copernicus' activities during his stay of more than four years in Allenstein (now Olsztyn).

688. Des Kopernikus Beziehungen zu Schlesien. *Archiv für schlesische Kirchengeschichte*, 1955, *12*: 138–156. Corrects some errors committed in Erice Rigoni's publication of a legal document written in Copernicus' own handwriting. Interprets Copernicus' departure from his uncle's episcopal residence some time before the bishop died as a personal decision not to follow in his uncle's footsteps in quest of wealth through the accumulation of multiple church benefices, but instead to pursue a career of devotion to astronomy.
Reviewed **43, 278, 578, 833.**

SCHMEIDLER, FELIX
689. Das Lebenswerk des Nikolaus Kopernikus. *Sternenwelt*, 1951, *3*: 176–178. Reflections on the Kubach-Zeller ed. of Copernicus (118–119).
Reviewed **121.**

SCHNEIDER, ERICH
690. *Von Kopernikus zur Kobaltwolke: unser Weltbild und seine dreihundertjährige Geschichte* (Berlin and Munich, 1955) p. 416. The discussion of Copernicus (pp. 11–16) contains the misstatements that he studied astronomy at Vienna, that nothing is known about his observational instruments, and that the cost of publishing the *Revolutions* was paid by Cardinal Schönberg.

SCHNELLER, HERIBERT
691. Nikolaus Kopernikus. *Die Sterne*, 1943. *23*: 49–59. The author errs in saying that the Gregorian calendar was based on the heliocentric astronomy. Why does he assert (p. 54) that Copernicus wrote the *Commentariolus* "under the pressure of his friends"?
Reviewed **119, 125.**

SCHOENBORN, HANS JOACHIM
692. Coppernicus der Deutsche, at pp. 17–23 in 278.

ANNOTATED COPERNICUS BIBLIOGRAPHY

SCHOUTEN, WILLEM JOHANNES ADRIAAN

693. *Grote Sterrenkundigen van Ptolemaeus tot De Sitter* (Rijswijk, 1950) vii + 285 p. The chapter on Copernicus (pp. 19–36) misapplies his expression "our dear fatherland" to Germany, and mistakenly puts in Copernicus' room a bit of Latin poetry which was first connected with him some forty years after his death.

SCHULTZE-PFAELZER, G.

694. Zum 400. Todestage von Nikolaus Kopernikus. Der Mann und sein Weg. *Koralle*, 1943, *11*: 212–213. Mistakenly asserts that Ptolemy considered the earth to be a convex disk ("gewölbte Scheibe") at whose boundaries one could fall down into bottomless space; that Copernicus went out and got Rheticus to be his assistant; and that his death was due to excitement caused by Osiander s preface.

SCHUMACHER, JOSEPH

695. Die griechischen und deutschen Elemente im kopernikanischen Denken, at pp. 110–115 in Diergart, ed. (151). German thought is complete, not one-sided (like non-German thought). Therefore Copernicus is a German.

696. Die theoria der Griechen und die kopernikanische Umwälzung. *Die Tatwelt*, 1942, *18*: 182–190. Based on a lecture delivered at Aachen on April 4.

697. Deutsches Denken und Forschen bei Nikolaus Kopernikus. *Europäischer Wissenschafts-Dienst*, 1943, *3*: no. 5, pp. 2–4. Calls Copernicus a compatriot ("Landsmann") of Melanchthon, who himself called Copernicus a "Sarmatian."

SCHWARTZ, GEORGE and BISHOP, PHILIP W., edd.

698. *Moments of Discovery*, 2 vols. (New York, 1958), xvii, xi + 1005 p. Excerpts from Book I of the *Revolutions* at pp. 220–231. Why do the editors say (pp. 218–219) that Oresme's "works were included in Copernicus's curriculum," and that Osiander intended his preface "to placate the Protestants"?

SCHWARZ, FRIEDRICH

699. Kopernikus-Bildnisse, at pp. 143–171 in 370. Proposes a genealogical table, as it were, of all the known portraits of Copernicus.

700. Wie sah Kopernikus aus?, at pp. 110–132 in Georgens, ed. Argues that the most faithful portrait of Copernicus (the one executed by Tobias Stimmer in the cathedral of Strasbourg) was not based on a self-portrait painted by the astronomer himself, but on a portrait for which he posed at some time between 1503 and 1510. Schwarz is confident that he can recognize the features of a man in whose veins North German blood flows.

SCHWITALLA, ALPHONSE M.

701. Quadricentennial of Copernicus. *Historical Bulletin*, 1942–1943, *21*: 79–80, 90. Why does Schwitalla say that Copernicus "must have been recognized as a capable and violent opponent of Luther"?

Science, 1943, 97

702. The Copernican Quadricentennial (p. 259);

703. The Copernican Quadricentennial Celebration (pp. 417–418);

704. The Four Hundredth Anniversary of the Death of Copernicus (p. 460);

705. The Copernican Quadricentennial (p. 504).

Science News Letter, 1942, *42*: 330

706. Copernicus Honored. An announcement of the quadricentennial celebrations of Copernicus.

SCOTT, DOUGLAS

707. The Acceptance of the Copernican System in England. *Astronomical Society of the Pacific*, Leaflet no. 304 (1954, September) p. 8. This paper by a student at the University of California, Los Angeles, confuses solar parallax with stellar parallax in explaining the resistance to Copernicanism.

ANNOTATED COPERNICUS BIBLIOGRAPHY

SCZANIECKI, MICHAŁ
708. Mikołaj Kopernik, tytan epoki Odrodzenia. *Kwartalnik historyczny*, 1953, *60*: no. 3, pp. 44–56 (summary in French at pp. 7–9). Expansion of a lecture on "Nicholas Copernicus, Titan of the Renaissance Epoch," which was delivered on February 19, 1953 at the opening of the Copernicus Exhibition in Poznań.

SELLE, GÖTZ VON
709. *Ostdeutsche Biographien* (Würzburg, 1955) no pagination. The 134th of these biographies deals with Copernicus, who is brought back from Italy two years too late, and whose astronomy is mistakenly said to have "shocked nobody for a long time." See Lichtenberg.

SEMENOV, V.
710. Blestiashchaia stranitsa polskoi historii. *Slaviane*, 1953, no. 12, pp. 21–24. This "brilliant page in Polish history" concerns Copernicus in his Renaissance setting.

Sesja kopernikowska 15–16 IX. 1953 (Warsaw, 1955) p. 482
711. A handsome volume reporting the Copernicus celebration held at Warsaw on September 15–16, 1953 under the auspices of the Polish Academy of Science. All the addresses as well as the discussions are published in at least four languages: Polish, Russian, French, and English. The principal participants are Jan Dembowski, president of the Polish Academy of Science; Stefan Żolkiewski, secretary of the Academy; Józef Witkowski, "The Reform of Copernicus"; Vladimir Aleksandrovich Fok, "The Systems of Copernicus and Ptolemy from the Standpoint of the Contemporary Theory of Gravitation," with discussion by Eugeniusz Rybka; Bogusław Leśnodorski, "Copernicus the Humanist"; Antonio Banfi, "Copernicus and Italian Culture"; with discussion by Georg Klaus on the relation between Copernicus and Rheticus, by Szczepan Szczeniowski on Copernicus' influence on physics, by Marcel Minnaert on Fok's paper, with a reply by Fok; by Roman Stanisław In-

garden on Fok's paper; by Antonio Signorini, Felice Gioelli, and Jan Mukarovsky; by Paul Libois on Witkowski's paper; by Sava Tzolov-Ganovsky; by Evry Schatzman on Minnaert's remarks; by Laslo Kalmar; by Alfons Kauffeldt, Gheorghe Demetrescu, Nikola Bonev Ivanov, Chou Pei-Yuan, Henryk Dunajewski, Bolesław Hryniewiecki, and Aleksander Birkenmajer.

SHAPLEY, HARLOW
712. In the name of Copernicus. *American Scientist*, 1943, *31*: 177–178. Reprinted in *The Humanist*, 1943–1944, *3*: 68–69. Reproduces the portrait of Copernicus by Arthur Szyk, and explains some of its symbols. See Mizwa, ed. (493).

SHAPLEY, HARLOW *et al.*, edd.
713. *Readings in the Physical Sciences* (New York, 1948) xiii + 501 p. Excerpts from Book I of the *Revolutions* at pp. 75–79.

SHCHEGLOV, V. P.
714. *Nicholas Copernicus, the Great Reformer of Science* (Tashkent, 1954) p. 31 (in Uzbek).

SIEHM, PETER
715. Ein grosser Sohn des polnischen Volkes. *Jenseits der Oder*, 1952, *3*: no. 3, pp. 12–13. In summarizing the reaction of Polish scientists to the Copernican astronomy, Siehm mistakenly asserts that Copernicus himself never wrote in German.

SIEVEKING, GERHART
716. Das Siebengestirn. *Atlantis*, 1943, *15*: 234–237. A new translation into German of the *Septem sidera*, a religious Latin poem allegedly composed by Copernicus, or copied by him, or chosen by him as his epitaph.

SILVA, GIOVANNI
717. Nicolò Copernico astronomo, at pp. 7–18 in *Copernico* (Padua, 1944) p. 69; *Opuscoli accademici*, Facoltà di lettere e filosofia, Università di Padova, serie

liviana, no. 8 (see Troilo, Erminio). A lecture delivered at a Copernicus celebration at Ferrara in 1943.

SINGER, CHARLES
718. *A Short History of Science to the Nineteenth Century* (Oxford, 1941) p. 399; reprinted with corrections in 1943. Deals with Copernicus at pp. 179–182, 256. See Dingle (167), and Price (576).

SKARŻYŃSKI, BOLESŁAW
719. Kopernik lekarzem. *Problemy*, 1953, *9*: 665–668. Copernicus as a doctor.

SKAZKIN, SERGEI DANIELOVICH
720. Kopernik i Vozrozdenie. *Istoricheskii Zurnal*, 1943, *10*: 60–63. Epocha Kopernika i jejo liudi, at pp. 43–63 in Mikhailov, ed. (483).

SKIMINA, STANISŁAW
721. *Twórczość poetycka Jana Dantyszka* (Kraków, 1948) vi + 199 p.; Polska akademia umiejętności, rozprawy wydziału filologicznego, *68*: no. 1. This study of the poetic production of John Dantiscus at pp. 76–77 analyzes his poem in praise of Copernicus.
722. *Ioannis Dantisci . . . carmina* (Kraków, 1950) xxxiv + 324 p. At pp. 208–209 reprints the Latin text of Dantiscus' poem *In Copernici libellum epigramma*, which was printed in Copernicus' *De lateribus et angulis triangulorum*.

Sky and Telescope, 1942–1943, *2*: no. 9, p. 19
723. Copernican Celebrations.

Slaviane, 1953, no. 3, pp. 61–62
724. Jubilei velikogo polskogo astronoma. A report on Copernicus celebrations in the U.S.S.R. in the Copernicus Year, 1953.
725. Sobranie pamiati N. Kopernika v Akademia nauk SSSR. A report on the meeting in honor of Copernicus under the auspices of the Soviet Academy of Science.

ŚLIZIŃSKI, JERZY
726. Kopernikiana czeskie. *Problemy*, 1953, *9*: 669. The interest of the Czechs in Copernicus.

SLOUKA, HUBERT
727. Kopernikova cesta k sluneční soustavě. *Říše hvězd*, 1953, *34*: 108–113. Copernicus and the solar system.

ŚNIADECKI, JAN
728. *O Koperniku* (Warsaw, 1953) p. 108. Reviewed by Jan Zygmunt Jakubowski, *Życie szkoły wyższej*, 1953, *1*: no. 11, p. 148, and by Jan Gadomski, *Urania* (Kraków), 1956, *27*: 25. A reprint of Śniadecki's classic essay on Copernicus, with a preface by Andrzej Nowicki.
729. Another reprint (Wrocław, 1955) with commentary by Mirosława Chamcówna (Biblioteka narodowa, series I, no. 159; lxxvii + 240 p.).

SNYDER, LOUIS L.
730. *The Age of Reason* (New York, 1955) p. 185. The treatment of Copernicus (pp. 18–19), although brief, abounds in errors. The misstatement that Osiander's unsigned foreword to the *Revolutions* was included in Copernicus' dedication shows that Snyder has not so much as looked at the book he is discussing. The quality of his scholarship is indicated by the fact that in his Table of Contents (p. 3) he does not even cite the title of Bishop Butler's famous work correctly.

SOLERI, GIACOMO
731. Copernico, at Vol. I, col. 1236–1238, in *Enciclopedia filosofica* (Venice and Rome, 1957) 4 vols. Besides committing errors of lesser importance, this article mistakenly declares that Copernicus presented his system only as a hypothesis.

SOMERVILLE, JOHN
732. *The Way of Science* (New York, 1953) p. 172. Copernicus is discussed at pp. 57–68 in this attractive little book for young people.

SOMMER, J.
733. Kopernikus und die Weltsysteme, at pp. 57–109 in Georgens, ed. Keeps Copernicus in Italy three years too long.

ANNOTATED COPERNICUS BIBLIOGRAPHY

SOMMER, RICHARD

734. Eine Coppernicus-Gesamtausgabe. *Das Weltall*, 1942, *42*: 141. An announcement of the forthcoming *Nikolaus Kopernikus Gesamtausgabe*.

735. Neue Ergebnisse der Coppernicusforschung. *Op. cit.*, 1943, *43*: 64–65. Summarizes the results of Ramsauer's investigation of Copernicana in Sweden.

736. Eine Sonnenuhr von Coppernicus. *Op. cit.*, p. 65. Summarizes Zinner's study of the sundials attributed to Copernicus.

737. Coppernicus im Film. *Op. cit.*, p. 66. Quotes a report that a film company in Czechoslovakia was making a movie about Copernicus.

738. Die Schreibweise des Namens Coppernicus. *Op. cit.*, p. 80. Reports an official decision of the German government that the astronomer's name is to be written "Nikolaus Coppernicus" (later changed to "Kopernikus"; see *op. cit.*, p. 98). In criticizing an American astronomer, Sommer ignorantly says that "Nicholas" is the Polish form of that given name.

739. Koppernikus-Feiern 1943. *Op. cit.*, pp. 95–98. A report of the German celebrations of Copernicus in 1943. Reviewed **89, 555, 602, 863.**

SONNEDECKER, GLENN

740. He Spun the Earth. *Science News Letter*, 1943, *43*: 330–331. This article confuses two documents written by Copernicus. Osiander suppressed, not "the dedicatory letter prepared by Copernicus," but his introduction to Book I of the *Revolutions*.

Spectator, 1939, *162*: 350

741. Coppernicus, Kant, Rosenberg. An unnamed correspondent reflects (in German) on Alfred Rosenberg's discussion of Copernicus and Kant at the celebration held at the University of Königsberg (now Kaliningrad).

SPONSEL, HEINZ

742. *Alles dreht sich um die Sonne* (Schloss Bleckede an der Elbe, 1949) p. 167; 2d ed., 1952, p. 159, under the title *Kopernikus*

(Meissners Jugendbücher, Band 2). Reviewed by A. Robl, *Sternenwelt*, 1952, *4*: 156.

A lively fictionalized biography of Copernicus for children.

STAHL, WILLIAM H.

See Dreyer.

STEARNS, RAYMOND PHINEAS, ed.

743. *Pageant of Europe* (New York, 1948) xxix + 1032 p. An excerpt from the *Commentariolus* at pp. 64–65. Why does the editor say that Osiander was a pupil of Copernicus?

STEBBINS, JOEL

744. Copernicus and Modern Revolutions. *Popular Astronomy*, 1943, *51*: 291–296. A new Copernicus is needed to solve contemporary astronomy's problems.

STEGEMANN, VIKTOR

745. Der griechische Philosoph und Astronom Hiketas von Syrakus als Nicetas (-us) bei Kopernikus und Giordano Bruno, at pp. 97–99 in Diergart, ed. (151). The faulty transmission of a passage in Cicero caused Copernicus to alter the name of the Greek philosopher Hicetas to "Nicetas."

STEIN, JOHAN W.

746. La parte dell' Osiander nell' edizione dell' opera di Copernico. *Coelum*, 1941, *11*: 71–73. In discussing Osiander's part in the publication of the *Revolutions*, the late director of the Vatican Observatory somehow failed to recognize that Kepler's interlocutor was Pierre de la Ramée (Petrus Ramus). See Vocca (818).

747. Copernico era sacerdote? *Memorie della società astronomica italiana*, 1945, *17*: 3. Translated into French as Copernic était-il prêtre? *Specola astronomica vaticana. Miscellanea astronomica*, 1950, *3*: 88–89 (no. 103).

Calls attention to Brachvogel's acceptance of Sighinolfi's contention that Copernicus was ordained a priest.

STEINER, RUDOLF

748. Kopernikus und seine Zeit, at no. 12 (pp. 36) in *Menschengeschichte im Lichte* *der Geistesforschung* (Basel, 1946). A lecture delivered at Berlin on February 15, 1912; the author did not supervise this draft, which was edited by Marie Steiner. Ancient man could see things more deeply than his mere senses could perceive and his reason understand. Human effort can be explained only by assuming repeated earthly existences. The rejection of this truth will have to be withdrawn, just as the Roman Catholic church had to withdraw its rejection of Copernicanism.

STENZ, EDWARD

749. *Nicholas Copernicus, the Founder of Modern Astronomy* (Kabul, 1943) p. 7. Expansion of a lecture delivered at a Copernican celebration in Kabul in 1943.

750. Kopernik o atmosferze ziemskiej. *Urania* (Kraków), 1953, *24*: 193–197. Copernicus' view of the earth's atmosphere.

751. The Upper East Wind of Copernicus. *Acta geophysica polonica*, 1953, *1*: 75–81. Copernicus divided the earth's atmosphere into two layers, of which the lower participates in the earth's rotation, whereas the upper layer does not.

An obituary notice of Stenz (+ February 21, 1956) was published by Jan Gadomski, *Urania* (Kraków), 1956, 27: 190–191.

Sterne, 1942, 22: 120–121
752. Kopernikus-Gesamtausgabe in Vorbereitung. An announcement of the forthcoming *Nikolaus Kopernikus Gesamtausgabe*.

STETSON, HARLAN T.

753. Copernicus and Science; from Yesterday until Tomorrow. *Popular Astronomy*, 1943, *51*: 425–433. The Federal Government should not control the development of science.

STEVERS, MARTIN D.

754. *Mind Through the Ages* (New York, 1940) xii + 521 p. Why does the author assert (p. 401) that Copernicus estimated the distance from the earth to the stars as a million miles?

STICKER, BERNHARD

755. Ursprung und Vollzug der kopernikanischen Wende, at pp. 121–126 in Diergart, ed. (151). Contends that neoplatonism in general, and Marsilio Ficino in particular, supplied the foundations of Copernicus' thought.

756. *Die geschichtliche Bedeutung der kopernikanischen Wende* (Bonn, 1943) p. 20; Kriegsvorträge der Rheinischen Friedrich-Wilhelms-Universität, Bonn, no. 79. Sticker says that Copernicus studied at Rome, where in fact he lectured.

757. 1543–1643. *Die Himmelswelt*, 1943, *53*: 1–2. The century between Copernicus' death and Newton's birth witnessed decisive developments in astronomy. Reviewed **89, 125, 151, 500, 863**.

STILLFRIED, E.

758. Im Schatten des Kopernikus. *Das Vorfeld*, 1943, *3*: 91–94. Imaginative reconstruction of important episodes in Copernicus' life.

STÖRIG, HANS JOACHIM

759. *Kleine Weltgeschichte der Wissenschaft* (Stuttgart, 1954) xviii + 778 p. Pp. 230–232 deal with Copernicus, whose references to his ancient predecessors are erroneously said to have been suppressed when Osiander replaced Copernicus' introduction by the false preface.

STOKLEY, JAMES

760. War Prevents Celebration of Copernicus Anniversary. *Science News Letter*, 1940, *38*: 108; reprinted in *Science Digest*, 1940, *8*: no. 4, p. 62.

761. Astronomy's Birthday. *Science News Letter*, 1943, *43*: 282–283. Misdates the beginning of the writing of the *Revolutions* "around 1530."

STRAUSS, FRANZ

762. Nikolaus Kopernikus ein Deutscher und Schöpfer eines neuen Lehrgebäudes in der Astronomie. *Nationalsozialistisches Bildungswesen*, 1942, 7: 200–208.

Strauss' ludicrous errors are too numerous to be listed or corrected here. Two, however, are funny enough to be noted: Strauss transfers Cardinal Schönberg from Capua to Padua, and has Leopold Prowe answer Polish arguments in 1922, thirty-five years after his death.

763. Nikolaus Kopernikus, der Deutsche. *Geographischer Anzeiger*, 1944, *45*: 83–91. Generously grants a Master of Arts degree to Copernicus, who did not receive this degree in his lifetime. Attributes German verses to Copernicus, who wrote no poetry. Says that Luther wrote about Copernicus what he is only reported to have said in conversation. Ascribes the decisive confirmation of the Copernican astronomy to Tycho Brahe, who always steadfastly rejected it.

STRONSKI, STANISLAW

764. Germany and Copernicus. *Times* (of London), 1943 (May 24), p 5. A letter maintaining that Copernicus was a Pole.

STRUVE, OTTO

765. The Work of Copernicus and the Structure of the Universe. *Bulletin of the Polish Institute of Arts and Sciences in America*, 1942–1943, *1*: 731–738. In formulating his astronomy, Copernicus had to reject arguments based on common sense. In like manner, recent astrophysics has had to refute common-sense arguments in order to find the true position of our solar system in galactic space. But the galactic revolution was accomplished more easily than the Copernican, because blind reverence for supposed authority has vanished from scientific discussion, and because the Roman Catholic church no longer has the power to interfere with the development of astronomical thinking. Reviewed **491, 620.**

ŠTYCH, JAROSLAV

766. Mikuláš Koperník. *Říše hvězd*, 1953, *34*: 101–103. An article in Czech commemorating the 410th anniversary of Copernicus.

SZANCER, JAN MARCIN

767. *Kopernik* (Warsaw, 1953). Sixteen original paintings, accompanied by explanatory text (20 pp.) written by Stefan Flukowski.

Reviewed by Konrad Rudnicki, *Urania* (Kraków), 1954, *25*: 191–193, and by Alfons Triller, *Zeitschrift für die Geschichte und Altertumskunde Ermlands*, 1956–1957, *29*: 155–156.

SZCZENIOWSKI, SZCZEPAN

768. Wpływ Kopernika na rozwój fizyki. *Studia i materiały z dziejów nauki polskiej*, 1954, *2*: 55–91. Copernicus' influence on the study of the optics of systems in motion, and on the theory of relativity.

769. Wpływ idei Kopernika na rozwój fizyki. *Postępy fizyki*, 1954, *5*: 239–266. The impact of Copernicus' ideas on the development of physics.

See *Sesja kopernikowska.*

SZCZĘŚNIAK, BOLESŁAW

770. Notes on the Development of Astronomy in the Far East. *Polish Science and Learning*, 1943 (June), no. 3, pp. 39–47. The spread of Copernicanism to China and Japan.

771. The Penetration of the Copernican Theory into Feudal Japan. *Journal of the Royal Asiatic Society*, 1944, pp. 52–61.

772. Notes on the Penetration of the Copernican Theory into China (Seventeenth–Nineteenth Centuries). *Op. cit.*, 1945, pp. 30–38.

773. The Penetration of the Copernican Theory into China and Japan (XVII–XIX Centuries). *Bulletin of the Polish Institute of Arts and Sciences in America*, 1945, *3*: 699–717.

774. Notes on Kepler's *Tabulae Rudolphinae* in the Library of Pei-t'ang in Pekin. *Isis*, 1949, *40*: 344–347.

SZYC, JAN

775. Chronologia dzieła i życia Kopernika. *Urania* (Kraków), 1954, *25*: 239–244. Important dates in Copernicus' life and work, from his ancestors' arrival in Kraków in 1367 to the Copernicus Year in 1953.

See Baranowski (28).

SZYMAŃSKI, STANISŁAW
776. Wieża Kopernika we Fromborku. *Ochrona zabytków*, 1953, *6*: 57–63. Copernicus' tower at Frombork.

SZYMAŃSKI, STANISŁAW and WEGNER, CZESŁAW
777. Frombork, mieszkanie i pracownia Mikołaja Kopernika. *Urania* (Kraków), 1953, *24*: 1–9. Frombork, where Copernicus lived and worked. See Gadomski (238).

TATON, RENÉ, ed.
778. *Histoire générale des sciences*, Vol. II: *La Science moderne* (Paris, 1958) vii + 800 p. Excellent discussion of Copernicus (pp. 57–67) and of the spread of his ideas (pp. 67–75).

TAUBENSCHLAG, RAPHAEL
779. The University of Cracow in the Times of Copernicus. *Bulletin of the Polish Institute of Arts and Sciences in America*, 1942–1943, *1*: 748–751.

TAUBES, JACOB
780. Dialectic and Analogy. *Journal of Religion*, 1954, *34*: 111–119. Because Copernicus destroyed the theist's cosmological basis, this Israeli political theologian argues that "the Catholic church was right in attacking the Copernican theory."

TAYLOR, FRANK SHERWOOD
781. *The March of Mind—a Short History of Science* (New York, 1939) p. 320 (issued previously in London as *A Short History of Science*). Mistakenly asserts that before publishing the *Revolutions* Copernicus "sounded the opinion of the Pope and . . . the Church's view was favourable" (p. 122).
782. Nicolaus Copernicus—His Reputation after Four Hundred Years. *Tablet* (London), 1943, May 22, *181*: no. 5376, pp. 245–246. Falsely describes Copernicus as a priest. Why did Taylor say that Copernicus' works were prohibited "as of Polish origin" and that "Germany has abandoned her claim" to him?

783. An Illustrated History of Science (London and New York, 1955), p. 178. These lectures, delivered at the Royal Institution in 1952, deal with Copernicus at pp. 30, 65–68.

TAYLOR, JACK
See Copernicus, translations (130). Reviewed **439.**

TESKE, ARNIM
784. Koncepcje Kopernika a nowa era w fizyce. *Problemy*, 1952, *8*: 306–313. Copernicus' conception and the new era in physics.

THIEL, RUDOLF
785. *Und es ward Licht* (Hamburg, 1956) p. 395. Translated from German into English by Richard and Clara Winston under the title *And There Was Light* (New York, 1957) xv + 415 p. Pp. 73–90 deal with Copernicus.

THORAK, JOSEF
786. A photograph of this sculptor's statue of Copernicus, which was unveiled in Thorn (now Toruń) on May 24, 1943, was published in *Westermanns Monatshefte*, 1942–1943, *87*: 524, and on the cover of *Die Himmelswelt*, 1943, *53*: no. 10–12.

THORNDIKE, LYNN
787. *A History of Magic and Experimental Science* (New York, 1923–1958) 8 vols. "The Copernican Theory" is discussed in vol. V (1941), pp. 406–429, and "Post-Copernican Astronomy" in vol. VI (1941), pp. 3–66.
788. Pre-Copernican Astronomical Activity. *Proceedings of the American Philosophical Society*, 1950, *94*: 321–326. Assembles evidence of astronomical activity in the 14th and 15th centuries, as revealed by unpublished manuscripts in several European libraries.

THÜRING, BRUNO
789. Nikolaus Kopernikus—der grosse deutsche Astronom, at pp. 212–232 in Kubach, ed. (393). Discusses Copernicus' method and its influence on later astronomers.

TIEGHEM, PAUL VAN
790. *La littérature latine de la Renaissance* (Paris, 1944) p. 254; offprinted from *Bibliothèque d'humanisme et renaissance*, 1944, *4*: 177–418. Still attributes the *Septem sidera* to Copernicus (p. 54 = p. 224).

Times (of London), 1943 (May 24), p. 5
791. Copernicus. A quadricentennial editorial.
792. On September 7, 1948 (p. 5) the (unnamed) Warsaw correspondent reported on the extensive measures being taken by the Polish government to preserve and restore the sites where Copernicus lived and worked.

Times Literary Supplement (London), 1943, May 29, pp. 258, 262
793. Two Forerunners—Our Debt to Copernicus and Vesalius.

TIMPANARO, SEBASTIANO
794. Galileo e Copernico. *Sapere*, 1943, *17–18*: 371-373; reprinted in Timpanaro's *Scritti di storia e critica della scienza* (Florence, 1952), pp. 91–101. Those who say that Copernicus devised a mathematical system equivalent to Ptolemy's, only simpler, overlook Copernicus' contribution to the subsequent development of mechanics and physics.

TOKARSKI, ZBIGNIEW
795. Uroczystości kopernikowskie w Polsce. *Nauka polska*, 1953, *1*: no. 3, pp. 150–151. A report on the Copernican celebrations in Poland.
796. Sesja kopernikowska polskiej akademii nauk. *Życie szkoły wyższej*, 1953, *1*: no. 11, pp. 102–109. A report on the Copernicus celebrations under the auspices of the Polish Academy of Science.

TOLSTOI, IVAN IVANOVICH
797. Kopernik i ego latinski perevod Pisem Teofilakta Simokatti, at pp. 64–83 in Mikhailov, ed. (483).
See 123.

TORCOLETTI, LUIGI MARIA
798. *Il processo di Galileo; clero ed astronomia* (Monza, 1956) p. 375. In Chap-

ter IV, pp. 41–48, devoted to Copernicus, the author mistakenly accepts the supposed proof that Copernicus was a priest.

TORWIRT, LEONARD
799. Zagadnienie autentyczności portretu Mikołaja Kopernika, znajdującego się w Muzeum pomorskim w Toruniu. *Ochrona zabytków*, 1953, *6*: 40–46. Reviewed by Bohdan Marconi, *Biuletyn historii sztuki*, 1954, *16*: 277–279.
Believes that the Torun portrait of Copernicus was painted while the astronomer was still alive.

TOYNBEE, ARNOLD J.
800. *A Study of History.* In Vol. IX (London, 1954), p. 47, Copernicus' universe is correctly said not to have been infinite.

TROILO, ERMINIO
801. Copernico dal punto di vista filosofico e umanistico. *Sapere*, 1943, *17–18*: 368–370. Accepts the *Septem sidera* as authentic.
802. Copernico dal punto di vista filosofico e umanistico, at pp. 21–69 in *Copernico* (Padua, 1944) p. 69; Opuscoli accademici, Facoltà di lettere e filosofia, Università di Padova, serie liviana, no. 8 (see Silva, Giovanni). Amplification of an address delivered at a Copernicus celebration at Ferrara in 1943.
803. La dialettica copernicana. *Atti della Accademia nazionale dei lincei*, Rendiconti, classe di scienze morali, 1953, *8*: 453–465. In this lecture, which was delivered at a meeting of the Lincei on November 14, 1953, Troilo mistakenly asserted that the *Revolutions* was published on the day of Copernicus' death.

TURSKI, STANISŁAW
804. Znaczenie odkrycia Kopernika dla rozwoju myśli matematycznej. *Studia i materiałj z dziejów nauki polskiej*, 1954, *2*: 93–101. This lecture, delivered to the Eighth Congress of Polish Mathematicians on September 6, 1953, discusses the significance of Copernicus' discovery for the development of mathematical thought.

UEMOV, A. I.

805. Geliotsentricheskaia sistema Kopernika i teoriia otnositelnosti, at pp. 299–331 in *Filosofskie voprosy sovremennoi fiziki* (Philosophical Problems of Contemporary Physics) Moscow, 1952, p. 576, ed. Aleksandr Aleksandrovich Maksimov et al. Translated into Polish under the title System heliocentryczny Kopernika a teoria względności, *Myśl filozoficzna*, 1953, no. 1, pp. 144–175, and into German by Gertraude Zahn under the title Das heliozentrische System des Kopernikus und die Relativitätstheorie, *Deutsche Zeitschrift für Philosophie*, 1954, 2: 418–445.

Vigorously denies that the relativity theory, properly understood, supports the theses that the Copernican system is equivalent to the Ptolemaic system, either kinematically or dynamically, and that Copernicus merely simplified the Ptolemaic theory or devised a more convenient fiction.

Urania (Kraków), 1953, *24*: 145–147

806. 480-lecie urodzin Mikołaja Kopernika. Polish commemorations of the 480th birthday of Copernicus.

Op. cit., 1954, *25*: 84

807. Tablica pamiątkowa ku czci Kopernika we Włocławku. The commemorative plaque in honor of Copernicus at Włocław.

Op. cit., 1955, *26*: 379

808. Kopernik jako Kopicki. A curious transmutation of Copernicus' surname.

Op. cit., 1956, *27*: 18

809. Calls attention to a postage stamp issued by the U.S.S.R. in honor of Copernicus with Jan Matejko's well-known painting of the astronomer reproduced on the face of the stamp.

VAUCOULEURS, GÉRARD DE

810. *Discovery of the Universe* (New York, 1957) p. 328. Pp. 42–46 deal with Copernicus, whose universe is mistakenly described as small.

VELASCO DE MILLÁS, ISOLINA DE

811. *Nicolás Copérnico, su vida y su obra* (61 pp.). A lecture delivered at Havana on June 21, 1943.

VERENIGING NEDERLAND-POLEN, brochure no. 8

812. *Mikołaj Kopernik*, p. 47 (s.l.s.a.). Contains the following items: Siegfried van Praag, "Kopernik en Nederland" (pp. 3–5); H. Zanstra, "De Betekenis van Kopernik voor de Ontwikkeling van de Wetenschappen" (p. 6); "Het Kopernikjaar" (p. 7); Marcel Minnaert, a lecture delivered at the University of Amsterdam (pp. 10–21); Jan Gadomski, "De Lotgevallen van de Manuscripten van Mikołaj Kopernik" (pp. 22–32); Józef Cyrankiewicz, excerpts from a speech delivered at the opening of the Copernicus Year (pp. 33–34); "Tijdperk en Geboortestreek van M. Kopernik" (pp. 34–35); "Kopernik als Econoom" (p. 36); "Eerste astronomisch Geschrift van Kopernik" (pp. 37–38); "Het Kopernik-Museum te Frombork" (pp. 39–40); "Het Kopernik-Monument te Warszawa" (p. 41); "Ansichtkaarten ter Ere van Kopernik" (p. 43); "Kopernik-Literatuur" (p. 43); and Marcel Minnaert, "De Beoefening van de Astronomie in het Polen van heden" (pp. 44–47).

Reviewed by Eduard Jan Dijksterhuis, *Hemel en Dampkring*, 1955, *53*: 16.

Vestnik akademii nauk SSSR, 1943, *13*: 111–118

813. Torzhestvennoe zasedanie Akademii Nauk SSSR, posviashchennoe 400-letiiu so dnia smerti Nikolaia Kopernika. A report of the quadricentennial celebration of Copernicus under the auspices of the Soviet Academy of Science.

Op. cit., 1953, *23*: no. 6, pp. 25–31

814. Torzhestvennoe sobranie v Akademii Nauk SSSR. A report on the ceremonial meeting in the Soviet Academy of Science to commemorate the 410th anniversary of Copernicus' death.

VETTER, QUIDO

815. Osud koperníkových rukopisů. *Národní listy*, 1939, December 9, p. 2. Traces the history of Copernicus' autograph manuscript of the *Revolutions* until its re-discovery in the mid-nineteenth century.

VICTORIA UNIVERSITY COLLEGE
816. *Nicholas Copernicus: Quadricentennial Addresses, 1543–1943* (Wellington, New Zealand, 1943) p. 26. Contains the following three addresses, delivered at Victoria University College in Wellington, New Zealand, on May 24, 1943: J. C. Beaglehole, "Copernicus and His Times" (pp. 3–13); F. F. Miles, "Copernicus and the Development of Astronomy" (pp. 14–21); Kazimierz A. Wodzicki, "Copernicus and the Spirit of Poland" (pp. 22–26). Reviewed in *Observatory*, 1943–1944, *65*: 204.

VOCCA, PAOLO
817. Il quarto centenario di una fondamentale svolta nel pensiero astronomico. *Coelum*, 1941, *11*: 35–37. Mistakenly asserts that Rheticus addressed the *Narratio prima* to the Cardinal of Capua.
818. *Coelum*, 1941, *11*: 73. A reply to an article by Johan W. Stein.

VOISÉ, WALDEMAR
819. Droga kopernikowskiego odkrycia, at pp. 109–111 in *Odrodzenia w Polsce*, vol. 2, part 2 (Warsaw, 1956). The path that led to Copernicus' discovery.

VOLEDI, I.
820. *N. Copernic* (Bucharest, 1954) p. 115.

Vorfeld, Das 1943, *3*: 89–90
821. Reports a commemorative speech delivered on May 24, 1943 by Hans Frank, Nazi head of the Generalgouvernement, who pointed to Copernicus as justifying German, rather than Polish, rule over the valley of the Vistula.

WAAGE, E.
822. Zur Geschichte des Planetenproblems. *Die Sterne*, 1939, *19*: 222–227, 235–242. Copernicus' orbit for Venus is discussed at p. 238.
823. Venusbahn und Weltsysteme. *Zeitschrift für mathematischen und naturwissenschaftlichen Unterricht*, 1939, *70*: 67–75, 117–120. Copernicus' construction for Venus is treated at pp. 117–118.

824. Die Bestimmung der Marsbahn nach Ptolemäus und Kopernikus. *Op. cit.*, 1941, *72*: 47–57, 81–82. The fourth section deals with Copernicus' determination of the orbit of Mars.

Wacht, Die 1939, *5*: 95
825. Kopernikus. Mistakenly asserts that the epitaph on Copernicus' grave "was chosen by himself."

WACHUŁKA, ADAM
See Dianni, Jadwiga (147).

WAERDEN, B. L. VAN DER
826. Die Vorgänger des Kopernikus im Altertum, at pp. 100–104 in Diergart, ed. (151). The development of heliocentrism runs from the Pythagoreans through Heraclides and Aristarchus to Copernicus.

WAKEFIELD, MARION
827. Pioneer of Intellectual Freedom. *Christian Science Monitor*, Weekly Magazine Section, 1943, May 22, p. 14. A quadricentennial salute to Copernicus.

WALLIS, CHARLES GLENN
See Copernicus, translations: *On the Revolutions of the Heavenly Spheres* (124).

WALLIS, MIECZYSŁAW
828. Trzy przyczynki do historii autoportretu w Polsce. *Łódzkie towarzystwo naukowe, Sprawozdania z czynności i posiedzeń*, 1953, *8*: 57–60. Of the three additions to the history of self-portraiture in Poland which are discussed in this article, the first concerns Copernicus (pp. 57–58).
829. Kopernik a malarstwo. *Wiedza i życie*, 1954, *21*: 268–269. Copernicus and the art of painting.

WANKOWICZ, MARTA
830. Nicolaus Copernicus. *Commonweal*, 1943, *38*: 94–96. Of Galileo, who was condemned for heresy and sentenced to life imprisonment, we are told that he "was warned by the authorities at Rome against the Copernican system." This understatement would be incredible, even in a Catholic magazine, were it not accompanied by such ludicrous blunders as

references to Giese as "Archbishop of Cuhn" (instead of bishop of Kulm), and to Andreas "Ossianus" (instead of "Osiander").

831. A condensed, and uncorrected, version in *Catholic Digest*, 1942–1943, 7: no. 9, pp. 25–29.

WASCHINSKI, EMIL
832. Die Mitarbeit des Astronomen Nicolaus Coppernicus an der preussischen Münz- und Währungsreform des 16. Jahrhunderts. *Deutsche Münzblätter*, 1940, *60*: 173–177. Briefly surveys Copernicus' writings and activities in connection with a projected reform of the currency.

833. Des Astronomen Nikolaus Coppernicus Denkschrift zur preussischen Münz- und Währungsreform 1519–1528. *Elbinger Jahrbuch*, 1941, *16*: 1–40. Reviewed by Hans Schmauch, *Zeitschrift für die Geschichte und Altertumskunde Ermlands*, 1939–1942, *27*: 628–629, and by Franz Buchholz, *Jomsburg*, 1942, *6*: 143–148. A careful examination of Copernicus' essay on currency.
Reviewed **673**.

WATTENBERG, DIETRICH
834. Nikolaus Kopernikus—Leben und Werk. *Natur und Kultur*, 1943, *40*: 61–65. "In general, all of Copernicus' scientific work is distinguished by its Aryan German nature." Wattenberg errs in saying that when Luther called Copernicus a fool, he was "evidently impelled to do so by the *Narratio prima*," which was first published in 1540, a year after Luther's denunciation of Copernicus. Nor did the Roman Catholic church condemn Copernicanism for the first time in 1633.

835. Nikolaus Coppernicus und sein Werk. *Das Weltall*, 1943, *43*: 55–61. Mistranslates as "Prussian" Melanchthon's description of Copernicus as a "Sarmatian."
Reviewed **370**.

WAVRE, ROLIN
836. A propos de Copernic. *Revue de théologie et de philosophie*, 1944, *32*: 75–91.

Why does Wavre speak of Copernicus' "pupils" in the plural (p. 83)?

WĘDKIEWICZ, STANISŁAW
837. Etudes coperniciennes. *Académie polonaise des sciences et des lettres, Centre polonais de recherches scientifiques de Paris, Bulletin*, 1955–1957, no. 13–16, vii + 315 pp. Reviewed by Edward Rosen, *Isis* (forthcoming).

A valuable survey of both the older and the recent literature concerning various aspects of Copernicus' biography and his activities as humanist, economist, physician, geographer, painter, and engineer. French translations of Copernican studies by other Polish scholars make this volume especially useful to those readers who know no Polish but would nevertheless like to know what is being said by Polish students of Copernicus.

WEGNER, CZESŁAW
See Szymański, Stanisław (777).

WEIGLE, FRITZ
838. Deutsche Studenten in Italien. *Die Mittelstelle*, 1943 (May), *2*: no. 19, pp. 38–42. German student life at Italian universities in the later Middle Ages and the Renaissance.

WENTSCHER, ERICH
839. Blutslinien um Nikolaus Koppernik (Coppernicus). *Archiv für Sippenforschung*, 1944, *21*: 21–29, 51–58. Copernicus viewed as a product of the symbiosis of two neighboring populations.

WERNER, HELMUT
840. Das ptolemäische und kopernikanische Weltsystem und das Zeiss-Planetarium. *Naturwissenschaftliche Rundschau*, 1948, *1*: 161–166. An excellent description of the way in which the Zeiss Planetarium vividly exhibits the motions of the planets in both the Ptolemaic and Copernican systems.

WETZEL, FRANZ
841. Das kopernikanische Weltbild und der Index. *Natur und Kultur*, 1940, *37*:

ANNOTATED COPERNICUS BIBLIOGRAPHY

344. Objects to the view expressed by Wolfgang-Günther Dahlenkamp (*op. cit.*, p. 280) that it was necessary for the Roman Catholic church to oppose Copernicanism.

WIELEITNER, HEINRICH
842. *Geschichte der Mathematik:* I, *Von den ältesten Zeiten bis zur Wende des 17. Jahrhunderts* (Berlin, 1939); reprinted from the original ed. of 1922. Copernicus' astronomy and trigonometry are briefly mentioned at pp. 94 and 97.

WIESE, LEOPOLD VON
843. Das Selbstbewusstsein des Menschen und das kopernikanische Weltbild, at pp. 126–130 in Diergart, ed. (151). Copernicanism brought with it a modest appraisal of man's place in the universe, neither the exaggerated self-importance that accompanied geocentrism and anthropocentrism, nor the excessive self-abasement that went with the view that man is but a worm.

WILSON, GROVE
844. *Great Men of Science* (New York, 1942). A reprint of the volume originally published as *The Human Side of Science* (New York, 1929) and reissued as *Great Men of Science* (Garden City, 1932). Chapter 12 (pp. 86–92) deals with Copernicus.

Wissenschaftliche Annalen, 1954, *3*: 57
845. Festsitzung der polnischen Akademie der Wissenschaften zum Kopernikus-Jahr. An account of the Polish Academy of Science's Copernicus celebration on September 15–16, 1953.

WITKOWSKI, JÓZEF
846. Reforma Kopernika. *Urania* (Kraków), 1951, *22*: 225–232. Copernicus' reformation of astronomy.
847. Reforma Kopernika. *Nauka polska*, 1953, *1*: no. 4, pp. 70–93. Sections of this speech, delivered on September 15, 1953 before the Polish Academy of Science. were translated into French by Allan Kosko at pp. 115–119 in Wędkiewicz.
848. Kopernikańska teoria ruchu planet na tle antycznych systemów. *Postępy astronomii*, 1953, *1*: 5–12. The

Copernican theory of planetary motion against the background of the ancient systems.
See *Sesja kopernikowska.*

WODZICKI, KAZIMIERZ A.
See Victoria University College.

WOLF, ABRAHAM
849. *A History of Science, Technology, and Philosophy in the 16th and 17th Centuries*, 2d ed. (London, 1950) xvii + 692 p. The 2d ed., prepared by Douglas McKie, of a now classic work, originally published in 1935 and reissued in 1938. Chapter II (pp. 11–26) deals with "The Copernican Revolution."

WOLF, FRANZ
850. *Von der Welt des Kopernikus bis in die Fernen der Spiralnebel* (Karlsruhe, 1943) p. 23; Karlsruher akademische Reden, no. 22: Kopernikus-Gedenkstunde zum 400. Todestag des Schöpfers unseres Weltbildes, Technische Hochschule, Karlsruhe, 1943, July 10. By moving the earth out of the center of the universe and assigning it an annual revolution around the sun, Copernicus induced observers to search for a corresponding parallax in the fixed stars, and thereby provided the impulse for the creation of stellar astronomy.

YAROVOI, MIKHAIL
851. God Kopernika v Polshe. *Ogonek*, 1953, *31*: no. 24, p. 11. From Warsaw this Russian correspondent describes the celebration of the Copernicus year in Poland.

YOUNG, WILLIAM LINDSAY
852. The Greatness of Copernicus. *Quarterly Review of Higher Education among Negroes*, 1943, *11*: no. 3, pp. 25–28. This commencement address to the 1943 graduates of Johnson C. Smith University mistakenly asserts that "the Reverend Mr. Copernicus," as it calls him, was a Catholic priest.

ZAGAR, FRANCESCO
853. Per il IV centenario della morte di Nicolò Copernico: il soggiorno del grande astronomo in Italia. *Gli annali della*

Università italiana, 1942–1943, *4*: 396–402. A review of the benefits derived by Copernicus from his years of study in Italy. Why does Zagar say that Copernicus was associated with the University of Rome?

854. Manifestazioni per il IV centenario della morte di Nicolò Copernico. *Coelum*, 1943, *13*: 27–30. Describes the quatercentennial Copernican ceremonies at the universities of Ferrara and Koenigsberg.

855. *Nicolò Copernico e il sistema eliocentrico del mondo* (Bologna, 1943) p. 19. A lecture delivered on May 15, 1943 at a Copernican celebration in Bologna.

ZANSTRA, H.
See Vereniging Nederland-Polen.

ZAUNICK, RUDOLF

856. Koppernicks Grabinschrift. *Mitteilungen zur Geschichte der Medizin, der Naturwissenschaften und der Technik*, 1940, *39*: 211. The pious quatrain on Copernicus' tomb was lifted from a poem by Aeneas Sylvius Piccolomini and placed on the astronomer's grave, decades after his death, by Melchior Pyrnesius.

ZAWACKI, EDMUND

857. Copernicus, the Man and His Times. *Bulletin of the Polish Institute of Arts and Sciences in America*, 1942–1943, *1*: 738–747.

ZAWILEC

858. Czym Kopernik dla Polski, at pp. 14–18 in 367, 2d ed.

Zeitschrift für die gesamte Naturwissenschaft

859. 1938–1939, *4*: 465–479. Das kopernikanische Weltbild. Copernicus' astronomy is analyzed by an (unidentified) associate of the University of Königsberg (now Kaliningrad).

860. 1942, *8*: 198–199. Kopernikus-Gesamtausgabe in Vorbereitung. An announcement of the plans for the *Nikolaus Kopernikus Gesamtausgabe*.

ŻELAZOWSKI, BRONISŁAW

861. O wielkości Kopernika jako astronoma, at pp. 9–14 in 367, 2d ed. The greatness of Copernicus as an astronomer.

ZELLER, FRANZ
See Copernicus, editions: *Nikolaus Kopernikus Gesamtausgabe* (119).

ZELLER, KARL

862. Der Forschungsweg des Nikolaus Kopernikus, at pp. 104–110 in Diergart, ed. (151). Repeats Brachvogel's contention that Copernicus did not find his starting point in ancient Greek thinkers, even though Copernicus says he did.

863. *Des Georg Joachim Rheticus Erster Bericht über die 6 Bücher des Kopernikus von den Kreisbewegungen der Himmelsbahnen* (Munich and Berlin, 1943) xii + 196 p. Reviewed by Bernhard Sticker, *Die Himmelswelt*, 1943, *53*: 72; by Max Caspar, *Kant-Studien*, 1943, *43*: 475; by Herman von Schelling, *Zentralblatt für Mathematik und ihrer Grenzgebiete*, 1943–1944, *28*: 2; by Paul ten Bruggencate, *Vierteljahrsschrift der astronomischen Gesellschaft*, 1944, *79*: 99; and by Richard Sommer, *Das Weltall*, 1944, *44*: 30–31.

The first complete German translation of Rheticus' *Narratio prima*, with an introduction and notes by Karl Zeller.

864. Zum vierhundertsten Todestag des Nikolaus Kopernikus. *Zeitschrift für die gesamte Naturwissenschaft*, 1943, *9*: 97–101. Claims that Copernicus did not begin to write the holograph manuscript of the *Revolutions* until 1529, and finished it before 1536.

See Copernicus, editions: *Nikolaus Kopernikus Gesamtausgabe* (118–119), and Copernicus, translations: Zeller, Karl (127).

ZELLER, MARY CLAUDIA

865. Copernicus Bibliography in the University of Michigan Library. *Bulletin of the Polish Institute of Arts and Sciences in America*, 1942–1943, *1*: 695–704.

866. *The Development of Trigonometry from Regiomontanus to Pitiscus* (Ann Arbor, 1946) vi + 119 p.; Ph.D. dissertation,

ANNOTATED COPERNICUS BIBLIOGRAPHY

University of Michigan. Copernicus' trigonometry, and its relation to Regiomontanus', are discussed at pp. 40–56.

ZICH, OTAKAR

867. Ke 410. výročí smrti Mikuláše Koperníka. *Časopis pro pěstovaní matematiky*, 1953, *78*: 297–304. An article in Czech on the 410th anniversary of Copernicus' death.

ZIEMSEN, I.

See Copernicus, bibliography: Kossmann and Ziemsen (113).

ZILSEL, EDGAR

868. Copernicus and Mechanics. *Journal of the History of Ideas*, 1940, *1*: 113–118 (reprinted in *Roots of Scientific Thought*, edd. Philip P. Wiener and Aaron Noland, New York, 1957, pp. 276–280). Emphasizes the big gap between modern mechanics and Copernicus' teleological, metaphysical, even animistic explanations of phenomena now treated in mechanical terms.

ZINNER, ERNST

869. Nikolaus Koppernick als Schöpfer der modernen Sternforschung. *Deutsche Monatshefte in Polen*, 1939, *6*: 15–26. Copernicus viewed as one of a long line of German astronomers who fought against oriental astrology and for empirical observation and technical progress.

870. Die Sonnenuhren des Nikolaus Coppernicus. *Forschungen und Fortschritte*, 1942, *18*: 183. Translated into Spanish under the title Los relojes de sol de Nicolás Copérnico, *Investigación y progreso*, 1943, *14*: 172–174.

Reviewed by Harald Geppert, *Zentralblatt für Mathematik und ihre Grenzgebiete*, 1943, *27*: 3.

See Richard Sommer (736).

The various sundials attributed to Copernicus were made in the 17th and 18th centuries, except the one at Allenstein.

871. *Entstehung und Ausbreitung der coppernicanischen Lehre* (Erlangen, 1943;

Sitzungsberichte der physikalisch-medizinischen Sozietät zu Erlangen, 74) xii + 594 p. Reviewed in *Astronomische Nachrichten*, 1943, *273*: 295; by J. Weber, *Die Himmelswelt*, 1943, *53*: 71–72; by N. V. E. Nordenmark, *Populär astronomisk tidskrift*, 1943, *24*: 151–152; by Max Caspar, *Kant-Studien*, 1943, *43*: 475; by Henrik Sandblad, *Lychnos*, 1943, pp. 369–372; by Joseph Ehrenfried Hofmann, *Zentralblatt für Mathematik und ihre Grenzgebiete*, 1944, *28*: 194; by Paul ten Bruggencate, *Vierteljahrsschrift der astronomischen Gesellschaft*, 1944, *79*: 100; by Elis Strömgren, *Nordisk astronomisk tidsskrift*, 1944, p. 34; by August Kopff, *Astronomischer Jahresbericht* for 1943–1946, *45*: 25; and by Edward Rosen, *Isis*, 1946, *36*: 261–266.

See Bornkamm.

The origin and spread of the Copernican theory.

872. Zur 400-Jahrfeier der coppernicanischen Lehre. *Die Himmelswelt*, 1943, *53*: 33–40. Through an unfortunate misprint (p. 39) the title of Reinhold's *Tabulae prutenicae* becomes almost unrecognizable.

873. Die Allensteiner Sonnenuhr des Nikolaus Coppernicus. *Naturforschende Gesellschaft in Bamberg*, 1946, Bericht no. 29, pp. 28–29. A description of the sundial at Allenstein which Zinner believes was executed by Copernicus.

874. War Coppernicus ein Sarmate oder Pole? *Naturforschende Gesellschaft in Bamberg*, 1950, Bericht no. 32, pp. 55–58; reprinted in *Kleine Veröffentlichungen der Remeis-Sternwarte*, no. 4, pp. 55–58. When Melanchthon called Copernicus that "Sarmatian astronomer," Melanchthon was not well acquainted with the region in question, and besides he wrote in anger.

See Copernicus, translations: Zinner, *Astronomie: Geschichte ihrer Probleme* (128). Reviewed **118–119**.

ZONN, WŁODZIMIERZ

875. Mikołaj Kopernik twórca nowego światopoglądu. *Wiedza i życie*, 1953, *20*: 234–240. Copernicus, founder of the new conception of the universe.

876. Une nouvelle conception du

ANNOTATED COPERNICUS BIBLIOGRAPHY

ZONN (cont.)

monde. *Démocratie nouvelle*, 1953, 7: 469–471. When Zonn, the director of the Warsaw Observatory, was attending an international congress of astrophysicists at Paris, 875 was translated into French. See Kurdybacha, Łukasz.

ZULUETA

ZULUETA, LUIS DE
877. Las tres lecciones del sabio, at pp. 25–27 in 108. From Copernicus we learn these three lessons: to love the truth, to know the real world, and to be humble in face of it.

ADDENDA

2a. Le rivoluzioni dei mondi celesti, at vol. 6, pp. 329–330, in *Dizionario letterario Bompiani delle opere* (Milan, 1947–1950). The German translation (1879) of the *Revolutions* does not form part of the 1873 ed. This error was repeated in the French translation, *Laffont-Bompiani Dictionnaire des oeuvres* (Paris, 1952–1955), vol. 4, pp. 261–262.

2b. Copernico, Niccolò, at vol. 1, pp. 543–548, in *Dizionario letterario Bompiani degli autori* (Milan, 1956–1957). Copernicus stayed in Italy seven (not ten) years. He took possession of his canonry by proxy from Bologna (not in person) in 1497 (not in 1499). He circulated the *Commentariolus* long before 1530.

50. See 376.
69. See 288, 397, and 862.
84. See 270.
124. See 106.
125. See 378.
126. See 502.
135. Revised 2d ed. issued as a paperback in 2 vols. (Garden City, 1959) under the title *Medieval and Early Modern Science.*
277. See 416.
305. See 178.

KELLY, ERIC PHILBROOK
349a. *From Star to Star* (New York, 1944) xi+239 p. This story for youngsters about life in Kraków in 1493 is highly imaginative, and so is the author's remark that "German claims to Copernicus have definitely ended" (p. 233).

KOSSMANN, EUGEN OSKAR
(using the pseudonym "Dr. K. Müller)
373a. *Nikolaus Coppernicus* (Berlin, 1939) p. 16; Bund Deutscher Osten, *12.* A Nazi propaganda pamphlet denying that Copernicus was a Pole. See 113.
374. The second entry should bear a separate number, which was omitted by an unintentional oversight.
416. See 277.
429. See 74.

PÂRVULESCU, CONSTANTIN
549a. *Copernic* (Bucharest, 1943) p. 135. A popular work by a Rumanian astronomer attached to the observatory of the University of Cluj.
589. See 98–99 and 642.
603. Reissued as a paperback (New York, 1948).
659. See 469.
715. See 323.
720. The second entry should bear a separate number, which was omitted by an unintentional oversight.
812. Published at Amsterdam in 1955. See 488.
833. Republished in an enlarged and corrected version, *Zeitschrift für die Geschichte und Altertumskunde Ermlands,* 1956–1958, *29:* 389–427.
837. Reviewed by Edward Rosen, *Isis,* 1959, *50:* 177–178.
869. See 184, 249, 392, and 835.

INDEX

A CATALOG OF SELECTED
DOVER BOOKS
IN SCIENCE AND MATHEMATICS

A CATALOG OF SELECTED

DOVER BOOKS

IN SCIENCE AND MATHEMATICS

Astronomy

BURNHAM'S CELESTIAL HANDBOOK, Robert Burnham, Jr. Thorough guide to the stars beyond our solar system. Exhaustive treatment. Alphabetical by constellation: Andromeda to Cetus in Vol. 1; Chamaeleon to Orion in Vol. 2; and Pavo to Vulpecula in Vol. 3. Hundreds of illustrations. Index in Vol. 3. 2,000pp. 6⅛ x 9¼.
23567-X, 23568-8, 23673-0 Three-vol. set

THE EXTRATERRESTRIAL LIFE DEBATE, 1750–1900, Michael J. Crowe. First detailed, scholarly study in English of the many ideas that developed from 1750 to 1900 regarding the existence of intelligent extraterrestrial life. Examines ideas of Kant, Herschel, Voltaire, Percival Lowell, many other scientists and thinkers. 16 illustrations. 704pp. 5⅜ x 8½. 40675-X

A HISTORY OF ASTRONOMY, A. Pannekoek. Well-balanced, carefully reasoned study covers such topics as Ptolemaic theory, work of Copernicus, Kepler, Newton, Eddington's work on stars, much more. Illustrated. References. 521pp. 5⅜ x 8½.
65994-1

AMATEUR ASTRONOMER'S HANDBOOK, J. B. Sidgwick. Timeless, comprehensive coverage of telescopes, mirrors, lenses, mountings, telescope drives, micrometers, spectroscopes, more. 189 illustrations. 576pp. 5⅝ x 8¼. (Available in U.S. only.)
24034-7

STARS AND RELATIVITY, Ya. B. Zel'dovich and I. D. Novikov. Vol. 1 of *Relativistic Astrophysics* by famed Russian scientists. General relativity, properties of matter under astrophysical conditions, stars, and stellar systems. Deep physical insights, clear presentation. 1971 edition. References. 544pp. 5⅝ x 8¼. 69424-0

Chemistry

CHEMICAL MAGIC, Leonard A. Ford. Second Edition, Revised by E. Winston Grundmeier. Over 100 unusual stunts demonstrating cold fire, dust explosions, much more. Text explains scientific principles and stresses safety precautions. 128pp. 5⅜ x 8½. 67628-5

THE DEVELOPMENT OF MODERN CHEMISTRY, Aaron J. Ihde. Authoritative history of chemistry from ancient Greek theory to 20th-century innovation. Covers major chemists and their discoveries. 209 illustrations. 14 tables. Bibliographies. Indices. Appendices. 851pp. 5⅜ x 8½. 64235-6

CATALYSIS IN CHEMISTRY AND ENZYMOLOGY, William P. Jencks. Exceptionally clear coverage of mechanisms for catalysis, forces in aqueous solution, carbonyl- and acyl-group reactions, practical kinetics, more. 864pp. 5⅜ x 8½.
65460-5

Physics

OPTICAL RESONANCE AND TWO-LEVEL ATOMS, L. Allen and J. H. Eberly. Clear, comprehensive introduction to basic principles behind all quantum optical resonance phenomena. 53 illustrations. Preface. Index. 256pp. 5⅜ x 8½. 65533-4

ULTRASONIC ABSORPTION: An Introduction to the Theory of Sound Absorption and Dispersion in Gases, Liquids and Solids, A. B. Bhatia. Standard reference in the field provides a clear, systematically organized introductory review of fundamental concepts for advanced graduate students, research workers. Numerous diagrams. Bibliography. 440pp. 5⅜ x 8½. 64917-2

QUANTUM THEORY, David Bohm. This advanced undergraduate-level text presents the quantum theory in terms of qualitative and imaginative concepts, followed by specific applications worked out in mathematical detail. Preface. Index. 655pp. 5⅜ x 8½. 65969-0

ATOMIC PHYSICS (8th edition), Max Born. Nobel laureate's lucid treatment of kinetic theory of gases, elementary particles, nuclear atom, wave-corpuscles, atomic structure and spectral lines, much more. Over 40 appendices, bibliography. 495pp. 5⅜ x 8½. 65984-4

AN INTRODUCTION TO HAMILTONIAN OPTICS, H. A. Buchdahl. Detailed account of the Hamiltonian treatment of aberration theory in geometrical optics. Many classes of optical systems defined in terms of the symmetries they possess. Problems with detailed solutions. 1970 edition. xv + 360pp. 5⅜ x 8½. 67597-1

THIRTY YEARS THAT SHOOK PHYSICS: The Story of Quantum Theory, George Gamow. Lucid, accessible introduction to influential theory of energy and matter. Careful explanations of Dirac's anti-particles, Bohr's model of the atom, much more. 12 plates. Numerous drawings. 240pp. 5⅜ x 8½. 24895-X

ELECTRONIC STRUCTURE AND THE PROPERTIES OF SOLIDS: The Physics of the Chemical Bond, Walter A. Harrison. Innovative text offers basic understanding of the electronic structure of covalent and ionic solids, simple metals, transition metals and their compounds. Problems. 1980 edition. 582pp. 6⅛ x 9¼. 66021-4

HYDRODYNAMIC AND HYDROMAGNETIC STABILITY, S. Chandrasekhar. Lucid examination of the Rayleigh-Benard problem; clear coverage of the theory of instabilities causing convection. 704pp. 5⅜ x 8¼. 64071-X

INVESTIGATIONS ON THE THEORY OF THE BROWNIAN MOVEMENT, Albert Einstein. Five papers (1905–8) investigating dynamics of Brownian motion and evolving elementary theory. Notes by R. Fürth. 122pp. 5⅜ x 8½. 60304-0

THE PHYSICS OF WAVES, William C. Elmore and Mark A. Heald. Unique overview of classical wave theory. Acoustics, optics, electromagnetic radiation, more. Ideal as classroom text or for self-study. Problems. 477pp. 5⅜ x 8½. 64926-1

HYDRODYNAMIC AND HYDROMAGNETIC STABILITY, S. Chandrasekhar. Lucid examination of the Rayleigh-Benard problem; clear coverage of the theory of instabilities causing convection. 704pp. 5⅜ x 8¼. 64071-X

INVESTIGATIONS ON THE THEORY OF THE BROWNIAN MOVEMENT, Albert Einstein. Five papers (1905–8) investigating dynamics of Brownian motion and evolving elementary theory. Notes by R. Fürth. 122pp. 5⅜ x 8½. 60304-0

THE PHYSICS OF WAVES, William C. Elmore and Mark A. Heald. Unique overview of classical wave theory. Acoustics, optics, electromagnetic radiation, more. Ideal as classroom text or for self-study. Problems. 477pp. 5⅜ x 8½. 64926-1

PHYSICAL PRINCIPLES OF THE QUANTUM THEORY, Werner Heisenberg. Nobel Laureate discusses quantum theory, uncertainty, wave mechanics, work of Dirac, Schroedinger, Compton, Wilson, Einstein, etc. 184pp. 5⅜ x 8½. 60113-7

ATOMIC SPECTRA AND ATOMIC STRUCTURE, Gerhard Herzberg. One of best introductions; especially for specialist in other fields. Treatment is physical rather than mathematical. 80 illustrations. 257pp. 5⅜ x 8½. 60115-3

AN INTRODUCTION TO STATISTICAL THERMODYNAMICS, Terrell L. Hill. Excellent basic text offers wide-ranging coverage of quantum statistical mechanics, systems of interacting molecules, quantum statistics, more. 523pp. 5⅜ x 8½. 65242-4

THEORETICAL PHYSICS, Georg Joos, with Ira M. Freeman. Classic overview covers essential math, mechanics, electromagnetic theory, thermodynamics, quantum mechanics, nuclear physics, other topics. xxiii+885pp. 5⅜ x 8½. 65227-0

PROBLEMS AND SOLUTIONS IN QUANTUM CHEMISTRY AND PHYSICS, Charles S. Johnson, Jr. and Lee G. Pedersen. Unusually varied problems, detailed solutions in coverage of quantum mechanics, wave mechanics, angular momentum, molecular spectroscopy, more. 280 problems, 139 supplementary exercises. 430pp. 6½ x 9¼. 65236-X

THEORETICAL SOLID STATE PHYSICS, Vol. I: Perfect Lattices in Equilibrium; Vol. II: Non-Equilibrium and Disorder, William Jones and Norman H. March. Monumental reference work covers fundamental theory of equilibrium properties of perfect crystalline solids, non-equilibrium properties, defects and disordered systems. Total of 1,301pp. 5⅜ x 8½. Vol. I: 65015-4 Vol. II: 65016-2

WHAT IS RELATIVITY? L. D. Landau and G. B. Rumer. Written by a Nobel Prize physicist and his distinguished colleague, this compelling book explains the special theory of relativity to readers with no scientific background, using such familiar objects as trains, rulers, and clocks. 1960 ed. vi+72pp. 23 b/w illustrations. 5⅜ x 8½. 42806-0 $6.95

A TREATISE ON ELECTRICITY AND MAGNETISM, James Clerk Maxwell. Important foundation work of modern physics. Brings to final form Maxwell's theory of electromagnetism and rigorously derives his general equations of field theory. 1,084pp. 5⅜ x 8½. Two-vol. set. Vol. I: 60636-8 Vol. II: 60637-6

METHODS OF THERMODYNAMICS, Howard Reiss. Outstanding text focuses on physical technique of thermodynamics, typical problem areas of understanding, and significance and use of thermodynamic potential. 1965 edition. 238pp. 5⅜ x 8½. 69445-3

TENSOR ANALYSIS FOR PHYSICISTS, J. A. Schouten. Concise exposition of the mathematical basis of tensor analysis, integrated with well-chosen physical examples of the theory. Exercises. Index. Bibliography. 289pp. 5⅜ x 8½. 65582-2

RELATIVITY IN ILLUSTRATIONS, Jacob T. Schwartz. Clear nontechnical treatment makes relativity more accessible than ever before. Over 60 drawings illustrate concepts more clearly than text alone. Only high school geometry needed. Bibliography. 128pp. 6⅛ x 9¼. 25965-X

THE ELECTROMAGNETIC FIELD, Albert Shadowitz. Comprehensive undergraduate text covers basics of electric and magnetic fields, builds up to electromagnetic theory. Also related topics, including relativity. Over 900 problems. 768pp. 5⅝ x 8¼. 65660-8

GREAT EXPERIMENTS IN PHYSICS: Firsthand Accounts from Galileo to Einstein, edited by Morris H. Shamos. 25 crucial discoveries: Newton's laws of motion, Chadwick's study of the neutron, Hertz on electromagnetic waves, more. Original accounts clearly annotated. 370pp. 5⅜ x 8½. 25346-5

RELATIVITY, THERMODYNAMICS AND COSMOLOGY, Richard C. Tolman. Landmark study extends thermodynamics to special, general relativity; also applications of relativistic mechanics, thermodynamics to cosmological models. 501pp. 5⅜ x 8½. 65383-8

LIGHT SCATTERING BY SMALL PARTICLES, H. C. van de Hulst. Comprehensive treatment including full range of useful approximation methods for researchers in chemistry, meteorology and astronomy. 44 illustrations. 470pp. 5⅜ x 8½. 64228-3

STATISTICAL PHYSICS, Gregory H. Wannier. Classic text combines thermodynamics, statistical mechanics and kinetic theory in one unified presentation of thermal physics. Problems with solutions. Bibliography. 532pp. 5⅜ x 8½. 65401-X

Paperbound unless otherwise indicated. Available at your book dealer, online at **www.doverpublications.com**, or by writing to Dept. GI, Dover Publications, Inc., 31 East 2nd Street, Mineola, NY 11501. For current price information or for free catalogues (please indicate field of interest), write to Dover Publications or log on to **www.doverpublications.com** and see every Dover book in print. Dover publishes more than 500 books each year on science, elementary and advanced mathematics, biology, music, art, literary history, social sciences, and other areas.